D0306775

AS & A2 Chemistry

Exam Board: AQA

Complete Revision and Practice

AS-Level Contents

A2-Level Contents

Published by CGP.

Editors:
Amy Boutal, Mary Falkner, Sarah Hilton, Paul Jordin, Sharon Keeley, Simon Little, Andy Park, Michael Southorn, Julie Wakeling.

Contributors:
Mike Bossart, Rob Clarke, Vikki Cunningham, Ian H. Davis, John Duffy, Max Fishel, Emma Grimwood, Richard Harwood, Lucy Muncaster, Jane Simoni, Derek Swain, Paul Warren, Chris Workman.

Proofreaders:
Barrie Crowther, Paul Jordin, Glenn Rogers, Julie Wakeling.

ISBN: 978 1 84762 420 8

With thanks to Laura Jakubowski for the copyright research.

With thanks to Science Photo Library for permission to reproduce the photograph used on page 62.

Groovy website: www.cgpbooks.co.uk
Jolly bits of clipart from CorelDRAW®
Printed by Elanders Ltd, Newcastle upon Tyne.

Based on the classic CGP style created by Richard Parsons.

Photocopying — it's dull, grey and sometimes a bit naughty. Luckily, it's dead cheap, easy and quick to order more copies of this book from CGP — just call us on 0870 750 1242. Phew!

Text, design, layout and original illustrations © Coordination Group Publications Ltd. (CGP) 2009
All rights reserved.

AS-Level
Chemistry

Exam Board: AQA

The Scientific Process

'How Science Works' is all about the scientific process — how we develop and test scientific ideas.
It's what scientists do all day, every day (well except at coffee time — never come between scientists and their coffee).

Scientists Come Up with **Theories** — Then **Test Them**...

Science tries to explain **how** and **why** things happen. It's all about seeking and gaining **knowledge** about the world around us. Scientists do this by **asking** questions and **suggesting** answers and then **testing** them, to see if they're correct — this is the **scientific process**.

1) **Ask** a question — make an **observation** and ask **why or how** whatever you've observed happens.
   E.g. Why does sodium chloride dissolve in water?

2) **Suggest** an answer, or part of an answer, by forming a **theory** or a **model** (a possible **explanation** of the observations or a description of what you think is happening actually happening).
   E.g. Sodium chloride is made up of charged particles which are pulled apart by the polar water molecules.

3) Make a **prediction** or hypothesis — a **specific testable statement**, based on the theory, about what will happen in a test situation.
   E.g. A solution of sodium chloride will conduct electricity much better than water does.

4) Carry out **tests** — to provide **evidence** that will support the prediction or refute it.
   E.g. Measure the conductivity of water and of sodium chloride solution.

The evidence supported Quentin's Theory of Flammable Burps.

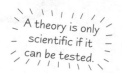

A theory is only scientific if it can be tested.

...Then They **Tell** Everyone About Their **Results**...

The results are **published** — scientists need to let others know about their work. Scientists publish their results in **scientific journals**. These are just like normal magazines, only they contain **scientific reports** (called papers) instead of the latest celebrity gossip.

1) Scientific reports are similar to the **lab write-ups** you do in school. And just as a lab write-up is **reviewed** (marked) by your teacher, reports in scientific journals undergo **peer review** before they're published.

   Scientists use standard terminology when writing their reports. This way they know that other scientists will understand them. For instance, there are internationally agreed rules for naming organic compounds, so that scientists across the world will know exactly what substance is being referred to. See page 36.

2) The report is sent out to **peers** — other scientists who are experts in the **same area**. They go through it bit by bit, examining the methods and data, and checking it's all clear and logical. When the report is approved, it's **published**. This makes sure that work published in scientific journals is of a **good standard**.

3) But peer review **can't guarantee** the science is **correct** — other scientists still need to **reproduce** it.

4) Sometimes **mistakes** are made and bad work is published. Peer review **isn't perfect** but it's probably the best way for scientists to self-regulate their work and to publish **quality reports**.

...Then **Other Scientists** Will **Test** the Theory Too

1) Other scientists read the published theories and results, and try to **test the theory** themselves. This involves:
   - Repeating the **exact same experiments**.
   - Using the theory to make **new predictions** and then testing them with **new experiments**.

2) If all the experiments in the world provide evidence to back it up, the theory is thought of as **scientific 'fact'** (for now).

3) If **new evidence** comes to light that **conflicts** with the current evidence the theory is questioned all over again. More rounds of **testing** will be carried out to try to find out where the theory **falls down**.

   This is how the scientific process works — evidence supports a theory, loads of other scientists read it and test it for themselves, eventually all the scientists in the world agree with it and then bingo, you get to learn it.

   This is exactly how scientists arrived at the structure of the atom (see pages 6-7) — and how they came to the conclusion that electrons are arranged in shells and orbitals (see page 10). It took years and years for these models to be developed and accepted — this is often the case with the scientific process.

The Scientific Process

If the **Evidence** Supports a Theory, It's **Accepted** — for Now

Our currently accepted theories have survived this '**trial by evidence**'. They've been tested **over and over again** and each time the results have backed them up. **BUT**, and this is a big but (teehee), they never become totally indisputable fact. Scientific **breakthroughs or advances** could provide new ways to question and test the theory, which could lead to **changes and challenges** to it. Then the testing starts all over again...

And this, my friend, is the **tentative nature of scientific knowledge** — it's always **changing** and **evolving**.

When CFCs were first used in fridges in the 1930s, scientists thought they were problem-free — well, why not? There was no evidence to say otherwise. It was decades before anyone found out that CFCs were actually making a whopping great hole in the ozone layer. See page 67.

Evidence Comes From **Lab Experiments**...

1) Results from **controlled experiments** in **laboratories** are **great**.
2) A lab is the easiest place to **control variables** so that they're all **kept constant** (except for the one you're investigating).
3) This means you can draw meaningful **conclusions**.

For example, if you're investigating how temperature affects the rate of a reaction you need to keep everything but the temperature constant, e.g. the pH of the solution, the concentration of the solution, etc.

...But You **Can't** Always do a Lab Experiment

There are things you **can't** study in a lab. And outside the lab controlling the variables is tricky, if not impossible.

- *Are increasing CO_2 emissions causing climate change?*
  There are other variables which may have an effect, such as changes in solar activity. You can't easily rule out every possibility. Also, climate change is a very **gradual process**. Scientists won't be able to tell if their predictions are correct for donkey's years.

See pages 42-43 for more on climate change.

- *Does drinking chlorinated tap water increase the risk of developing certain cancers?*
  There are always differences between groups of people. The best you can do is to have a **well-designed study** using **matched groups** — **choose two groups** of people (those who drink tap water and those who don't) which are **as similar as possible** (same mix of ages, same mix of diets etc.). But you still can't rule out every possibility. Taking newborn identical twins and treating them identically, except for making one drink gallons of tap water and the other only pure water, might be a fairer test, but it would present huge **ethical problems**.

Samantha thought her study was very well designed — especially the fitted bookshelf.

Science Helps to Inform **Decision-Making**

Lots of scientific work eventually leads to **important discoveries** that **could** benefit humankind — but there are often **risks** attached (and almost always **financial costs**).

Society (that's you, me and everyone else) must weigh up the information in order to **make decisions** — about the way we live, what we eat, what we drive, and so on. Information is also be used by **politicians** to devise policies and laws.

- **Chlorine** is added to water in **small quantities** to disinfect it. Some studies link drinking chlorinated water with certain types of cancer (see page 57). But the risks from drinking water contaminated by nasty bacteria are far, far greater. There are other ways to get rid of bacteria in water, but they're heaps **more expensive**.
- Scientific advances mean that **non-polluting hydrogen-fuelled cars** can be made. They're better for the environment, but are really expensive. Also, it'd cost a fortune to adapt the existing filling stations to store hydrogen.
- Pharmaceutical drugs are really expensive to develop, and drug companies want to make money. So they put most of their efforts into developing drugs that they can sell for a good price. Society has to consider the **cost** of buying new drugs — the **NHS** can't afford the most expensive drugs without **sacrificing** something else.

So there you have it — how science works...

Hopefully these pages have given you a nice intro to how science works, e.g. what scientists do to provide you with 'facts'. You need to understand this, as you're expected to know how science works yourselves — for the exam and for life.

The Atom

This stuff about atoms and elements should be ingrained in your brain from GCSE. You do need to know it perfectly though if you are to negotiate your way through the field of man-eating tigers which is AS Chemistry.

Atoms are made up of **Protons**, **Neutrons** and **Electrons**

All elements are made of **atoms**. Atoms are made up of 3 types of particle — **protons**, **neutrons** and **electrons**.

Electrons
1) Electrons have **–1** charge.
2) They whizz around the nucleus in **orbitals**. The orbitals take up most of the **volume** of the atom.

Nucleus
1) Most of the **mass** of the atom is concentrated in the nucleus.
2) The **diameter** of the nucleus is rather titchy compared to the whole atom.
3) The nucleus is where you find the **protons** and **neutrons**.

The mass and charge of these subatomic particles is **really small**, so **relative mass** and **relative charge** are used instead.

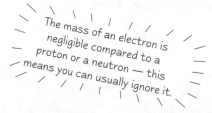
The mass of an electron is negligible compared to a proton or a neutron — this means you can usually ignore it.

Subatomic particle	Relative mass	Relative charge
Proton	1	+1
Neutron	1	0
Electron, e^-	$\frac{1}{2000}$	–1

Nuclear Symbols Show Numbers of **Subatomic Particles**

You can figure out the **number** of protons, neutrons and electrons from the **nuclear symbol**.

Mass number
This tells you the **total** number of **protons** and **neutrons** in the nucleus.

Element symbol

$$^A_Z X$$

Atomic (proton) number
1) This is the number of **protons** in the nucleus — it identifies the element.
2) **All** atoms of the same element have the **same** number of protons.

Sometimes the atomic number is left out of the nuclear symbol, e.g. ^7Li. You don't really need it because the element's symbol tells you its value.

1) For **neutral** atoms, which have no overall charge, the number of electrons is **the same as** the number of protons.
2) The number of neutrons is just **mass number minus atomic number**, i.e. 'top minus bottom' in the nuclear symbol.

Nuclear symbol	Atomic number, Z	Mass number, A	Protons	Electrons	Neutrons
^7_3Li	3	7	3	3	7 – 3 = **4**
$^{80}_{35}Br$	35	80	35	35	80 – 35 = **45**
$^{24}_{12}Mg$	12	24	12	12	24 – 12 = **12**

"Hello, I'm Newt Ron..."

Ions have **Different** Numbers of **Protons** and **Electrons**

Negative ions have **more electrons** than protons...
E.g.

Br^- The negative charge means that there's 1 more electron than there are protons. Br has 35 protons (see table above), so Br^- must have 36 electrons. The overall charge = + 35 – 36 = –1.

...and **positive** ions have **fewer electrons** than protons. It kind of makes sense if you think about it.
E.g.

Mg^{2+} The 2+ charge means that there's 2 fewer electrons than there are protons. Mg has 12 protons (see table above), so Mg^{2+} must have 10 electrons. The overall charge = +12 – 10 = +2.

The Atom

Isotopes are Atoms of the Same Element with Different Numbers of Neutrons

Make sure you **learn** this definition and totally **understand** what it means —

Isotopes of an element are atoms with the same number of protons but different numbers of neutrons.

Chlorine-35 and chlorine-37 are examples of isotopes.

Different mass numbers mean different numbers of neutrons.

$35 - 17 = 18$ neutrons ⟵

⟶ $37 - 17 = 20$ neutrons

$$^{35}_{17}\text{Cl}$$

The **atomic numbers** are the same. **Both** isotopes have 17 protons and 17 electrons.

$$^{37}_{17}\text{Cl}$$

1) It's the **number** and **arrangement** of electrons that decides the **chemical properties** of an element. Isotopes have the **same configuration of electrons**, so they've got the **same** chemical properties.

2) Isotopes of an element do have slightly different **physical properties** though, such as different densities, rates of diffusion, etc. This is because **physical properties** tend to depend more on the **mass** of the atom.

Here's another example — naturally occurring **magnesium** consists of 3 isotopes.

^{24}Mg (79%)	^{25}Mg (10%)	^{26}Mg (11%)
12 protons	12 protons	12 protons
12 neutrons	**13** neutrons	**14** neutrons
12 electrons	12 electrons	12 electrons

The Periodic Table gives the atomic number for each element. The other number isn't the mass number, it's the relative atomic mass (see page 8). They're a bit different, but you can often assume they're equal — it doesn't matter unless you're doing really accurate work.

Practice Questions

Q1 Draw a diagram showing the structure of an atom, labelling each part.

Q2 Define the term 'isotope' and give an example.

Q3 Draw a table showing the relative charge and relative mass of the three subatomic particles found in atoms.

Q4 Using an example, explain the terms 'atomic number' and 'mass number'.

Q5 Where is the mass concentrated in an atom, and what makes up most of the volume of an atom?

Exam Questions

Q1 Hydrogen, deuterium and tritium are all isotopes of each other.
 a) Identify one similarity and one difference between these isotopes. [2 marks]
 b) Deuterium can be written as ^{2}H. Determine the number of protons, neutrons and electrons in a deuterium atom. [3 marks]
 c) Write the nuclear symbol for tritium, given that it has 2 neutrons. [1 mark]

Q2 This question relates to the atoms or ions A to D: A. ^{32}S^{2-},   B. ^{40}Ar,   C. ^{30}S,   D. ^{42}Ca.
 a) Identify the similarity for each of the following pairs, justifying your answer in each case.
  (i)   A and B. [2 marks]
  (ii) A and C. [2 marks]
  (iii) B and D. [2 marks]
 b) Which two of the atoms or ions are isotopes of each other? Explain your reasoning. [2 marks]

Got it learned yet? — Isotope so...

This is a nice straightforward page just to ease you in to things. Remember that positive ions have fewer electrons than protons, and negative ions have more electrons than protons. Get that straight in your mind or you'll end up in a right mess. There's nowt too hard about isotopes neither. They're just the same element with different numbers of neutrons.

Atomic Models

The model of the atom on the previous pages is darn useful for understanding loads of ideas in chemistry.
You can picture what's happening in your mind really well. But it is just a model. So it's not completely like that really.

The **Accepted Model** of the **Atom** Has **Changed** Throughout History

1) The model of the atom you're expected to know (the one on page 4) is one of the **currently accepted** ones.
   But in the past, **completely different** models were accepted, because they fitted the evidence available at the time.

2) As scientists did more experiments, **new evidence** was found and the models were **modified** to fit it.

3) At the start of the 19th century John Dalton described atoms as **solid spheres**,
   and said that different spheres made up the different elements.

4) In 1897 **J J Thompson** concluded from his experiments that atoms **weren't**
   solid and indivisible. His measurements of **charge** and **mass** showed that an
   atom must contain even smaller, negatively charged particles — **electrons**.
   The 'solid sphere' idea of atomic structure had to be changed.
   The new model was known as the '**plum pudding model**'.

delicious pudding

positively charged 'pudding'

electrons

Rutherford Showed that the **Plum Pudding** Model Was **Wrong**

1) In 1909 Ernest Rutherford and his students Hans Geiger and Ernest Marsden conducted the famous **gold foil
   experiment**. They fired **alpha particles** (which are positively charged) at an extremely thin sheet of gold.

2) From the plum pudding model, they were expecting **most** of the alpha particles to be deflected **very slightly** by the
   positive 'pudding' that made up most of an atom. In fact, most of the alpha particles passed **straight through** the
   gold atoms, and a very small number were deflected **backwards**. So the plum pudding model **couldn't be right**.

3) So Rutherford came up with a model that **could** explain
   this new evidence — the **nuclear model** of the atom.
   In this, there's a **tiny, positively charged nucleus** at the
   centre, surrounded by a '**cloud**' of **negative electrons** —
   most of the atom is **empty space**.

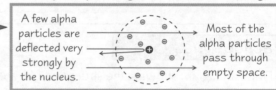

A few alpha
particles are
deflected very
strongly by
the nucleus.

Most of the
alpha particles
pass through
empty space.

> This is nearly always the way scientific knowledge develops — **new evidence** prompts people
> to come up with **new, improved ideas**. Then other people go through each new, improved
> idea with a fine-tooth comb as well — modern '**peer review**' (see p2) is part of this process.

The **Refined Bohr Model** Explains a Lot...

1) There were quite a few other modifications to the model before we got to our currently accepted versions.
   Niels Bohr got pretty close though.

2) Scientists realised that electrons in a '**cloud**' around the nucleus of an atom, as Rutherford described, would quickly
   spiral down into the nucleus, causing the atom to **collapse**. Niels Bohr proposed a new model of the atom with four
   basic principles:

> 1) Electrons can only exist in **fixed orbits**, or **shells**, and not anywhere in between.
>
> 2) Each shell has a **fixed energy**.
>
> 3) When an electron moves between shells **electromagnetic radiation** is **emitted** or **absorbed**.
>
> 4) Because the energy of shells is fixed, the radiation will have a **fixed frequency**.

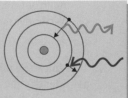

3) The frequencies of radiation emitted and absorbed by atoms were already known from experiments.
   The Bohr model **fitted these observations** — it looked good.

> One of the things that makes a theory **scientific** is that it's 'falsifiable' —
> you can **make predictions** using the theory, then if you test the predictions
> and they turn out to be **wrong**, you know that the **theory's wrong**.

4) Scientists discovered that not all the electrons in a shell had the same energy.
   This meant that the Bohr model wasn't quite right. So, they **refined** it to include **subshells**.

Atomic Models

The **Bohr Model** Explained Why Some Gases are **Inert**

1) The Bohr model also explained why some elements (the noble gases) are **inert**.

2) Bohr said that the shells of an atom can only hold **fixed numbers of electrons**, and that an element's reactivity is due to its electrons. So, when an atom has **full shells** of electrons it's **stable** and does not react.

3) Loads of observations fitted in with the **Bohr model**, and the refined Bohr model was even better. But...

There's **More Than One** Model of Atomic Structure in Use Today

1) We now know that the refined Bohr model is **not perfect** — but it's still widely used to describe atoms because it is simple and explains many observations from experiments, like bonding and ionisation energy trends.

2) The most accurate model we have today involves complicated quantum mechanics. Basically, you can never know where an electron is or which direction it's going in at any moment, but you can say **how likely** it is to be at a certain point in the atom. Oh, and electrons can act as **waves** as well as particles. But you don't need to worry about that.

3) It might be **more accurate**, but it's a lot harder to get your head round and visualise. It **does** explain some observations that can't be accounted for by the Bohr model though.

4) So scientists use whichever model is most relevant to whatever they're investigating.

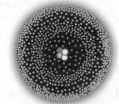

The quantum model of an atom with two shells of electrons. The denser the dots, the more likely an electron is to be there.

Practice Questions

Q1 Who developed the 'nuclear' model of the atom? What evidence did they have for it?

Q2 Describe the Bohr model of an atom. How was it later refined?

Q3 What predictions were made from the Bohr model that turned out to be correct?

Q4 Is there only one accepted model of the atom today?

Exam Questions

Q1 Read the passage below and then answer the questions that follow.

In 1911 Ernest Rutherford proposed a new model of the atom based on observations of the behaviour of atoms made by his students two years earlier. He said that atoms consisted of a small, positively charged nucleus around which negative electrons orbited. Scientists predicted that as electrons orbited the nucleus they would emit radiation with a continuous range of frequencies. This was tested and it was found that atoms emitted only certain fixed frequencies of radiation. In 1915 Niels Bohr proposed a model of the atom in which electrons were constrained to fixed orbits and could not exist anywhere between.

a) Why did Rutherford think that a new model of the atom was needed? [1 mark]

b) Why is Bohr's model thought to be a truer description of the atom than Rutherford's? [2 marks]

c) More accurate models of the atom have been developed since the Bohr model. Explain why the Bohr model is still used today. [1 mark]

Q2 The ion Ca^{2+} has the same electronic configuration as an argon atom, whilst the Ca^+ ion has an electronic configuration identical to that of a potassium atom.

a) Which of these two ions is most stable? [1 mark]

b) How do atomic models help to explain the relative stability of these ions? [2 marks]

These models are tiny — even smaller than size zero, I reckon...

The process of developing a model to fit the evidence available, looking for more evidence to show if it's correct or not, then revising the model if necessary is really important. It happens with all new scientific ideas. Remember, scientific 'facts' are only accepted as true because no one's proved yet that they aren't. It _might_ all be bunkum.

Relative Mass

Relative mass...What? Eh?...Read on...

Relative Masses are Masses of Atoms Compared to Carbon-12

The actual mass of an atom is **very**, **very tiny**. Don't worry about exactly how tiny for now, but it's far **too small** to weigh. So, the mass of one atom is compared to the mass of a different atom. This is its **relative mass**. Here are some definitions to learn:

Relative atomic mass is an average, so it's not usually a whole number. Relative isotopic mass is always a whole number (at AS level anyway). E.g. a natural sample of chlorine contains a mixture of ^{35}Cl (75%) and ^{37}Cl (25%), so the relative isotopic masses are 35 and 37. But its relative atomic mass is 35.5.

The **relative atomic mass**, A_r, is the **average mass** of an atom of an element on a scale where an atom of **carbon-12** is 12.

Relative isotopic mass is the mass of an atom of an **isotope** of an element on a scale where an atom of **carbon-12** is 12.

The **relative molecular mass** (or **relative formula mass**), M_r, is the average mass of a **molecule** or **formula unit** on a scale where an atom of **carbon-12** is 12.

To find the relative molecular mass, just add up the relative atomic mass values of all the atoms in the molecule, e.g. $M_r(C_2H_6O) = (2 \times 12) + (6 \times 1) + 16 = 46$.

Relative formula mass is used for compounds that are ionic (or giant covalent, such as SiO_2). To find the relative formula mass, just add up the relative atomic masses (A_r) of all the ions in the formula unit. (A_r of ion = A_r of atom. The electrons make no difference to the mass.) E.g. $M_r(CaF_2) = 40 + (2 \times 19) = 78$.

Relative Masses can be Measured Using a Mass Spectrometer

You can use a **mass spectrometer** to find out loads of stuff. It can tell you the **relative atomic mass**, **relative molecular mass**, **relative isotopic abundance**, **molecular structure** and your **horoscope** for the next fortnight.

There are **5** things that happen when a sample is squirted into a mass spectrometer.

① **Vaporisation** — the sample is turned into **gas** (**vaporised**) using an electrical heater.

② **Ionisation** — the gas particles are bombarded with **high-energy electrons** to ionise them. Electrons are knocked off the particles, leaving **positive ions**.

③ **Acceleration** — the positive ions are accelerated by an **electric field**.

④ **Deflection** — The positive ions' paths are altered with a **magnetic field**. **Lighter ions** have less momentum and are deflected **more** than heavier ions. For a given magnetic field, **only** ions with a particular **mass/charge ratio** make it to the detector.

⑤ **Detection** — the magnetic field strength is **slowly increased**. As this happens, different ions (ones with a higher mass/charge ratio) can reach the detector. A **mass spectrum** is produced.

A Mass Spectrum

The y-axis gives the **abundance of ions**, often as a percentage. For an element, the height of each peak gives the **relative isotopic abundance**, e.g. 75.5% are the ^{35}Cl isotope.

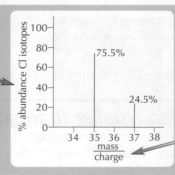

If the sample is an **element**, each line will represent a **different isotope** of the element.

The x-axis units are given as a 'mass/charge' ratio. Since the charge on the ions is mostly **+1**, you can often assume the x-axis is simply the **relative isotopic mass**.

Relative Mass

A_r and *Relative Isotopic Abundance* can be *Worked Out* from a *Mass Spectrum*

You need to know how to calculate the **relative atomic mass** (A_r) of an element from the **mass spectrum**.

Here's how to calculate A_r for magnesium, using the mass spectrum below —

Step 1: For each peak, read the **% relative isotopic abundance** from the y-axis and the **relative isotopic mass** from the x-axis. **Multiply** them together to get the total mass for each isotope. $79 \times 24 = 1896$; $10 \times 25 = 250$; $11 \times 26 = 286$

Step 2: **Add** up these totals. $1896 + 250 + 286 = 2432$

Step 3: **Divide by 100** (since percentages were used). $A_r(Mg) = \dfrac{2432}{100} = 24.32 \approx \underline{\textbf{24.3}}$

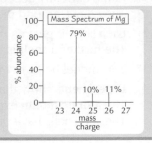

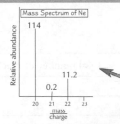

If the relative abundance is **not** given as a percentage, the total abundance may not add up to 100. In this case, don't panic. Just do steps 1 and 2 as above, but then divide by the **total relative abundance** instead of 100 — like this:

$$A_r(Ne) = \frac{(114 \times 20) + (0.2 \times 21) + (11.2 \times 22)}{114 + 0.2 + 11.2} \approx 20.18$$

Mass spectrometry is a good way to identify elements and molecules (it's kind of like fingerprinting). For instance, small mass spectrometers have been used in probes to find out what the Martian atmosphere is made of.

Mass Spectrometry can be used to *Find Out* M_r

You can also get a mass spectrum for a **molecular sample**, such as ethanol (CH_3CH_2OH).

1) A **molecular ion**, $M^+_{(g)}$, is formed when the bombarding electrons remove 1 electron from the molecule. This gives the peak in the spectrum with the **highest mass** (furthest to the right, ignoring isotopes). The mass of M^+ gives M_r for the molecule, e.g. $CH_3CH_2OH^+$ has $M_r = 46$.

2) But it's not that simple — bombarding with electrons makes some molecules break up into fragments. These all show up on the mass spectrum, making a **fragmentation pattern**. For ethanol, the fragments you get include: CH_3^+ ($M_r = 15$), $CH_3CH_2^+$ ($M_r = 29$) and CH_2OH^+ ($M_r = 31$). Fragmentation patterns are actually pretty cool because you can use them to identify **molecules** and even their **structure**.
There's more about fragmentation patterns on p80.

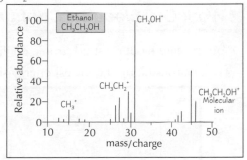

Practice Questions

Q1 Explain what relative atomic mass (A_r) and relative isotopic mass mean.

Q2 Explain the difference between relative molecular mass and relative formula mass.

Q3 Describe how a mass spectrometer works.

Exam Questions

Q1 Copper, Cu, exists in two main isotopic forms, ^{63}Cu and ^{65}Cu.
   a) Calculate the relative atomic mass of Cu using the information from the mass spectrum. [2 marks]
   b) Explain why the relative atomic mass of copper is not a whole number. [2 marks]

Q2 The percentage make-up of naturally occurring potassium is 93.11% ^{39}K, 0.12% ^{40}K and 6.77% ^{41}K.
   a) What method is used to determine the mass and abundance of each isotope? [1 mark]
   b) Use the information to determine the relative atomic mass of potassium. [2 marks]

You can't pick your relatives — you just have to learn them...

Working out M_r is dead easy — and using a calculator makes it even easier. It'll really help if you know the mass numbers for the first 20 elements or so, or you'll spend half your time looking back at the periodic table. I hope you've done the practice and exam questions, cos they pretty much cover the rest of the stuff, and if you can get them right, you've nailed it.

Electronic Structure

Those little electrons prancing about like mini bunnies decide what'll react with what — it's what chemistry's all about.

Electron Shells are Made Up of Sub-Shells and Orbitals

1) In the currently accepted model of the atom, electrons have **fixed energies**.
   They move around the nucleus in certain regions of the atom called **shells** or **energy levels**.

2) Each shell is given a number called the **principal quantum number**.
   The **further** a shell is from the nucleus, the **higher** its energy and the **larger** its principal quantum number.

3) This model helps to explain why electrons are **attracted** to the nucleus, but are not **drawn into it** and destroyed.

4) **Experiments** show that not all the electrons in a shell have exactly the same energy.
   The **atomic model** explains this — shells are divided up into **sub-shells** that have slightly different energies.
   The sub-shells have different numbers of **orbitals** which can each hold up to **2 electrons**.

This table shows the number of electrons that fit in each type of sub-shell.

Sub-shell	Number of orbitals	Maximum electrons
s	1	$1 \times 2 = 2$
p	3	$3 \times 2 = 6$
d	5	$5 \times 2 = 10$
f	7	$7 \times 2 = 14$

And this one shows the sub-shells and electrons in the first four energy levels.

Shell	Sub-shells	Total number or electrons	
1st	$1s$	2	= 2
2nd	$2s$ $2p$	$2 + (3 \times 2)$	= 8
3rd	$3s$ $3p$ $3d$	$2 + (3 \times 2) + (5 \times 2)$	= 18
4th	$4s$ $4p$ $4d$ $4f$	$2 + (3 \times 2) + (5 \times 2) + (7 \times 2)$	= 32

5) The two electrons in each orbital spin in **opposite directions**.

Work Out Electron Configurations by Filling the Lowest Energy Levels First

You can figure out most electronic configurations pretty easily, so long as you know a few simple rules —

1) Electrons fill up the **lowest** energy sub-shells first.

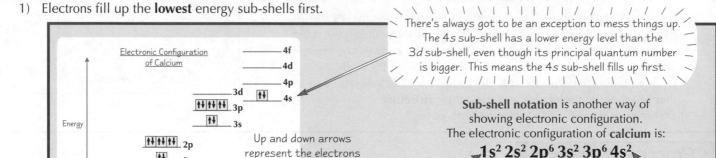

There's always got to be an exception to mess things up. The 4s sub-shell has a lower energy level than the 3d sub-shell, even though its principal quantum number is bigger. This means the 4s sub-shell fills up first.

Up and down arrows represent the electrons spinning in opposite directions.

Sub-shell notation is another way of showing electronic configuration.
The electronic configuration of **calcium** is:

$$1s^2\ 2s^2\ 2p^6\ 3s^2\ 3p^6\ 4s^2$$

Energy level / shell (principal quantum number)   Sub-shell   Number of electrons

2) Electrons fill orbitals **singly** before they start sharing.

	1s	2s	2p		
Nitrogen	↑↓	↑↓	↑	↑	↑

	1s	2s	2p		
Oxygen	↑↓	↑↓	↑↓	↑	↑

3) For the configuration of **ions** from the **s** and **p** blocks of the periodic table, just **remove or add** the electrons to or from the highest energy occupied sub-shell.
   E.g. $Mg^{2+} = 1s^2\ 2s^2\ 2p^6$, $Cl^- = 1s^2\ 2s^2\ 2p^6\ 3s^2\ 3p^6$

See the next page for more on the s and p block.

Watch out — **noble gas symbols**, like that of argon (Ar), are sometimes used in electron configurations.
For example, calcium ($1s^2\ 2s^2\ 2p^6\ 3s^2\ 3p^6\ 4s^2$) can be written as $[Ar]4s^2$, where $[Ar] = 1s^2\ 2s^2\ 2p^6\ 3s^2\ 3p^6$.

Electronic Structure

Transition Metals Behave Unusually

1) **Chromium** (Cr) and **copper** (Cu) are badly behaved. They donate one of their **4s** electrons to the **3d sub-shell**. It's because they're happier with a **more stable** full or half-full d sub-shell.

   Cr atom (24 e$^-$): $1s^2\ 2s^2\ 2p^6\ 3s^2\ 3p^6\ 3d^5\ 4s^1$ Cu atom (29 e$^-$): $1s^2\ 2s^2\ 2p^6\ 3s^2\ 3p^6\ 3d^{10}\ 4s^1$

2) And here's another weird thing about transition metals — when they become **ions**, they lose their **4s** electrons **before** their 3d electrons.

   Fe atom (26 e$^-$): $1s^2\ 2s^2\ 2p^6\ 3s^2\ 3p^6\ 3d^6\ 4s^2$ → Fe^{3+} ion (23 e$^-$): $1s^2\ 2s^2\ 2p^6\ 3s^2\ 3p^6\ 3d^5$

Electronic Structure Decides the Chemical Properties of an Element

The number of **outer shell electrons** decides the chemical properties of an element.

1) The **s block** elements (Groups 1 and 2) have 1 or 2 outer shell electrons. These are easily **lost** to form positive ions with an **inert gas configuration**. E.g. Na — $1s^2\ 2s^2\ 2p^6\ 3s^1$ → Na$^+$ — $1s^2\ 2s^2\ 2p^6$ (the electronic configuration of neon).

2) The elements in Groups 5, 6 and 7 (in the p block) can **gain** 1, 2 or 3 electrons to form negative ions with an **inert gas configuration**. E.g. O — $1s^2\ 2s^2\ 2p^4$ → O^{2-} — $1s^2\ 2s^2\ 2p^6$. Groups 4 to 7 can also **share** electrons when they form covalent bonds.

3) Group 0 (the inert gases) have **completely filled** s and p sub-shells and don't need to bother gaining, losing or sharing electrons — their full sub-shells make them **inert**.

4) The **d block** elements (transition metals) tend to **lose** s and d electrons to form positive ions.

Practice Questions

Q1 Write down the sub-shells in order of increasing energy up to 4p.

Q2 How many electrons would full s, p and d sub-shells contain?

Q3 Chromium and copper don't fill up their shells in the same way as other atoms. Explain the differences.

Q4 Which groups of the Periodic Table tend to gain electrons to form negative ions?

Exam Questions

Q1 Potassium reacts with oxygen to form potassium oxide, K$_2$O.

   a) Give the electron configurations of the K atom and K$^+$ ion. [2 marks]

   b) Using arrow-in-box notation, give the electron configuration of the oxygen atom. [2 marks]

   c) Explain why it is the outer shell electrons, not those in the inner shells, which determine the chemistry of potassium and oxygen. [2 marks]

Q2 This question concerns the electron configurations in atoms and ions.

   a) What is the electron configuration of a manganese atom? [1 mark]

   b) Using arrow-in-box notation, give the electron configuration of the Al^{3+} ion. [2 marks]

   c) Identify the element with the 4th shell configuration $4s^2 4p^2$. [1 mark]

   d) Suggest the identity of an atom, a positive ion and a negative ion with the configuration $1s^2\ 2s^2\ 2p^6\ 3s^2\ 3p^6$. [3 marks]

She shells sub-sells on the shesore...

The way electrons fill up the orbitals is kind of like how strangers fill up seats on a bus. Everyone tends to sit in their own seat till they're forced to share. Except for the huge, scary, smelly man who comes and sits next to you. Make sure you learn the order the sub-shells are filled up, so you can write electron configurations for any atom or ion they throw at you.

Ionisation Energies

This page gets a trifle brain-boggling, so I hope you've got a few aspirin handy...

Ionisation is the Removal of One or More Electrons

When electrons have been removed from an atom or molecule, it's been **ionised**.
The energy you need to remove the first electron is called the **first ionisation energy** (or often just ionisation energy).

> The **first ionisation energy** is the energy needed to remove 1 electron from **each atom** in **1 mole** of **gaseous** atoms to form 1 mole of gaseous 1+ ions.

You can write **equations** for this process — here's the equation for the **first ionisation of oxygen** :

$$O_{(g)} \rightarrow O^+_{(g)} + e^- \quad \text{1st ionisation energy} = +1314 \text{ kJ mol}^{-1}$$

Here are a few rather important points about ionisation energies:

1) You **must** use the gas state symbol, **(g)**, because ionisation energies are measured for gaseous atoms.

2) Always refer to **1 mole** of atoms, as stated in the definition, rather than to a single atom.

3) The **lower** the ionisation energy, the **easier** it is to form an ion.

The Factors Affecting Ionisation Energy are...

 **Nuclear Charge** The **more protons** there are in the nucleus, the more positively charged the nucleus is and the **stronger the attraction** for the electrons.

 Distance from Nucleus Attraction falls off very **rapidly with distance**. An electron **close** to the nucleus will be **much more** strongly attracted than one further away.

Shielding As the number of electrons **between** the outer electrons and the nucleus **increases**, the outer electrons feel less attraction towards the nuclear charge. This lessening of the pull of the nucleus by inner shells of electrons is called **shielding (or screening)**.

A **high ionisation energy** means there's a **high attraction** between the **electron** and the **nucleus**.

Ionisation Energy Decreases Down Group 2

1) This provides **evidence** that electron shells **REALLY DO EXIST**.

2) If each element down Group 2 has an **extra electron shell** compared to the one above, the extra inner shells will **shield** the outer electrons from the attraction of the nucleus.

3) Also, the extra shell means that the outer electrons are **further away** from the nucleus, so the nucleus's attraction will be greatly reduced.

It makes sense that both of these factors will make it **easier** to remove outer electrons, resulting in a **lower ionisation energy**.

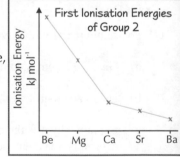

Ionisation Energy Increases Across a Period

The graph below shows the first ionisation energies of the elements in **Period 3**.

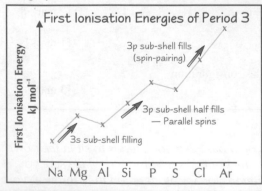

1) As you **move across** a period, the general trend is for the ionisation energies to **increase** — i.e. it gets harder to remove the outer electrons.

2) This can be explained because the number of protons is increasing, which means a stronger **nuclear attraction**.

3) All the extra electrons are at **roughly the same** energy level, even if the outer electrons are in different orbital types.

4) This means there's generally little **extra shielding** effect or **extra distance** to lessen the attraction from the nucleus.

5) But, there are **small drops** between Groups 2 and 3, and 5 and 6. Tell me more, I hear you cry. Well, alright then...

Ionisation Energies

The Drop between Groups 2 and 3 Shows **Sub-Shell Structure**

Mg	$1s^2\,2s^2\,2p^6\,3s^2$	1st ionisation energy = 738 kJ mol^{-1}
Al	$1s^2\,2s^2\,2p^6\,3s^2\,3p^1$	1st ionisation energy = 578 kJ mol^{-1}

1) Aluminium's outer electron is in a **3p orbital** rather than a 3s. The 3p orbital has a **slightly higher** energy than the 3s orbital, so the electron is, on average, to be found **further** from the nucleus.

2) The 3p orbital has additional shielding provided by the **$3s^2$ electrons**.

3) Both these factors together are strong enough to **override** the effect of the increased nuclear charge, resulting in the ionisation energy **dropping** slightly.

4) This pattern in ionisation energies provides **evidence** for the theory of electron sub-shells.

The Drop between Groups 5 and 6 is due to **Electron Repulsion**

P	$1s^2\,2s^2\,2p^6\,3s^2\,3p^3$	1st ionisation energy = 1012 kJ mol^{-1}
S	$1s^2\,2s^2\,2p^6\,3s^2\,3p^4$	1st ionisation energy = 1000 kJ mol^{-1}

1) The **shielding is identical** in the phosphorus and sulfur atoms, and the electron is being removed from an identical orbital.

2) In phosphorus's case, the electron is being removed from a **singly-occupied** orbital. But in sulfur, the **electron** is being **removed** from an orbital containing two electrons.

Phosphorus: (Ne) 3s ⇅ 3p ↑ ↑ ↑   Sulfur: (Ne) 3s ⇅ 3p ⇅ ↑ ↑

The **repulsion** between two electrons in an orbital means that electrons are **easier to remove** from shared orbitals.

3) Yup, yet more **evidence** for the electronic structure model.

Practice Questions

Q1 Define first ionisation energy and give an equation as an example.

Q2 Describe the three main factors that affect ionisation energies.

Q3 When an atom is ionised, does it release or absorb energy?

Q4 Do ionisation energies increase or decrease as you go down Group 2?

Exam Questions

Q1 The first ionisation energies of the elements lithium to neon are given below in kJ mol^{-1}:

Li	Be	B	C	N	O	F	Ne
519	900	799	1090	1400	1310	1680	2080

a) Write an equation, including state symbols, to represent the first ionisation energy of lithium. [2 marks]

b) Explain why the ionisation energies show an overall tendency to increase across the period. [3 marks]

c) Explain the irregularities in this trend for:
   (i) boron
   (ii) oxygen [4 marks]

Q2 First ionisation energy decreases down Group 2.

Explain how this trend provides evidence for the arrangement of electrons in levels. [3 marks]

Shirt crumpled — ionise it...

When you're talking about ionisation energies in exams, always use the three main factors — shielding, nuclear charge and distance from nucleus. Make sure you understand how ionisation energies provide evidence that electron shells and subshells DO exist. They don't prove the model is right, but they do make the scientific community think it's a good'un.

The Mole

It'd be handy to be able to count out atoms — but they're way too tiny. You can't even see them, never mind get hold of them with tweezers. But not to worry — using the idea of relative mass, you can figure out how many atoms you've got.

A **Mole** is Just a (Very Large) **Number of Particles**

1) Amount of substance is measured using a unit called the **mole** (**mol** for short) and given the symbol **n**.

2) One mole is roughly **6×10^{23} particles** (**Avogadro's constant**).

3) It **doesn't matter** what the particles are. They can be atoms, molecules, electrons, ions, penguins — **anything**.

> In the reaction $C + O_2 \rightarrow CO_2$:
> **1 atom** of carbon reacts with **1 molecule** of oxygen to make **1 molecule** of carbon dioxide,
> so **1 mole** of carbon reacts with **1 mole** of oxygen to make **1 mole** of carbon dioxide.

> In the reaction $2Mg + O_2 \rightarrow 2MgO$:
> **2 moles** of magnesium react with **1 mole** of oxygen molecules to make **2 moles** of magnesium oxide.

Molar Mass is the Mass of *One Mole*

Molar mass, **M**, is the mass of **one mole** of something.

But the main thing to remember is:

> **Molar mass is just the same as the relative molecular mass, M_r**
> (or relative formula mass)

That's why the mole is such a ridiculous number of particles (6×10^{23}) — it's the number of particles for which the weight in g is the same as the relative molecular mass.

The only difference is you stick a 'g mol^{-1}' for grams per mole on the end...

> **Example:** Find the molar mass of $CaCO_3$.
> Relative formula mass, M_r, of $CaCO_3$ = $40 + 12 + (3 \times 16) = 100$
> So the molar mass, M, is **100 g mol^{-1}** — i.e. 1 mole of $CaCO_3$ weighs 100 g.

Here's another formula. This one's really important — you need it **all the time**:

> $$\text{Number of moles} = \frac{\text{mass of substance}}{\text{molar mass}}$$

> **Example:** How many moles of aluminium oxide are present in 5.1 g of Al_2O_3?
> Molar mass of Al_2O_3 $= (2 \times 27) + (3 \times 16)$
> $= 102$ g mol^{-1}
> Number of moles of Al_2O_3 $= \frac{5.1}{102} = $ **0.05 moles**

The **Concentration** of a Solution is Measured in **mol dm^{-3}**

1) The **concentration** of a solution is how many **moles** are dissolved per **1 dm^3** of solution. The units are **mol dm^{-3}** (or M).

2) Here's the formula to find the **number of moles**.

1 dm^3 = 1000 cm^3 = 1 litre

> $$\text{Number of moles} = \frac{\text{Concentration} \times \text{Volume (in cm}^3)}{1000}$$

or just

> **Number of moles = Concentration × Volume (in dm^3)**

> **Example:** What mass of sodium hydroxide needs to be dissolved in 50 cm^3 of water to make a 2 M solution?
> $$\text{Number of moles} = \frac{2 \times 50}{1000} = 0.1 \text{ moles of NaOH}$$
> Molar mass, M, of NaOH = $23 + 16 + 1 = 40$ g mol^{-1}
> Mass = number of moles × M = $0.1 \times 40 = $ **4 g**

The Mole

All Gases Take Up the **Same Volume** under the Same Conditions

If temperature and pressure stay the same, **one mole** of **any** gas always has the **same volume**.
At **room temperature and pressure** (r.t.p.), this happens to be **24 dm³**, (r.t.p is 298 K (25 °C) and 100 kPa).
Here are two formulas for working out the number of moles in a volume of gas. Don't forget — **ONLY** use them for r.t.p.

$$\text{Number of moles} = \frac{\text{Volume in dm}^3}{24} \qquad \text{OR} \qquad \text{Number of moles} = \frac{\text{Volume in cm}^3}{24\,000}$$

Example: How many moles are there in 6 dm³ of oxygen gas at r.t.p.?

$$\text{Number of moles} = \frac{6}{24} = \textbf{0.25 moles of oxygen molecules}$$

Ideal Gas equation — $pV = nRT$

In the real world (and AQA exam questions), it's not always room temperature and pressure.
The **ideal gas equation** lets you find the **number of moles** in a certain volume at **any temperature and pressure**.

$pV = nRT$ 　　　　Where:　p = pressure (Pa)
　　　　　　　　　　　　　　　V = volume (m³)
　　　　　　　　　　　　　　　n = number of moles
The gas constant. Don't worry ⟶ $R = 8.31$ J K⁻¹mol⁻¹
about what it means. Just learn it.　　T = temperature (K)

$1\ cm^3 = 1 \times 10^{-6}\ m^3$
$1\ dm^3 = 1 \times 10^{-3}\ m^3$
$K = °C + 273$

Example:

At a temperature of 60 °C and a pressure of 250 kPa, a gas occupied a volume of 1100 cm³ and had a mass of 1.6 g.
Find its relative molecular mass.

$$n = \frac{pV}{RT} = \left(\frac{(250 \times 10^3) \times (1.1 \times 10^{-3})}{8.31 \times 333} \right) = 0.1 \text{ moles}$$

$1100\ cm^3 = 1.1 \times 10^{-3}\ m^3$

If 0.1 moles is 1.6 g, then 1 mole $= \frac{1.6}{0.1} = 16$ g. So the relative molecular mass (M_r) is **16**.

Practice Questions

Q1 How many molecules are there in one mole of ethane molecules?

Q2 What volume does 1 mole of gas occupy at r.t.p.?

Q3 Write down the ideal gas equation.

Exam Questions

Q1 Calculate the mass of 0.36 moles of ethanoic acid, CH_3COOH. 　　　　　　　　　　　　[2 marks]

Q2 What mass of H_2SO_4 is needed to produce 60 cm³ of 0.25 M solution? 　　　　　　　　　[2 marks]

Q3 What volume will be occupied by 88 g of propane gas (C_3H_8)
　　a)　at r.t.p.? 　　　　　　　　　　　　　　　　　　　　　　　　　　　　　　　　　　[2 marks]
　　b)　at 35 °C and 100 kPa? 　　　　　　　　　　　　　　　　　　　　　　　　　　　[2 marks]

Put your back teeth on the scale and find out your molar mass...

You need this stuff for loads of the calculation questions you might get, so make sure you know it inside out. Before you start plugging numbers into formulas, make sure they're in the right units. If they're not, you need to know how to convert them or you'll be tossing marks out the window.   Learn all the definitions and formulas, then have a bash at the questions.

Equations and Calculations

Balancing equations'll cause you a few palpitations — as soon as you make one bit right, the rest goes pear-shaped.

Balanced Equations have **Equal Numbers** of each Atom on **Both Sides**

1) Balanced equations have the **same number** of each atom on **both** sides. They're.. well... you know... balanced.

2) You can only add more atoms by adding **whole compounds**. You do this by putting a number **in front** of a compound or changing one that's already there. You **can't** mess with formulas — ever.

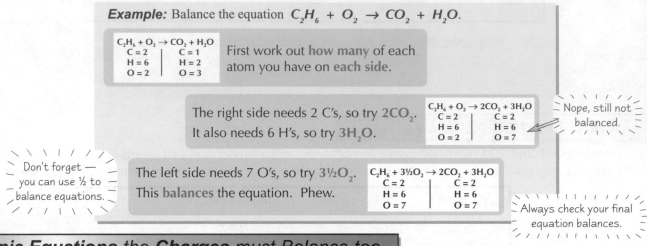

Example: Balance the equation $C_2H_6 + O_2 \rightarrow CO_2 + H_2O$.

$C_2H_6 + O_2 \rightarrow CO_2 + H_2O$

C = 2	C = 1
H = 6	H = 2
O = 2	O = 3

First work out **how many** of each atom you have on **each side**.

The right side needs 2 C's, so try $2CO_2$. It also needs 6 H's, so try $3H_2O$.

$C_2H_6 + O_2 \rightarrow 2CO_2 + 3H_2O$

C = 2	C = 2
H = 6	H = 6
O = 2	O = 7

Nope, still not balanced.

Don't forget — you can use ½ to balance equations.

The left side needs 7 O's, so try $3\frac{1}{2}O_2$. This **balances** the equation. Phew.

$C_2H_6 + 3\frac{1}{2}O_2 \rightarrow 2CO_2 + 3H_2O$

C = 2	C = 2
H = 6	H = 6
O = 7	O = 7

Always check your final equation balances.

In **Ionic Equations** the **Charges** must Balance too

In ionic equations, only the **reacting particles** are included. You don't have to worry about the rest of the stuff.

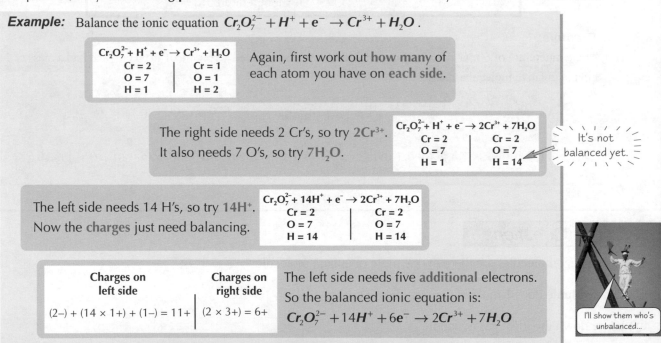

Example: Balance the ionic equation $Cr_2O_7^{2-} + H^+ + e^- \rightarrow Cr^{3+} + H_2O$.

$Cr_2O_7^{2-} + H^+ + e^- \rightarrow Cr^{3+} + H_2O$

Cr = 2	Cr = 1
O = 7	O = 1
H = 1	H = 2

Again, first work out **how many** of each atom you have on **each side**.

The right side needs 2 Cr's, so try $2Cr^{3+}$. It also needs 7 O's, so try $7H_2O$.

$Cr_2O_7^{2-} + H^+ + e^- \rightarrow 2Cr^{3+} + 7H_2O$

Cr = 2	Cr = 2
O = 7	O = 7
H = 1	H = 14

It's not balanced yet.

The left side needs 14 H's, so try $14H^+$. Now the **charges** just need balancing.

$Cr_2O_7^{2-} + 14H^+ + e^- \rightarrow 2Cr^{3+} + 7H_2O$

Cr = 2	Cr = 2
O = 7	O = 7
H = 14	H = 14

Charges on left side	Charges on right side
$(2-) + (14 \times 1+) + (1-) = 11+$	$(2 \times 3+) = 6+$

The left side needs five **additional** electrons. So the balanced ionic equation is:

$Cr_2O_7^{2-} + 14H^+ + 6e^- \rightarrow 2Cr^{3+} + 7H_2O$

I'll show them who's unbalanced...

Balanced Equations can be used to Work out Masses

Example: Calculate the mass of iron oxide produced if 28 g of iron is burnt in air.

$$2Fe + \frac{3}{2}O_2 \rightarrow Fe_2O_3$$

The molar mass, M, of Fe = 56 g mol⁻¹, so the number of moles in 28 g of Fe = $\frac{mass}{M} = \frac{28}{56} = 0.5$ moles

From the equation: 2 moles of Fe produces 1 mole of Fe_2O_3, so 0.5 moles of Fe produces 0.25 moles of Fe_2O_3.

Once you know the number of moles and the molar mass (M) of Fe_2O_3, it's easy to work out the mass.

M of $Fe_2O_3 = (2 \times 56) + (3 \times 16) = 160$ g mol⁻¹

Mass of Fe_2O_3 = no. of moles × M = 0.25 × 160 = **40 g**. And that's your answer.

Equations and Calculations

That's not all... *Balanced Equations* can be used to *Work Out Gas Volumes*

It's pretty handy to be able to work out **how much gas** a reaction will produce, so that you can use **large enough apparatus**. Or else there might be a rather large bang.

Example: How much gas is produced when 15 g of sodium is reacted with excess water at r.t.p.?

$$2Na_{(s)} + 2H_2O_{(l)} \rightarrow 2NaOH_{(aq)} + H_{2(g)}$$

M of Na = 23 g mol^{-1}, so number of moles in 15 g of Na = $\frac{15}{23}$ = 0.65 moles

From the equation, 2 moles Na produces 1 mole H$_2$,

so you know 0.65 moles Na produces $\frac{0.65}{2}$ = 0.326 moles H$_2$.

So the volume of H$_2$ = 0.326 × 24 = **7.8 dm³**

'Excess water' means you know all the sodium will react.

The reaction happens at room temperature and pressure, so you know 1 mole takes up 24 dm³.

State Symbols *Give a bit More Information about the Substances*

State symbols are put after each compound in an equation. They tell you what **state of matter** things are in.

s = solid
l = liquid
g = gas
aq = aqueous
   (solution in water)

To show you what I mean, here's an example —

$$CaCO_{3\,(s)} + 2HCl_{(aq)} \rightarrow CaCl_{2\,(aq)} + H_2O_{(l)} + CO_{2\,(g)}$$

solid   aqueous   aqueous   liquid   gas

Practice Questions

Q1 What is the state symbol for a solution of hydrochloric acid?

Q2 What is the difference between a full, balanced equation and an ionic equation?

Exam Questions

Q1 Calculate the mass of ethene required to produce 258 g of chloroethane, C$_2$H$_5$Cl.
$$C_2H_4 + HCl \rightarrow C_2H_5Cl$$
[4 marks]

Q2 15 g of calcium carbonate is heated strongly so that it fully decomposes. $CaCO_{3(s)} \rightarrow CaO_{(s)} + CO_{2(g)}$

a) Calculate the mass of calcium oxide produced. [3 marks]

b) Calculate the volume of gas produced. [3 marks]

Q3 Balance this equation: $KI + Pb(NO_3)_2 \rightarrow PbI_2 + 2KNO_3$ [1 mark]

Don't get in a state about equations...

Balancing equations is really, really important to the whole of AS Chemistry, so hang in there, and make sure you can do it. You can ONLY calculate reacting masses and gas volumes if you've got a balanced equation to work from. I've said it once, and I'll say it again — practise, practise, practise...it's the only road to salvation (by the way, where is salvation anyway?).

Titrations

*Titrations are used to find out the **concentration** of acid or alkali solutions. You're likely to have to do a **titration** for your **Practical Skills Assessment**, and even if you don't, you still need to know how to use titration results in **calculations**.*

Titrations need to be done **Accurately**

1) **Titrations** allow you to find out **exactly** how much acid is needed to **neutralise** a quantity of alkali.

2) You measure out some **alkali** using a pipette and put it in a flask, along with some **indicator**, e.g. **phenolphthalein**.

3) First of all, do a rough titration to get an idea where the **end point** is (the point where the alkali is **exactly neutralised** and the indicator changes colour). Add the **acid** to the alkali using a **burette** — giving the flask a regular **swirl**.

4) Now do an **accurate** titration. Run the acid in to within 2 cm³ of the end point, then add the acid **dropwise**. If you don't notice exactly when the solution changed colour you've **overshot** and your result won't be accurate.

5) **Record** the amount of acid used to **neutralise** the alkali. It's best to **repeat** this process a few times, making sure you get the same answer each time. This'll make sure your results are **reliable**.

Pipette
Pipettes measure only one volume of solution. Fill the pipette to just above the line, then take the pipette out of the solution. Now drop the level down carefully to the line.

Burette
Burettes measure different volumes and let you add the solution drop by drop.

acid

scale

alkali and indicator

You can also do titrations the other way round — adding alkali to acid.

You can Calculate **Concentrations** from Titrations

Now for the calculations...

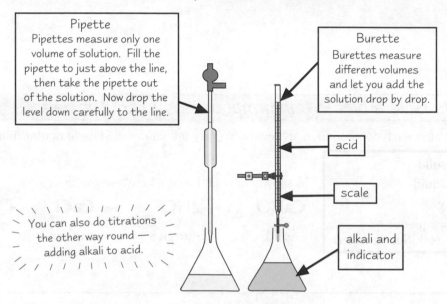

Example: 25 cm³ of 0.5 M HCl was used to neutralise 35 cm³ of NaOH solution.

Calculate the concentration of the sodium hydroxide solution in mol dm⁻³.

First write a **balanced equation** and decide **what you know** and what you **need to know**:

$$HCl + NaOH \rightarrow NaCl + H_2O$$

25 cm³   35 cm³
0.5 M   ?

It's just the formula from page 14.

Now work out how many **moles of HCl** you have:

$$\text{Number of moles HCl} = \frac{\text{concentration} \times \text{volume (cm}^3)}{1000} = \frac{0.5 \times 25}{1000} = 0.0125 \text{ moles}$$

From the equation, you know 1 mole of HCl neutralises 1 mole of NaOH.
So 0.0125 moles of HCl must neutralise **0.0125** moles of NaOH.

Now it's a doddle to work out the **concentration of NaOH**.

$$\text{Concentration of NaOH}_{(aq)} = \frac{\text{moles of NaOH} \times 1000}{\text{volume (cm}^3)} = \frac{0.0125 \times 1000}{35} = \textbf{0.36 mol dm}^{-3}$$

Titrations

You use a *Pretty Similar Method* to Calculate *Volumes* for Reactions

This is usually used for **planning experiments**.

You need to use this formula again, but this time **rearrange** it to find the volume.   ⟶ $$\text{number of moles} = \frac{\text{concentration} \times \text{volume (cm}^3)}{1000}$$

Example: 20.4 cm³ of a 0.5 M solution of sodium carbonate reacts with 1.5 M nitric acid. Calculate the volume of nitric acid required to neutralise the sodium carbonate.

Like before, first write a **balanced equation** for the reaction and decide **what you know** and what you **want to know**:

$$Na_2CO_3 + 2HNO_3 \rightarrow 2NaNO_3 + H_2O + CO_2$$

20.4 cm³   **?**

0.5 M   **1.5 M**

Now work out how many **moles** of Na_2CO_3 you've got.

$$\text{No. of moles of } Na_2CO_3 = \frac{\text{concentration} \times \text{volume (cm}^3)}{1000} = \frac{0.5 \times 20.4}{1000} = 0.0102 \text{ moles}$$

1 mole of Na_2CO_3 neutralises 2 moles of HNO_3, so 0.0102 moles of Na_2CO_3 neutralises **0.0204 moles of HNO_3**.

Now you know the number of moles of HNO_3 and the concentration, you can work out the **volume**:

$$\text{Volume of } HNO_3 = \frac{\text{number of moles} \times 1000}{\text{concentration}} = \frac{0.0204 \times 1000}{1.5} = \textbf{13.6 cm}^3$$

Practice Questions

Q1 Explain what a titration is.

Q2 Write down the formula for calculating number of moles from the concentration and volume of a solution.

Q3 Rearrange this formula so that you could use it to calculate concentration. Then do the same for volume.

Exam Questions

Q1 Calculate the concentration in mol dm⁻³ of a solution of ethanoic acid, CH_3COOH, if 25.4 cm³ of it is neutralised by 14.6 cm³ of 0.5 M sodium hydroxide solution. $CH_3COOH + NaOH \rightarrow CH_3COONa + H_2O$ [3 marks]

Q2 You are supplied with 0.75 g of calcium carbonate and a solution of 0.25 M sulfuric acid. What volume of acid will be needed to neutralise the calcium carbonate?
$CaCO_3 + H_2SO_4 \rightarrow CaSO_4 + H_2O + CO_2$ [4 marks]

Burettes and pipettes — big glass things, just waiting to be dropped...

Titrations are fiddly. But you do get to use big, impressive-looking equipment and feel like you're doing something important. Of course, then there's the results to do calculations with. The best way to start is always to write out the balanced equation and put what you know about each substance underneath it. Then think about what you're trying to find out.

Formulas, Yield and Atom Economy

Here's another page piled high with numbers — it's all just glorified maths really.

Empirical and Molecular Formulas are Ratios

You have to know what's what with empirical and molecular formulas, so here goes...

1) The **empirical formula** gives just the smallest whole number ratio of atoms in a compound.

2) The **molecular formula** gives the **actual** numbers of atoms in a molecule.

3) The molecular formula is made up of a whole **number** of empirical units.

Example: A molecule has an empirical formula of $C_4H_3O_2$, and a molecular mass of 166 g. Work out its molecular formula.

Empirical mass is just like the relative formula mass... (if that helps at all...).

First find the empirical mass — $(4 \times 12) + (3 \times 1) + (2 \times 16)$
$= 48 + 3 + 32 = 83$ g

Compare the empirical and molecular masses.

But the molecular mass is 166 g, so there are $\frac{166}{83} = 2$ empirical units in the molecule.

The molecular formula must be the empirical formula × 2, so the molecular formula = $C_8H_6O_4$.

Empirical Formulas can be Calculated from Percentage Composition

You need to know how to work out empirical formulas from the **percentages** of the different elements.

Example: A compound is found to have percentage composition 56.5% potassium, 8.7% carbon and 34.8% oxygen by mass. Calculate its empirical formula.

If you assume you've got 100 g of the compound, you can turn the % straight into mass, and then work out the number of moles as normal.

In **100 g** of compound there are:

 Use $n = \frac{mass}{M}$

$\frac{56.5}{39} = 1.449$ moles of K $\frac{8.7}{12} = 0.725$ moles of C $\frac{34.8}{16} = 2.175$ moles of O

Divide each number of moles by the **smallest number** — in this case it's 0.725.

K: $\frac{1.449}{0.725} = 2.0$ C: $\frac{0.725}{0.725} = 1.0$ O: $\frac{2.175}{0.725} = 3.0$

The ratio of K : C : O = 2 : 1 : 3. So you know the empirical formula's got to be K_2CO_3.

Percentage Yield Is Never 100%

1) The **theoretical yield** is the **mass of product** that **should** be formed in a chemical reaction. It assumes **no** chemicals are 'lost' in the process. You can use the **masses of reactants** and a **balanced equation** to calculate the theoretical yield for a reaction.

Example: 1.40 g of iron filings is reacted with ammonia and sulfuric acid to make hydrated ammonium iron(II) sulfate.
$$Fe_{(s)} + 2NH_{3\,(aq)} + 2H_2SO_{4\,(aq)} + 6H_2O_{(l)} \rightarrow (NH_4)_2Fe(SO_4)_2.6H_2O_{(s)} + H_{2\,(g)}$$
Calculate the theoretical yield.

Number of moles of **iron** ($A_r = 56$) reacted = mass ÷ molar mass = 1.40 ÷ 56 = **0.025 moles**.

From the equation, 'moles of iron : moles of ammonium iron(II) sulfate' is 1 : 1, so 0.025 moles of product should form.

Molar mass of $(NH_4)_2Fe(SO_4)_2.6H_2O_{(s)}$ = 392, so **theoretical yield** = 0.025 × 392 = **9.8 g**.

2) For any reaction, the **actual** mass of product (the **actual yield**) will always be **less** than the theoretical yield. There are many reasons for this. For example, sometimes not all the 'starting' chemicals react fully. And some chemicals are always 'lost', e.g. some solution gets left on filter paper, or is lost during transfers between containers.

3) Once you've found the **theoretical yield** and the **actual yield**, you can work out the **percentage yield**. $\Longrightarrow$

$$\text{Percentage Yield} = \frac{\text{Actual Yield}}{\text{Theoretical Yield}} \times 100$$

4) So, in the ammonium iron(II) sulfate example above, the theoretical yield was 9.8 g. Say you weighed the hydrated ammonium iron(II) sulfate crystals produced and found the actual yield was **5.2 g**. Then

Percentage yield = (5.2 ÷ 9.8) × 100 = **53%**

Formulas, Yield and Atom Economy

Atom Economy is a Measure of the Efficiency of a Reaction

1) The **efficiency** of a reaction is often measured by the **percentage yield**. This tells you how wasteful the **process** is — it's based on how much of the product is lost because of things like reactions not completing or losses during collection and purification.

2) But percentage yield doesn't measure how wasteful the **reaction** itself is. A reaction that has a 100% yield could still be very wasteful if a lot of the atoms from the **reactants** wind up in **by-products** rather than the **desired product**.

3) **Atom economy** is a measure of the proportion of reactant **atoms** that become part of the desired product (rather than by-products) in the **balanced** chemical equation. It's calculated using this formula:

$$\% \text{ atom economy} = \frac{\text{mass of desired product}}{\text{total mass of reactants}} \times 100$$

You can use the masses in grams, or their relative molecular masses.

Example: Bromomethane is reacted with sodium hydroxide to make methanol:

$$CH_3Br + NaOH \rightarrow CH_3OH + NaBr$$

Calculate the atom economy for this reaction.

Always make sure you're using a balanced equation.

$$\% \text{ atom economy} = \frac{\text{mass of desired product}}{\text{total mass of reactants}} \times 100$$

$$= \frac{(12+(3\times1)+16+1)}{(12+(3\times1)+80)+(23+16+1)} \times 100 = \frac{32\,g}{135\,g} \times 100 = \mathbf{23.7\%}$$

The relative molecular masses have been used here. You need to use the numbers of moles from the balanced equation.

Practice Questions

Q1 Define 'empirical formula'.

Q2 What is the difference between a molecular formula and an empirical formula?

Q3 Give two examples of how chemicals could be 'lost' during a reaction.

Q4 Write down the formula for calculating percentage yield.

Q5 What is the difference between percentage yield and atom economy?

Exam Questions

Q1 Hydrocarbon X has a molecular mass of 78 g. It is found to have 92.3% carbon and 7.7% hydrogen by mass. Calculate the empirical and molecular formulae of X. [3 marks]

Q2 Phosphorus trichloride (PCl_3) reacts with chlorine to give phosphorus pentachloride (PCl_5):

$$PCl_3 + Cl_2 \rightleftharpoons PCl_5$$

a) If 0.275 g of PCl_3 reacts with 0.142 g of chlorine, what is the theoretical yield of PCl_5? [2 marks]

b) When this reaction is performed 0.198 g of PCl_5 is collected. Calculate the percentage yield. [1 mark]

c) Changing conditions such as temperature and pressure will alter the percentage yield of this reaction. Will changing these conditions affect the atom economy? Explain your answer. [2 marks]

The Empirical Strikes Back...

With this stuff, it's not enough to learn a few facts parrot-fashion, to regurgitate in the exam — you've gotta know how to use them. The only way to do that is to practise. Go through all the examples on these two pages again, this time working the answers out for yourself. Then test yourself on the practice exam questions. It'll help you sleep at night — honest.

Ionic Bonding

Every atom's aim in life is to have a full outer shell of electrons. Once they've managed this, that's it — they're happy.

Compounds are Atoms of Different Elements Bonded Together

1) When different elements join or bond together, you get a **compound**.

2) There are two main types of bonding in compounds — **ionic** and **covalent**. You need to make sure you've got them **both** totally sussed.

E.g. when the elements hydrogen (H_2) and oxygen (O_2) combine, the compound water (H_2O) is formed.

Ionic Bonding is when Ions are Stuck Together by Electrostatic Attraction

1) Ions are formed when electrons are **transferred** from one atom to another.

2) The simplest ions are single atoms which have either lost or gained 1, 2 or 3 electrons so that they've got a **full outer shell**. Here are some examples of ions:

> A sodium atom (Na) **loses** 1 electron to form a sodium ion (Na^+) $Na \rightarrow Na^+ + e^-$
> A magnesium atom (Mg) **loses** 2 electrons to form a magnesium ion (Mg^{2+}) $Mg \rightarrow Mg^{2+} + 2e^-$
> A chlorine atom (Cl) **gains** 1 electron to form a chloride ion (Cl^-) $Cl + e^- \rightarrow Cl^-$
> An oxygen atom (O) **gains** 2 electrons to form an oxide ion (O^{2-}) $O + 2e^- \rightarrow O^{2-}$

3) You **don't** have to remember what ion **each element** forms — nope, for many of them you just look at the Periodic Table. Elements in the same **group** all have the same number of **outer electrons**. So they have to **lose or gain** the same number to get the full outer shell that they're aiming for. And this means that they form ions with the **same charges**.

4) **Electrostatic attraction** holds positive and negative ions together — it's **very** strong. When atoms are held together like this, it's called **ionic bonding**.

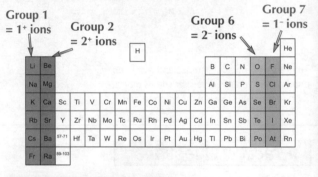

Sodium Chloride and Magnesium Oxide are Ionic Compounds

1) The formula of sodium chloride is **NaCl**. It just tells you that sodium chloride is made up of **Na^+ ions** and **Cl^- ions** (in a 1:1 ratio).

2) You can use '**dot-and-cross**' diagrams to show how ionic bonding works in sodium chloride —

Here, the dots represent the Na electrons and the crosses represent the Cl electrons (all electrons are really identical, but this is a good way of following their movement).

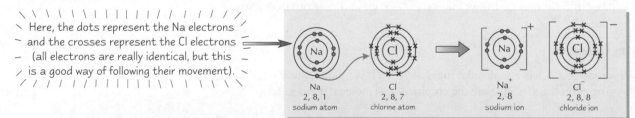

Na
2, 8, 1
sodium atom

Cl
2, 8, 7
chlorine atom

Na^+
2, 8
sodium ion

Cl^-
2, 8, 8
chloride ion

3) **Magnesium oxide**, MgO, is another good example:

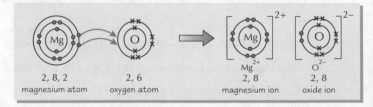

2, 8, 2
magnesium atom

2, 6
oxygen atom

Mg^{2+}
2, 8
magnesium ion

O^{2-}
2, 8
oxide ion

Dot (cross)

The positive charges in the compound **balance** the negative charges exactly — so the total overall charge is **zero**. This is a dead handy way of checking the formula.

- In **NaCl**, the single positive charge on the Na^+ ion balances the single negative charge on the Cl^- ion.

- In magnesium chloride, $MgCl_2$, the 2+ charge on the Mg^{2+} ion balances the two individual – charges on the two Cl^- ions.

Ionic Bonding

Sodium Chloride has a *Giant Ionic Lattice* Structure

1) Ionic crystals are giant lattices of ions. A **lattice** is just a **regular structure**.
2) The structure's called '**giant**' because it's made up of the same basic unit repeated over and over again.
3) In **sodium chloride**, the Na^+ and Cl^- ions are packed together. The sodium chloride lattice is **cube** shaped — different ionic compounds have different shaped structures, but they're all still giant lattices.

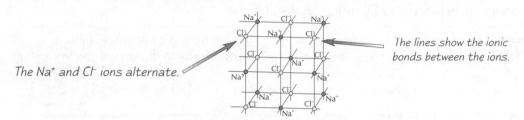

The Na⁺ and Cl⁻ ions alternate.

The lines show the ionic bonds between the ions.

The structure of ionic compounds decides their **physical properties**...

Ionic Structure Explains the *Behaviour* of *Ionic Compounds*

1) **Ionic compounds conduct electricity when they're molten or dissolved — but not when they're solid.**
   The ions in a liquid are free to move (and they carry a charge).
   In a solid they're fixed in position by the strong ionic bonds.

2) **Ionic compounds have high melting points.**
   The giant ionic lattices are held together by strong electrostatic forces. It takes loads of energy to overcome these forces, so melting points are very high (801 °C for sodium chloride).

3) **Ionic compounds tend to dissolve in water.**
   Water molecules are polar — part of the molecule has a small negative charge, and the other bits have small positive charges (see p28). The water molecules pull the ions away from the lattice and cause it to dissolve.

Practice Questions

Q1 What's a compound?

Q2 Draw a dot-and-cross diagram showing the bonding between magnesium and oxygen.

Q3 What type of force holds ionic substances together?

Q4 Do ionic compounds tend to dissolve in water? Why?

Exam Questions

Q1 a) Draw a labelled diagram to show the structure of sodium chloride. [3 marks]

b) What is the name of this type of structure? [1 mark]

c) Would you expect sodium chloride to have a high or a low melting point?
Explain your answer. [4 marks]

Q2 a) Ions can be formed by electron transfer. Explain this and give an example
of a positive and a negative ion. [3 marks]

b) Solid lead(II) bromide does not conduct electricity, but molten lead(II) bromide does.
Explain this with reference to ionic bonding. [3 marks]

Atom 1 says, "I think I lost an electron". Atom 2 replies, "are you positive?"...

Make sure that you can explain why ionic compounds do what they do. Their properties are all down to the fact that ionic crystals are made up of oppositely charged ions attracted to each other. Ionic bonding ONLY happens between a metal and a non-metal. If you've got two non-metal or two metals, they'll do different sorts of bonding — keep reading...

Covalent Bonding

And now for covalent bonding — this is when atoms share electrons with one another so they've all got full outer shells.

Molecules are Groups of Atoms Bonded Together

1) Molecules are the **smallest parts** of compounds that can take part in chemical reactions.
2) They're formed when **two or more** atoms bond together — it doesn't matter if the atoms are the **same** or **different**. Chlorine gas (Cl_2), carbon monoxide (CO), water (H_2O) and ethanol (C_2H_5OH) are all molecules.
3) Molecules are held together by strong **covalent bonds**.

In covalent bonding, two atoms **share** electrons, so they've **both** got **full outer shells** of electrons. Both the positive nuclei are attracted **electrostatically** to the shared electrons.

E.g. two iodine atoms bond covalently to form a molecule of iodine (I_2).

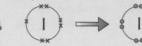

Covalent bonding happens between non-metals. Ionic bonding is between a metal and a non-metal.

Here's some more examples. These diagrams don't show all the electrons — just the ones in the **outer shells**:

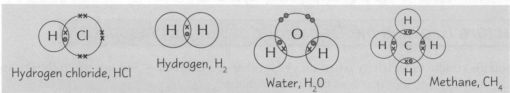

Hydrogen chloride, HCl — Hydrogen, H_2 — Water, H_2O — Methane, CH_4

There are Double and Triple Bonds Too

Atoms don't just form single bonds — **double** or even **triple covalent bonds** can form too.

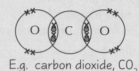

E.g. carbon dioxide, CO_2

Nitrogen, N_2
(nitrogen's a triple-bonder)

It's not over yet — the **typical** properties of simple covalent molecules are covered on page 32.

There are Giant Covalent Structures Too

1) **Giant covalent** structures have a huge network of **covalently** bonded atoms. (They're sometimes called **macromolecular structures**.)
2) **Carbon** atoms can form this type of structure because they can each form **four** strong, covalent bonds. There are two types of giant covalent carbon structure you need to know about, **graphite** and **diamond**.

Graphite — Sheets of Hexagons with Delocalised Electrons

The carbon atoms are arranged in sheets of flat hexagons covalently bonded with three bonds each. The fourth outer electron of each carbon atom is delocalised.

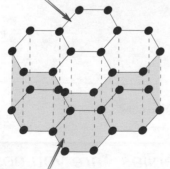

The sheets of hexagons are bonded together by weak van der Waals forces.
(see p29)

Graphite's **structure** means it has **certain properties**:

1) The weak bonds **between** the layers in graphite are easily broken, so the sheets can slide over each other — graphite feels **slippery** and is used as a **dry lubricant** and in **pencils**.
2) The '**delocalised**' electrons in graphite aren't attached to any particular carbon atoms and are **free to move** along the sheets, so an **electric current** can flow.
3) The layers are quite **far apart** compared to the length of the covalent bonds, so graphite has a **low density** and is used to make **strong**, **lightweight** sports equipment.
4) Because of the **strong covalent bonds** in the hexagon sheets, graphite has a **very high melting point** (it sublimes at over 3900 K).
5) Graphite is **insoluble** in any solvent. The covalent bonds in the sheets are **too difficult** to break.

'Sublimes' means it changes straight from a solid to a gas, skipping out the liquid stage.

Covalent Bonding

Diamond is the Hardest Known Substance

Diamond is also made up of **carbon atoms**. Each carbon atom is **covalently bonded** to **four** other carbon atoms. The atoms arrange themselves in a **tetrahedral** shape — its crystal lattice structure.

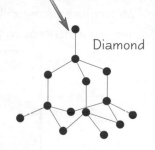

Diamond

Because of its **strong covalent** bonds:

1) Diamond has a **very high melting point** — it actually sublimes at over 3800 K.
2) Diamond is extremely **hard** — it's used in diamond-tipped drills and saws.
3) **Vibrations** travel easily through the stiff lattice, so it's a **good thermal conductor**.
4) It **can't conduct** electricity — all the outer electrons are held in localised bonds.
5) Like graphite, diamond won't dissolve in **any** solvent.

You can 'cut' diamond to form gemstones. Its structure makes it refract light a lot, which is why it sparkles.

Dative Covalent Bonding is where Both Electrons come from One Atom

The **ammonium ion** (NH_4^+) is formed by dative covalent (or coordinate) bonding — it's an example the examiners love. It forms when the nitrogen atom in an ammonia molecule **donates a pair of electrons** to a proton (H^+) —

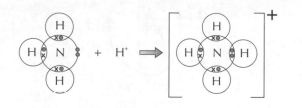 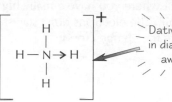

Dative covalent bonding is shown in diagrams by an arrow, pointing away from the 'donor' atom.

Practice Questions

Q1 Does covalent bonding occur between metal atoms or between non-metal atoms?

Q2 Describe how atoms are held together in covalent molecules.

Q3 Draw a dot-and-cross diagram to show the arrangement of the outer electrons in a molecule of iodine.

Q4 How are the carbon sheets in graphite held together?

Q5 In diamond, how many other carbons is each carbon atom bonded to?

Exam Questions

Q1 Methane, CH_4, is an organic molecule.
a) What type of bonding would you expect it to have? [1 mark]
b) Draw a dot-and-cross diagram to show the full electronic arrangement in a molecule of methane. [2 marks]

Q2 a) What type of bonding is present in the ammonium ion? [1 mark]
b) Explain how this type of bonding occurs. [2 marks]

Q3 Carbon can be found as diamond and as graphite.
a) What type of structure do diamond and graphite display? [1 mark]
b) Draw diagrams to illustrate the structures of diamond and graphite. [2 marks]
c) Compare and explain the electrical conductivities of diamond and graphite in terms of their structure and bonding. [4 marks]

Carbon is a girl's best friend...

More pretty diagrams to learn here folks — practise till you get every single dot and cross in the right place. It's totally amazing to think of these titchy little atoms sorting themselves out so they've got full outer shells of electrons. Remember — covalent bonding happens between two non-metals, whereas ionic bonding happens between a metal and a non-metal.

Shapes of Molecules

Chemistry would be heaps more simple if all molecules were flat. But they're not.

Molecular Shape depends on Electron Pairs around the Central Atom

Molecules and molecular ions come in loads of **different shapes**.

The shape depends on the **number of pairs** of electrons in the outer shell of the central atom.

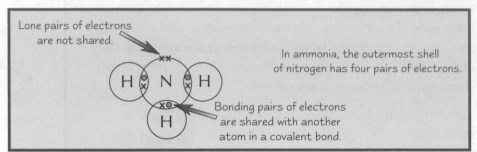

Lone pairs of electrons are not shared.

In ammonia, the outermost shell of nitrogen has four pairs of electrons.

Bonding pairs of electrons are shared with another atom in a covalent bond.

A lone pear

Electron Pairs exist as Charge Clouds

Bonding pairs and lone pairs of electrons exist as **charge clouds**.

A charge cloud is an area where you have a really **big chance** of finding an electron pair. The electrons don't stay still — they **whizz around** inside the charge cloud.

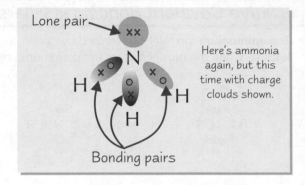

Lone pair

Here's ammonia again, but this time with charge clouds shown.

Bonding pairs

Electron Charge Clouds Repel Each Other

1) Electrons are all **negatively charged**, so it's pretty obvious that the charge clouds will **repel** each other as much as they can.

2) This sounds straightforward, but the **shape** of the charge cloud affects **how much** it repels other charge clouds. Lone-pair charge clouds repel **more** than bonding-pair charge clouds.

3) So, the **greatest** angles are between **lone pairs** of electrons, and bond angles between bonding pairs are often **reduced** because they are pushed together by lone-pair repulsion.

Lone-pair/lone-pair bond angles are the biggest.	*Lone-pair/bonding-pair bond angles are the second biggest.*	*Bonding-pair/bonding-pair bond angles are the smallest.*

4) This is known by the long-winded name '**Valence-Shell Electron-Pair Repulsion Theory**'.

The central atoms in these molecules all have **four pairs** of electrons in their outer shells, but they're all **different shapes**.

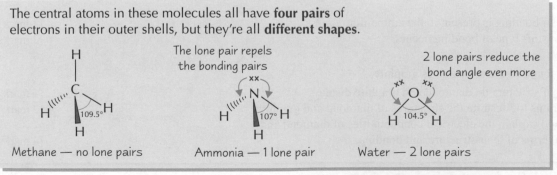

The lone pair repels the bonding pairs

2 lone pairs reduce the bond angle even more

Methane — no lone pairs   Ammonia — 1 lone pair   Water — 2 lone pairs

In a molecule diagram, use wedges to show that a bond sticks out of the page towards you, and a broken (or dotted) line to show a bond goes behind the page.

5) These rules mean that the **shapes and bond angles** of loads of molecules can be predicted.

Shapes of Molecules

Practise **Drawing** these Molecules

2 ELECTRON PAIRS ON CENTRAL ATOM —

Just treat double bonds the same as single bonds (even though there might be slightly more repulsion from a double bond).

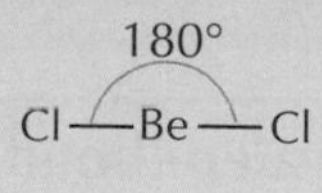

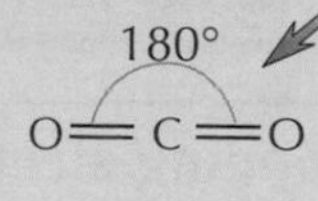

$BeCl_2$ Cl—Be—Cl   180°

CO_2 O=C=O   180°

Linear molecules

3 ELECTRON PAIRS ON CENTRAL ATOM —

BF₃
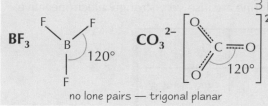
F, B, F, F   120°

no lone pairs — trigonal planar

CO₃²⁻
$$\left[\begin{array}{c} O \\ C = O \\ O \end{array} \right]^{2-}$$ 120°

(in CO_3^{2-} and NO_3^- the bonds are all midway between single and double bonds)

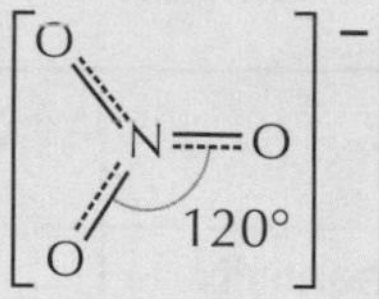

NO₃⁻
$$\left[\begin{array}{c} O \\ N = O \\ O \end{array} \right]^{-}$$ 120°

Here, the extra electron density in the double bonds cancels out the extra repulsion from the lone pair, so you still get 120°.

SO₂
O=S=O   xx   120°

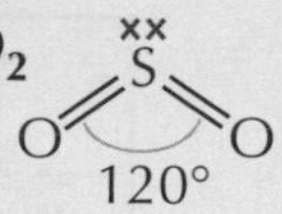

1 lone pair — non-linear or 'bent'

4 ELECTRON PAIRS ON CENTRAL ATOM —

NH₄⁺
H, N⁺, H, H, H   109.5°

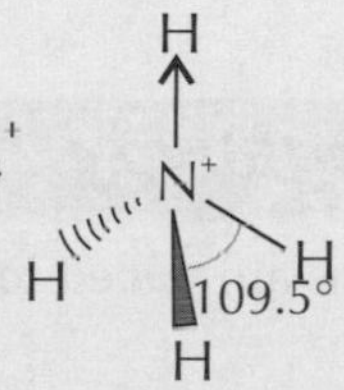

no lone pairs — tetrahedral

NH₃
H, N, H, H   xx   107°

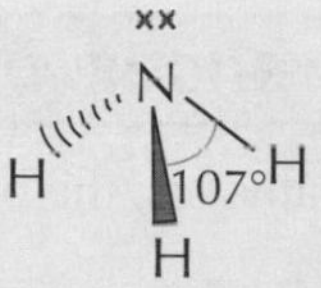

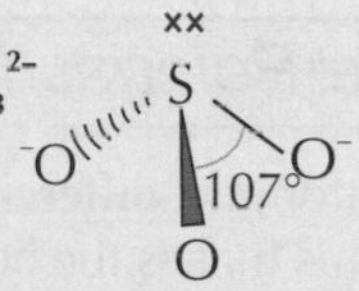

1 lone pair — trigonal pyramidal

SO₃²⁻
⁻O, S, O⁻, O   xx   107°

H₂O
H, O, H   xx xx   104.5°

2 lone pairs — non-linear or 'bent'

Some central atoms can use d orbitals and can 'expand the octet' — which means they can have more than eight bonding electrons. E.g. in PCl_5, phosphorus has 10 electrons in its outermost shell, while in SF_6, sulfur has 12.

5 ELECTRON PAIRS ON CENTRAL ATOM —

PCl₅
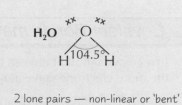
Cl, Cl, P, Cl, Cl, Cl   120°   90°

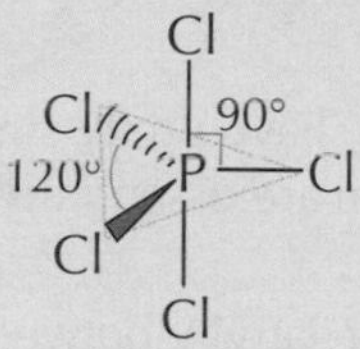

no lone pairs — trigonal bipyramidal

6 ELECTRON PAIRS ON CENTRAL ATOM —

SF₆
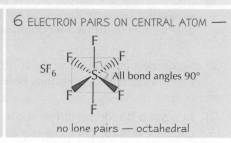
F, F, F, S, F, F, F   All bond angles 90°

no lone pairs — octahedral

Practice Questions

Q1 What is a lone pair of electrons?

Q2 What is a charge cloud?

Q3 Write down the order of the strength of repulsion between different kinds of electron pair.

Q4 Draw an example of a tetrahedral molecule.

Exam Question

Q1 Nitrogen and boron can form the chlorides NCl_3 and BCl_3.

a) Draw dot-and-cross diagrams to show the bonding in NCl_3 and BCl_3. [2 marks]

b) Draw the shapes of the molecules NCl_3 and BCl_3.
Show the approximate values of the bond angles on the diagrams and name each shape. [6 marks]

c) Explain why the shapes of NCl_3 and BCl_3 are different. [3 marks]

These molecules ain't square...

In the exam, those evil examiners might try to throw you by asking you to predict the shape of an unfamiliar molecule. Don't panic — it'll be just like one you do know, e.g. PH_3 is the same shape as NH_3. Make sure you can draw every single molecule on this page. Yep, that's right — from memory. And learn what the shapes are called too.

Polarisation and Intermolecular Forces

Intermolecular forces hold molecules together. They're pretty important, cos we'd all be gassy clouds without them. Some of these intermolecular forces are down to polarisation. So you best make sure you know about that first...

Some Atoms **Attract** Bonding Electrons More than Other Atoms

The ability to attract the bonding electrons in a covalent bond is called electronegativity.

Fluorine is the most electronegative element. Oxygen, nitrogen and chlorine are also very strongly electronegative.

Element	H	C	N	Cl	O	F
Electronegativity (Pauling Scale)	2.1	2.5	3.0	3.0	3.5	4.0

Covalent Bonds may be Polarised by **Differences** in **Electronegativity**

In a covalent bond between two atoms of **different** electronegativities, the bonding electrons are **pulled towards** the more electronegative atom. This makes the bond **polar**.

1) The covalent bonds in diatomic gases (e.g. H_2, Cl_2) are **non-polar** because the atoms have **equal** electronegativities and so the electrons are equally attracted to both nuclei.

2) Some elements, like carbon and hydrogen, have pretty **similar** electronegativities, so bonds between them are essentially **non-polar**.

3) In a **polar bond**, the difference in electronegativity between the two atoms causes a **dipole**. A dipole is a **difference in charge** between the two atoms caused by a shift in **electron density** in the bond.

$$\overset{\delta+}{H} \overset{\circ}{\underset{\times}{——}} \overset{\delta-}{Cl}$$

'δ' (delta) means 'slightly', so 'δ+' means 'slightly positive'.

Permanent polar bonding

4) So what you need to **remember** is that the greater the **difference** in electronegativity, the **more polar** the bond.

Polar Molecules have Permanent Dipole-Dipole Forces

The **δ+** and **δ–** charges on **polar molecules** cause **weak electrostatic forces** of attraction **between** molecules.

E.g. hydrogen chloride gas has polar molecules.

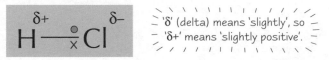

Now this bit's pretty cool:
If you put an **electrostatically charged rod** next to a jet of a polar liquid, like water, the liquid will **move** towards the rod. I wouldn't believe me either, but it's true. It's because **polar liquids** contain molecules with **permanent dipoles**. It doesn't matter if the rod is **positively** or **negatively** charged. The polar molecules in the liquid can **turn around** so the oppositely charged end is attracted towards the rod.

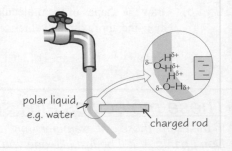

polar liquid, e.g. water

charged rod

Polarisation and Intermolecular Forces

Intermolecular Forces are **Very Weak**

Intermolecular forces are forces **between** molecules. They're much **weaker** than covalent, ionic or metallic bonds. There are three types you need to know about:

1) **Induced dipole-dipole** or **van der Waals** forces (this is the weakest type)
2) **Permanent dipole-dipole forces** (these are the ones that are caused by polar molecules — see the previous page)
3) **Hydrogen bonding** (this is the strongest type)

Van der Waals Forces are Found Between **All** Atoms and Molecules

Van der Waals forces cause **all** atoms and molecules to be **attracted** to each other.

1) **Electrons** in charge clouds are always **moving** really quickly. At any particular moment, the electrons in an atom are likely to be more to one side than the other. At this moment, the atom would have a **temporary dipole**.

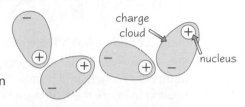

2) This dipole can cause **another** temporary dipole in the opposite direction on a neighbouring atom. The two dipoles are then **attracted** to each other.

3) The second dipole can cause yet another dipole in a **third atom**. It's kind of like a domino rally.

4) Because the electrons are constantly moving, the dipoles are being **created** and **destroyed** all the time. Even though the dipoles keep changing, the **overall effect** is for the atoms to be **attracted** to each another.

Van der Waals Forces Can Hold Molecules in a **Lattice**

Van der Waals forces are responsible for holding **iodine** molecules together in a **lattice**.

1) Iodine atoms are held together in pairs by **strong** covalent bonds to form molecules of I_2.

2) But the molecules are then held together in a **molecular lattice** arrangement by **weak** van der Waals attractions.

Stronger **Van der Waals Forces** mean **Higher Boiling Points**

1) Not all van der Waals forces are the same strength — larger molecules have **larger electron clouds**, meaning **stronger** van der Waals forces.

2) Molecules with greater **surface areas** also have stronger van der Waals forces because they have a **more exposed electron cloud**.

3) When you **boil** a liquid, you need to **overcome** the intermolecular forces, so that the particles can **escape** from the liquid surface. It stands to reason that you need **more energy** to overcome **stronger** intermolecular forces, so liquids with stronger van der Waals forces will have **higher boiling points**.

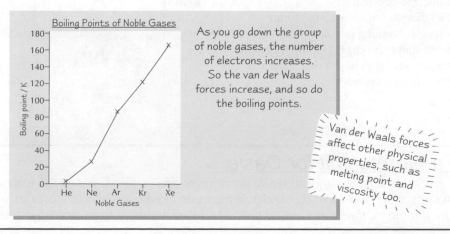

As you go down the group of noble gases, the number of electrons increases. So the van der Waals forces increase, and so do the boiling points.

Van der Waals forces affect other physical properties, such as melting point and viscosity too.

Polarisation and Intermolecular Forces

Hydrogen Bonding *is the* Strongest *Intermolecular Force*

1) Hydrogen bonding **only** happens when **hydrogen** is covalently bonded to **fluorine**, **nitrogen** or **oxygen**.

2) Fluorine, nitrogen and oxygen are very **electronegative**, so they draw the bonding electrons away from the hydrogen atom. The bond is so **polarised**, and hydrogen has such a **high charge density** because it's so small, that the hydrogen atoms form weak bonds with **lone pairs of electrons** on the fluorine, nitrogen or oxygen atoms of **other molecules**.

3) Molecules which have hydrogen bonding are usually **organic**, containing **-OH** or **-NH** groups. **Water** and **ammonia** both have hydrogen bonding.

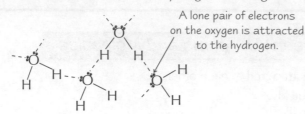

A lone pair of electrons on the oxygen is attracted to the hydrogen.

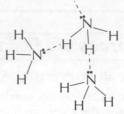

4) Hydrogen bonding has a **huge effect** on the properties of substances.

• Substances with hydrogen bonds have **higher boiling and melting points** than other similar molecules because of the **extra energy** needed to break the hydrogen bonds.

This is the case with **water**, and also **hydrogen fluoride**, which has a much **higher boiling point** than the other hydrogen halides.

• Ice has more hydrogen bonds than liquid water, and hydrogen bonds are relatively **long**. So the H_2O molecules in ice are further apart on average, making ice **less dense** than liquid water.

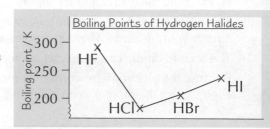

Practice Questions

Q1 What are the only bonds which can be purely non-polar?

Q2 What is the most electronegative element?

Q3 What is a dipole?

Q4 What's the strongest type of intermolecular force?

Q5 What is a hydrogen bond?

Exam Questions

Q1 Many covalent molecules have a permanent dipole, due to differences in electronegativities.
   a) Define the term electronegativity. [2 marks]
   b) Draw the shapes of the following molecules and mark any bond polarities clearly on your diagrams:
      (i) Br_2 (ii) H_2O (iii) NH_3 [5 marks]

Q2 a) Name three types of intermolecular force. [3 marks]
   b) Draw a clearly labelled diagram to show all the forms of intra- and intermolecular bonding in water. [4 marks]
   c) This graph shows the boiling points of the Group 6 hydrides. Explain why water's boiling point is higher than expected in comparison to other Group 6 hydrides? [2 marks]

Intra-molecular bonding is bonding inside molecules.

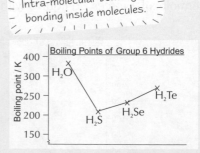

Van der Waal — a German hit for Oasis...

Just because intermolecular forces are a bit wimpy and weak, don't forget they're there. It'd all fall apart without them. Learn the three types — van der Waals, permanent dipole-dipole forces and hydrogen bonds. I bet fish are glad that water forms hydrogen bonds. If it didn't, their water would boil. (And they wouldn't have evolved in the first place.)

Metallic Bonding and Properties of Structures

Lots of this stuff you should already be able to recite in your sleep, but just in case it's fallen out of your brain, here it is...

Metals have Giant Structures

Metal elements exist as **giant metallic lattice structures**.

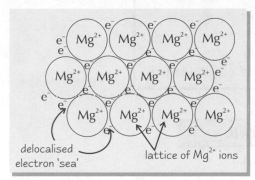

delocalised electron 'sea'

lattice of Mg^{2+} ions

1) The outermost shell of electrons of a metal atom is **delocalised** — the electrons are free to move about the metal. This leaves a **positive metal ion**, e.g. Na^+, Mg^{2+}, Al^{3+}.

2) The positive metal ions are **attracted** to the delocalised negative electrons. They form a lattice of closely packed positive ions in a **sea** of delocalised electrons — this is **metallic bonding**.

Metallic bonding explains why metals do what they do —

1) The **number of delocalised electrons per atom** affects the melting point. The **more** there are, the **stronger** the bonding will be and the **higher** the melting point. Mg^{2+} has **two** delocalised electrons per atom, so it's got a **higher melting point** than Na^+, which only has **one**. The **size** of the metal ion and the **lattice structure** also affect the melting point.

2) As there are **no bonds** holding specific ions together, the metal ions can slide over each another when the structure is pulled, so metals are **malleable** (can be shaped) and **ductile** (can be drawn into a wire).

3) The delocalised electrons can pass **kinetic energy** to each other, making metals **good thermal conductors**.

4) Metals are **good electrical conductors** because the **delocalised electrons** can carry a **current**.

5) Metals are **insoluble**, except in **liquid metals**, because of the **strength** of the metallic bonds.

The Physical Properties of Solids, Liquids and Gases Depend on Particles

1) A typical **solid** has its particles very **close** together. This gives it a high density and makes it **incompressible**. The particles **vibrate** about a **fixed point** and can't move about freely.

2) A typical **liquid** has a similar density to a solid and is virtually **incompressible**. The particles move about **freely** and **randomly** within the liquid, allowing it to flow.

3) In **gases**, the particles have **loads more** energy and are much **further apart**. So the density is generally pretty low and it's **very compressible**. The particles move about **freely**, with not a lot of attraction between them, so they'll quickly **diffuse** to fill a container.

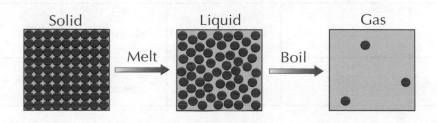

Solid — Melt → Liquid — Boil → Gas

The jelly state* occurs in solids when the particles start feeling a bit tired and achy.

*Don't write this in the exam, cos I just made it up, like...

Metallic Bonding and Properties of Structures

The Physical Properties of a **Solid** Depend on the **Nature** of its Particles

Here are some handy points that'll make AS chemistry a little less painful —

1) **Melting** and **boiling** points depend on **attraction** between particles.
2) The **closer** the particles, the **greater** the density.
3) If there are **charged** particles that are **free** to move, then it'll conduct electricity.
4) Solubility depends on the **type** of particles present.
   Water is a polar solvent and it tends to only dissolve other polar substances.
5) If a solid has a regular structure, it's called a **crystal**. The structure is a **crystal lattice**.

Covalent Bonds **Don't** Break during **Melting** and **Boiling***

This is something that confuses loads of people — prepare to be enlightened...

1) To **melt** or **boil** a simple covalent compound you only have to overcome the **van der Waals forces** or **hydrogen bonds** that hold the molecules together.
2) You **don't** need to break the much stronger covalent bonds that hold the atoms together in the molecules.
3) That's why simple covalent compounds have relatively **low melting** and **boiling points**. For example:

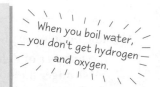
When you boil water, you don't get hydrogen and oxygen.

Chlorine, Cl_2, has **stronger** covalent bonds than bromine, Br_2.
But under normal conditions, chlorine is a **gas** and bromine a **liquid**.
Bromine has the higher boiling point because its molecules have **more electrons**, giving stronger van der Waals forces.

*Except for giant molecular substances, like diamond.

Learn the **Properties** of the Main Substance Types

Make sure you know this stuff like the back of your spam —

Bonding	Examples	Melting and boiling points	Typical state at STP	Does solid conduct electricity?	Does liquid conduct electricity?	Is it soluble in water?
Ionic	NaCl $MgCl_2$	High	Solid	No (ions are held firmly in place)	Yes (ions are free to move)	Yes
Simple molecular (covalent)	CO_2 I_2 H_2O	Low (have to overcome van der Waals forces or hydrogen bonds, not covalent bonds)	May be solid (like I_2), but usually liquid or gas (water is liquid because it has hydrogen bonds)	No	No	Depends on how polarised the molecule is
Giant molecular (covalent)	Diamond Graphite SiO_2	High	Solid	No (except graphite)	— (will generally sublime)	No
Metallic	Fe Mg Al	High	Solid	Yes (delocalised electrons)	Yes (delocalised electrons)	No

Metallic Bonding and Properties of Structures

Bonding Models Match Observations

Scientists develop **models** based on **experimental evidence** — they're an attempt to **explain observations**. Bonding models explain how substances behave.

E.g. the **physical properties** of ionic compounds provide evidence that supports the theory of ionic bonding.
1) They have **high melting points** — this tells you that the atoms are held together by a **strong attraction**. Positive and negative ions are strongly attracted, so the **model** fits the **evidence**.
2) They are often **soluble** in **water** but **not** in **non-polar solvents** — this tells you that the particles are **charged**. The ions are **pulled apart** by **polar molecules** like water, but **not** by **non-polar** molecules. Again, the **model** of ionic structures fits this evidence.

Models of Bonding Have Their Limitations

Like pretty much all models, bonding models aren't totally accurate.

1) **Dot-and-cross models** of ionic and covalent bonding are great for explaining what's happening nice and clearly. But like most things in life, it's not really quite as simple as that.
2) One important reason is that most bonds aren't **purely ionic** or **purely covalent** but somewhere in between. This is down to **bond polarisation** (see page 28). Most compounds end up with a **mixture** of ionic and covalent properties.

Practice Questions

Q1 Why can metals conduct electricity?

Q2 Why are metals malleable?

Q3 Describe the motion of particles in solids, liquids and gases.

Q4 Why do gases diffuse to fill the space available?

Q5 What is a solid with a regular structure called?

Q6 What types of bonds must be overcome in order for a substance to boil or melt?

Q7 Do ionic compounds conduct electricity?

Exam Questions

Q1 Illustrate with a suitable labelled diagram the structure of calcium and explain what is meant by metallic bonding. [4 marks]

Q2

Substance	Melting point	Electrical conductivity of solid	Electrical conductivity of liquid	Solubility in water
A	High	Poor	Good	Soluble
B	Low	Poor	Poor	Insoluble
C	High	Good	Good	Insoluble
D	Very High	Poor	Poor	Insoluble

a) Identify the type of structure present in each substance, A to D. [4 marks]

b) Which substance is most likely to be:
(i) diamond, (ii) aluminium, (iii) sodium chloride and (iv) iodine? [2 marks]

Q3 Explain the electrical conductivity of magnesium, sodium chloride and graphite.
In your answer you should consider the structure and bonding of each of these materials. [12 marks]

Gases — like flies in jam jars...

You need to learn the info in the table on the left. With a quick glance in my crystal ball, I can almost guarantee you'll need a bit of it in your exam...let me look a bit closer and tell you which bit....mmm....nah. It's clouded over. You'll have to learn the lot. Sorry. Tell you what — close the book and see how much of the table you can scribble out from memory.

Periodicity

Periodicity is one of those words you hear a lot in Chemistry without ever really knowing what it means.
Well it basically means trends that occur (in physical and chemical properties) as you move across the periods.
E.g. Metal to non-metal is a trend that occurs going left to right in each period... The trends repeat each period.

The **Periodic Table** arranges Elements by **Proton Number**

1) The periodic table is arranged into **periods** (rows) and **groups** (columns), by atomic (proton) number.

2) All the elements **within a period** have the same number of **electron shells** (if you don't worry about s and p sub-shells) E.g. the elements in Period 2 have 2 electron shells.

3) All the elements **within a group** have the **same number** of electrons in their **outer shell** — so they have **similar properties**.

4) The **group number** tells you the number of electrons in the outer shell, e.g. Group 1 elements have 1 electron in their outer shell, Group 4 elements have 4 electrons and so on...

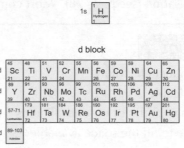

You can use the Periodic Table to work out **Electron Configurations**

The periodic table can be split into an **s block**, **d block** and **p block** like this: Doing this shows you which sub-shells all the electrons go into.

See page 10 if this sub-shell malarkey doesn't ring a bell.

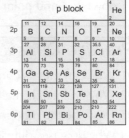

1) The **s-block** elements have an outer shell electron configuration of s^1 or s^2.

> **Examples** Lithium ($1s^2\ 2s^1$) and magnesium ($1s^2\ 2s^2\ 2p^6\ 3s^2$)

2) The **p-block** elements have an outer shell configuration of s^2p^1 to s^2p^6.

> **Example** Chlorine ($1s^2\ 2s^2\ 2p^6\ 3s^2\ 3p^5$)

3) The **d-block** elements have electron configurations in which d sub-shells are being filled.

> **Example** Cobalt ($1s^2\ 2s^2\ 2p^6\ 3s^2\ 3p^6\ 3d^7\ 4s^2$) ◄

Even though the 3d sub-shell fills last in cobalt, it's not written at the end of the line.

When you've got the periodic table **labelled** with the **shells** and **sub-shells** like the one up there, it's pretty easy to read off the electron structure of any element by starting at the top and working your way across and down until you get to your element.

A wee apology...
This bit's really hard to explain clearly in words. If you're confused, just look at the examples until you get it...

Example

Electron structure of phosphorus (P):

Period 1 — $1s^2$ ◄——— Complete sub-shells
Period 2 — $2s^2\ 2p^6$ ◄———
Period 3 — $3s^2\ 3p^3$ ◄——— Incomplete outer sub-shell

So the full electron structure of phosphorus is: $1s^2\ 2s^2\ 2p^6\ 3s^2\ 3p^3$

Atomic Radius **Decreases** across a Period

1) As the number of protons increases, the **positive charge** of the nucleus increases. This means electrons are **pulled closer** to the nucleus, making the atomic radius smaller.

2) The extra electrons that the elements gain across a period are added to the **outer energy level** so they don't really provide any extra shielding effect (shielding works with inner shells mainly).

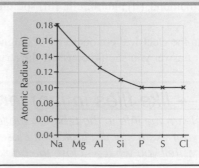

Periodicity

Melting and Boiling Points are linked to **Bond Strength** and **Structure**

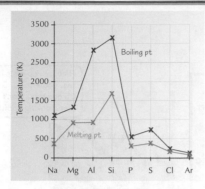

1) Sodium, magnesium and aluminium are **metals**. Their melting and boiling points **increase** across the period because the **metal-metal bonds** get stronger. The bonds get stronger because the metal ions have an increasing number of **delocalised electrons** and a decreasing **radius**. This leads to a higher **charge density**, which attracts the ions together more strongly.

2) Silicon is **macromolecular**, with a tetrahedral structure — **strong covalent bonds** link all its atoms together. **A lot** of energy is needed to break these bonds, so silicon has **high** melting and boiling points.

3) Phosphorus (P_4), sulfur (S_8) and chlorine (Cl_2) are all **molecular substances**. Their melting and boiling points depend upon the strength of the **van der Waals forces** (see page 29) between their molecules. Van der Waals forces are weak and easily overcome so these elements have **low** melting and boiling points.

4) More atoms in a molecule mean stronger van der Waals forces. Sulfur is the **biggest molecule** (S_8), so it's got higher melting and boiling points than phosphorus or chlorine.

5) Argon has **very low** melting and boiling points because it exists as **individual atoms** (they're monatomic) resulting in **very weak** van der Waals forces.

Sam is looking hot in the latest periodic trends.

Ionisation Energy Generally **Increases** across a Period

This is because of the **increasing attraction** between the outer shell electrons and the nucleus, due to the number of **protons** increasing (there are a few blips in the trend however — check back to pages 12-13 for more details).

Practice Questions

Q1 Which elements of Period 3 are found in the s block of the periodic table?

Q2 Write down the electronic configuration of sodium.

Q3 Which element in Period 3 has the largest atomic radius?

Q4 Which element in Period 3 has the highest melting point? Which has the highest boiling point?

Exam Questions

Q1 Explain why the melting point of magnesium is higher than that of sodium. [3 marks]

Q2 This table shows the melting points for the Period 3 elements.

Element	Na	Mg	Al	Si	P	S	Cl	Ar
Melting point / K	371	923	933	1680	317	392	172	84

In terms of structure and bonding explain why:

a) silicon has a high melting point.
b) the melting point of sulfur is higher than phosphorus.

Q3 State and explain the trend in atomic radius across Period 3. [4 marks]

Q4 Explain why the first ionisation energy of neon is greater than that of sodium. [2 marks]

Periodic trends — my mate Dom's always a decade behind...

He still thinks Oasis, Blur and REM are the best bands around. The sad muppet. But not me. Oh no sirree, I'm up with the times — April Lavigne... Linkin' Pork... Christina Agorrilla. I'm hip, I'm with it. Da ga da ga da ga da ga.. ooaarrr ooup *

<small>* Obscure reference to Austin Powers: International Man of Mystery. You should watch it — it's better than doing Chemistry.</small>

Basic Stuff

There are zillions of organic compounds. And loads of them are pretty similar, but with a slight, but crucial difference. There's no way you could memorise all their names, so chemists have devised a clever way of naming each of them.

IUPAC Rules Help Avoid Confusion

The **IUPAC system** for naming organic compounds is the agreed **international language** of chemistry. Years ago, organic compounds were given whatever names people fancied, such as acetic acid and ethylene. But these names caused **confusion** between different countries.

The IUPAC system means scientific ideas can be communicated **across the globe** more effectively. So it's easier for scientists to get on with testing each other's work, and either confirm or dispute new theories.

Fancy a cup of tea? Tea, you know... Brown stuff. Oh come on, don't just sit there looking confused...

Nomenclature is a Fancy Word for Naming Organic Compounds

You can name **any** organic compound using **rules** of nomenclature.

Here's how the rules are used to name a **branched alkane** (there's more on alkanes on page 40).

1) Count the carbon atoms in the **longest continuous chain** — this gives you the stem:

Number of carbons	1	2	3	4	5	6
Stem	meth-	eth-	prop-	but-	pent-	hex-

Don't forget — the longest carbon chain may be bent.

2) Decide what **type** of molecule you've got. This gives you crucial parts of the name — see the table on the right.

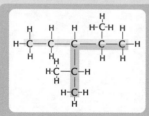

The longest chain is **5** carbons, so the stem is **pent-**

It's a branched alkane, so the name's a bit more complicated than just **pentane**.

Homologous series	Prefix or Suffix	Example
alkanes	-ane	Propane $CH_3CH_2CH_3$
branched alkanes	-yl	methylpropane $CH_3CH(CH_3)CH_3$
alkenes	-ene	propene $CH_3CH=CH_2$
haloalkanes (halogenoalkanes)	chloro- bromo- iodo-	chloroethane CH_3CH_2Cl

3) Number the carbons in the **longest** carbon chain. If there's more than one longest chain, pick the one with the **most side-chains**.

With other compounds, you number so that the most important functional group (see p38) is on the lowest number carbon.

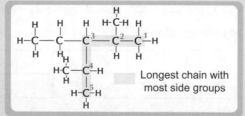

Longest chain with most side groups

4) Side-chains are added as prefixes at the start of the name. Put them in **alphabetical** order, with the **number** of the carbon atom each is attached to.

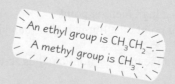

An ethyl group is CH_3CH_2–. A methyl group is CH_3–.

5) If there's more than one **identical** side-chain or functional group, use **di-** (2), **tri-** (3) or **tetra-** (4) before that part of the name — but ignore this when working out the alphabetical order.

There's an ethyl group on carbon-3, and methyl groups on carbon-2 and carbon-4, so it's **3-ethyl-2,4-dimethylpentane**.

Basic Stuff

Name *Haloalkanes* Using the *Same Rules*

Haloalkanes are just alkanes where one or more hydrogens have been swapped for a **halogen**.

dichloromethane   trichloromethane   2-iodopropane   2-bromo-2-chloro-1, 1, 1-trifluoroethane

Naming *Alkenes* — Look at the Position of the *Double Bond*

Alkenes have at least one **double bond** in their carbon chain.
For alkenes with more than three carbons, you need to say which carbon the double bond starts from.

Example — 1) The longest chain is **5** carbons, so the stem of the name is **pent-**.

2) The functional group is **C=C**, so it's **pentene**.

3) Number the carbons from right to left (so the double bond starts on the lowest possible number). The first carbon in the double bond is **carbon 2**. So this is **pent-2-ene**.

Here are some more examples:

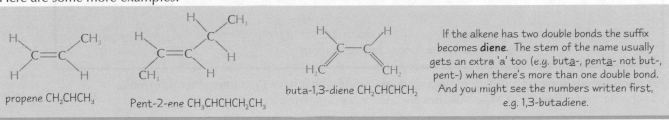

propene CH₂CHCH₃   Pent-2-ene CH₃CHCHCH₂CH₃   buta-1,3-diene CH₂CHCHCH₂

If the alkene has two double bonds the suffix becomes **diene**. The stem of the name usually gets an extra 'a' too (e.g. but<u>a</u>-, pent<u>a</u>- not but-, pent-) when there's more than one double bond. And you might see the numbers written first, e.g. 1,3-butadiene.

Practice Questions

Q1 What do you call an unbranched alkane with three carbon atoms?

Q2 In what order should prefixes be listed in the name of an organic compound?

Q3 What is a haloalkane?

Exam Questions

Q1 1-bromobutane is C_4H_9Br

a) Draw the structure of 1-bromobutane.   *Homologous series are "families" of organic chemicals, see the table on the left.*   [1 mark]

b) Which homologous series does 1-bromobutane belong to?   [1 mark]

c) 1-bromobutane can be made from the molecule shown on the right.
Name the molecule.   [2 marks]

Q2 Name the following molecules.

a)   [2 marks]

b)   [2 marks]

c)   [2 marks]

It's as easy as 1,2,3-trichloropent-2-ene...

The best thing to do now is find some random alkanes, alkenes and haloalkanes and work out their names using the rules. Then have a bash at it the other way around — read the name and draw the compound. It might seem a wee bit tedious now, but come the exam, you'll be thanking me. Doing the exam questions will give you some good practice too.

Formulas and Structural Isomerism

Isomers are fun — you can put the same atoms together in different ways to make completely different molecules. It's just like playing with plastic building bricks... But first some definitions to learn.

There are **Loads of Ways** of **Representing** Organic Compounds

TYPE OF FORMULA	WHAT IT SHOWS YOU	FORMULA FOR BUTANE
General formula	An algebraic formula that can describe **any member** of a family of compounds.	C_nH_{2n+2} (for all alkanes)
Empirical formula	The **simplest ratio** of atoms of each element in a compound (cancel the numbers down if possible). (So ethane, C_2H_6, has the empirical formula CH_3.)	C_2H_5
Molecular formula	The **actual** number of atoms of each element in a molecule.	$C_4H_{10}O$
Structural formula	Shows the atoms **carbon by carbon**, with the attached hydrogens and functional groups.	$CH_3CH_2CH_2CH_3$
Displayed formula	Shows how all the atoms are **arranged**, and all the bonds between them.	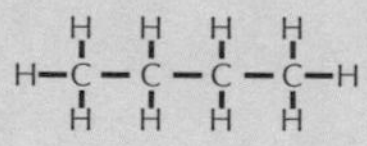

> A functional group is a reactive part of a molecule — it gives it most of its chemical properties. E.g. in an alcohol it's –OH.

A **homologous series** is a bunch of organic compounds which have the **same general formula**. Each member differs by $-CH_2-$. **Alkanes** are a homologous series.

Structural Isomers have different **Structural Arrangements** of Atoms

In structural isomers the atoms are **connected** in different ways. But they still have the **same molecular formula**. There are **three types** of structural isomers:

CHAIN ISOMERS

Chain isomers have different arrangements of the **carbon skeleton**. Some are **straight chains** and others **branched** in different ways.

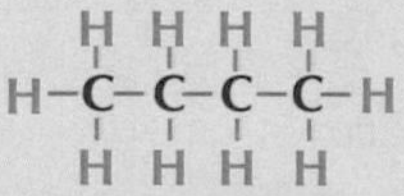

butane 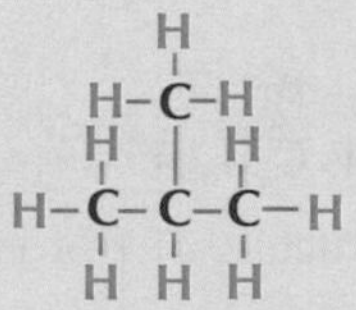   methylpropane

POSITIONAL ISOMERS

Positional isomers have the **same skeleton** and the **same atoms or groups of atoms** attached. The difference is that the atom or group of atoms is attached to a **different carbon atom**.

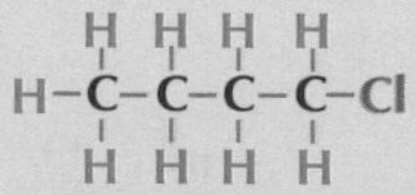

1-chlorobutane 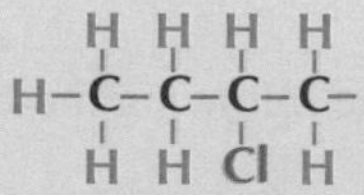   2-chlorobutane

FUNCTIONAL GROUP ISOMERS

Functional group isomers have the same atoms arranged into **different functional groups**.

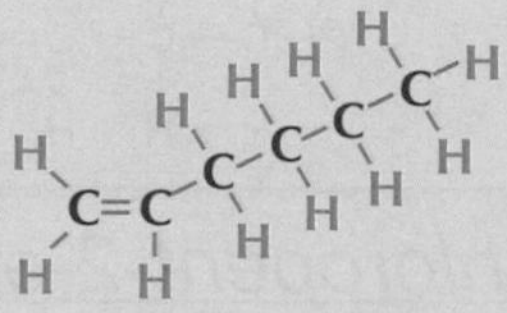

hex-1-ene 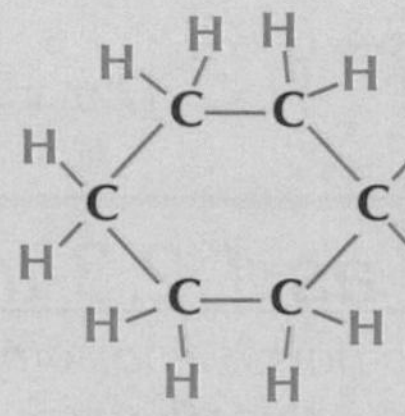   cyclohexane

Formulas and Structural Isomerism

Don't be Fooled — What Looks Like an Isomer Might **Not** Be

Atoms can rotate as much as they like around single **C–C bonds**. Remember this when you work out structural isomers — sometimes what looks like an isomer, isn't.

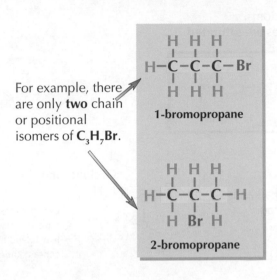

For example, there are only **two** chain or positional isomers of **C₃H₇Br**.

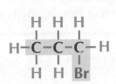

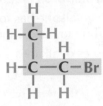

... and again
1-bromopropane

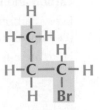

... and again
1-bromopropane

Practice Questions

Q1 Explain the difference between molecular formulas and structural formulas.

Q2 Draw the structural and displayed formulas of hexane.

Q3 What are structural isomers?

Q4 What is a positional isomer?

Exam Questions

Q1 There are four haloalkanes with the molecular formula C₄H₉Cl.
  a) Give the names of all four of these haloalkanes. [4 marks]
  b) Identify a pair of positional isomers from your answer to part a). [1 mark]
  c) Identify a pair of chain isomers from your answer to part a). [1 mark]

Q2 There are five chain isomers of the alkane C₆H₁₄.
  a) Draw and name all five isomers of C₆H₁₄. [10 marks]
  b) Alkanes are an example of a homologous series. What is a homologous series? [2 marks]
  c) (i) Write down the molecular formula for an alkane molecule that has 8 carbon atoms. [1 mark]
     (ii) Write out the full structural formula for the alkane molecule in part c)(i).
          Assume that it is an unbranched alkane. [1 mark]

Q3 The alkane with the molecular formula C₅H₁₂ has these chain isomers.
  a) Name these isomers. [3 marks]
  b) Explain what is meant by the term 'chain isomerism'. [2 marks]

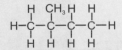

Human structural isomers...

Alkanes and Petroleum

Alkanes are the first set of organic chemicals you need to know about. They're what petroleum's mainly made of.

Alkanes are **Saturated Hydrocarbons**

1) Alkanes have the **general formula C_nH_{2n+2}**. They've only got **carbon** and **hydrogen** atoms, so they're **hydrocarbons**.

2) Every carbon atom in an alkane has **four single bonds** with other atoms. It's **impossible** for carbon to make more than four bonds, so alkanes are **saturated**.

   Here are a few examples of alkanes —

Methane	Ethane	Propane

3) You get **cycloalkanes** too. They have a ring of carbon atoms with two hydrogens attached to each carbon.

4) Cycloalkanes have a **different general formula** from that of normal alkanes (C_nH_{2n}, assuming they have only one ring), but they are still **saturated**.

Cyclohexane C_6H_{12} ← cycloalkanes have two fewer hydrogens than other alkanes

Crude Oil is Mainly **Alkanes**

1) **Petroleum** is just a **poncy word** for crude oil — the black, yukky stuff they get out of the ground with huge oil wells. It's mostly **alkanes**. They range from **smallish alkanes**, like pentane, to **massive alkanes** with more than 50 carbons.

2) Crude oil isn't very useful as it is, but you can **separate** it into more useful bits (or **fractions**) by **fractional distillation**.

Here's how fractional distillation works — don't try this at home.

1) First, the crude oil is **vaporised** at about 350 °C.

2) The vaporised crude oil goes into the **fractionating column** and rises up through the trays. The largest hydrocarbons don't **vaporise** at all, because their boiling points are too high — they just run to the bottom and form a gooey **residue**.

3) As the crude oil vapour goes up the fractionating column, it gets **cooler**. Because of the different chain lengths, each fraction **condenses** at a different temperature. The fractions are **drawn off** at different levels in the column.

4) The hydrocarbons with the **lowest boiling points** don't condense. They're drawn off as **gases** at the top of the column.

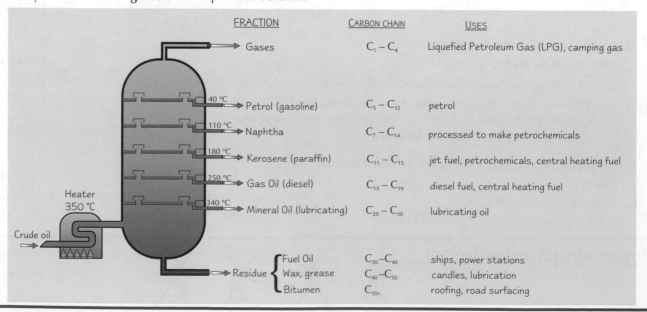

FRACTION	CARBON CHAIN	USES
Gases	$C_1 - C_4$	Liquefied Petroleum Gas (LPG), camping gas
Petrol (gasoline)	$C_5 - C_{12}$	petrol
Naphtha	$C_7 - C_{14}$	processed to make petrochemicals
Kerosene (paraffin)	$C_{11} - C_{15}$	jet fuel, petrochemicals, central heating fuel
Gas Oil (diesel)	$C_{15} - C_{19}$	diesel fuel, central heating fuel
Mineral Oil (lubricating)	$C_{20} - C_{30}$	lubricating oil
Fuel Oil	$C_{30} - C_{40}$	ships, power stations
Wax, grease	$C_{40} - C_{50}$	candles, lubrication
Bitumen	C_{50+}	roofing, road surfacing

Alkanes and Petroleum

Heavy Fractions can be 'Cracked' to Make Smaller Molecules

1) People want loads of the **light** fractions, like petrol and naphtha. They don't want so much of the **heavier** stuff like bitumen though. Stuff that's in high demand is much more **valuable** than the stuff that isn't.

2) To meet this demand, the less popular heavier fractions are **cracked**. Cracking is **breaking** long-chain alkanes into **smaller** hydrocarbons (which can include alkenes). It involves breaking the **C–C bonds**.
You could crack **decane** like this —

$$C_{10}H_{22} \rightarrow C_2H_4 + C_8H_{18}$$
decane   ethene   octane

There are **two types** of **cracking** you need to know about:

THERMAL CRACKING

- It takes place at **high temperature** (up to 1000 °C) and **high pressure** (up to 70 atm).
- It produces a lot of **alkenes**.
- These **alkenes** are used to make heaps of valuable products, like **polymers**. A good example is **poly(ethene)**, which is made from ethene (have a squiz at page 75 for more on polymers).

CATALYTIC CRACKING

- This makes mostly **motor fuels** and **aromatic** hydrocarbons. Aromatic compounds contain benzene rings. Benzene rings have six carbon atoms with three double bonds. They're pretty stable because the electrons are delocalised around the carbon ring.
- It uses something called a **zeolite catalyst (hydrated aluminosilicate)**, at a **slight pressure** and **high temperature** (about 450 °C).
- Using a catalyst **cuts costs**, because the reaction can be done at a **lower** temperature and pressure. The catalyst also **speeds** up the reaction, and time is money and all that.

Practice Questions

Q1 What is the general formula for alkanes?

Q2 Draw and name the first four alkanes.

Q3 What's the purpose of the fractional distillation of crude oil?

Q4 What is cracking?

Q5 What type of organic chemical does thermal cracking produce?

Exam Questions

Q1 Crude oil contains many different alkane molecules. These are separated using a process called fractional distillation.

a) Why do the components of crude oil need to be separated? [1 mark]

b) What physical property of the molecules is used to separate them? [1 mark]

c) A typical alkane found in the petrol (gasoline) fraction has 8 carbon atoms.

(i) Give the molecular formula for this alkane. [1 mark]

(ii) Would you find the petrol fraction near the top or bottom of the fractionating column? Explain your answer. [3 marks]

(iii) What is the molecular formula of a cycloalkane with 8 carbon atoms. [1 mark]

Q2 Crude oil is a source of fuels and petrochemicals. It's vaporised and separated into fractions using fractional distillation. Some heavier fractions are processed using cracking.

a) Describe one reason why cracking is carried out. [2 marks]

b) Write a possible equation for the thermal cracking of dodecane, $C_{12}H_{26}$. [1 mark]

Crude oil — not the kind of oil you could take home to meet your mother...

This ain't the most exciting page in the history of the known universe. Although in a galaxy far, far away there may be lots of pages on even more boring topics. But, that's neither here nor there, cos you've got to learn the stuff anyway. Get fractional distillation and cracking straight in your brain and make sure you know why people bother to do it.

Alkanes as Fuels

Alkanes are absolutely fantastic as fuels. Except for the fact that they produce loads of nasty pollutant gases.

Alkanes are Useful Fuels

1) If you burn (**oxidise**) alkanes with **plenty of oxygen**, you get **carbon dioxide** and water — it's a **combustion reaction**.

Here's the equation for the combustion of propane — $C_3H_{8(g)} + 5O_{2(g)} \rightarrow 3CO_{2(g)} + 4H_2O_{(g)}$

This is complete combustion. There's also incomplete combustion, which is really BAD. See below...

2) Alkanes make great fuels — burning just a small amount of **methane** releases a humungous amount **of energy**.

Carbon Monoxide is Formed if Alkanes Burn Incompletely

If there's not enough oxygen, hydrocarbons **combust incompletely**, and you get carbon monoxide gas instead of carbon dioxide. E.g.

$CH_{4(g)} + 1\frac{1}{2}O_{2(g)} \rightarrow CO_{(g)} + 2H_2O_{(g)}$ $C_8H_{18(g)} + 8\frac{1}{2}O_{2(g)} \rightarrow 8CO_{(g)} + 9H_2O_{(g)}$

This is bad news because carbon monoxide gas is poisonous. Carbon monoxide molecules bind to the same sites on **haemoglobin molecules** in red blood cells as oxygen molecules. So **oxygen** can't be carried around the body.

Luckily, carbon monoxide can be removed from exhaust gases by **catalytic converters** on cars.

And if that's Not Bad Enough... Burning Fuels Produces Other Pollutants Too

Unburnt Hydrocarbons and Oxides of Nitrogen (NOₓ) Contribute to Smog

1) Engines **don't burn** all the fuel molecules. Some of these come out as **unburnt hydrocarbons**.

2) **Oxides of nitrogen** (NO_x) are produced when the high pressure and temperature in a car engine cause the nitrogen and oxygen atoms in the air to react together.

3) The hydrocarbons and nitrogen oxides react in the presence of sunlight to form **ground-level ozone** (O_3), which is a major component of **smog**. **ground-level ozone** irritates people's eyes, aggravates respiratory problems and even causes lung damage (ozone isn't nice stuff, unless it is high up in the atmosphere as part of the ozone layer).

4) **Catalytic converters** on cars remove unburnt hydrocarbons and oxides of nitrogen from the exhaust.

Sulfur Dioxide

1) **Acid rain** is caused by burning fossil fuels that contain **sulfur**. The sulfur burns to produce **sulfur dioxide** gas which then enters the atmosphere, dissolves in the moisture, and is converted into **sulfuric acid**.
The same process occurs when nitrogen dioxide escapes into the atmosphere — nitric acid is produced.

2) Acid rain destroys trees and vegetation, as well as corroding buildings and statues and killing fish in lakes. Luckily, sulfur dioxide can be removed from power station flue gases using **calcium oxide**.

Yet More Bad News... Burning Fossil Fuels Contributes to Global Warming

1) The vast majority of scientists believe that **global warming** is caused by increased levels of **carbon dioxide** in the atmosphere due to burning **fossil fuels** (coal, oil and natural gas).

2) Not everyone agrees with this theory, but here's what is true:
- **Greenhouse gases** stop some of the heat from the Sun from escaping back into space.
  This is the greenhouse effect — it's what keeps the Earth warm enough for us to live here (see the next page).
- **Carbon dioxide** is a greenhouse gas.
- Burning **fossil fuels** produces carbon dioxide.
- The level of carbon dioxide in the atmosphere has **increased** in the last 50 years or so.
- The average **temperature** of the Earth has **increased dramatically** over the same period.
  This is **global warming**, and it's a big headache for the whole planet.

3) Most scientists have looked at all the **evidence** and agree that the rise in carbon dioxide levels is down to human activity, including burning fossil fuels. They also agree that the extra CO_2 is **enhancing** the greenhouse effect, and that this is the cause of global warming.

4) There are still a few scientists who think that there are **other explanations**, either for the rise in CO_2 levels, or for the cause of global warming. That's part of science — it can take a long time for everyone to accept a theory (and you never know when some new evidence might turn up to prove everyone wrong).

Alkanes as Fuels

Carbon Dioxide isn't the Only Greenhouse Gas

1) Some of the **electromagnetic radiation** from the Sun reaches the Earth and is **absorbed**. The Earth then **re-emits** it as **infrared radiation** (heat).

2) Various gases in the troposphere (the lowest layer of the atmosphere) **absorb** some of this infrared radiation... and **re-emit** it in **all directions** — including back towards Earth, keeping us warm. This is called the '**greenhouse effect**' (even though a real greenhouse doesn't actually work like this, annoyingly).

3) The three main greenhouse gases are **water vapour**, **carbon dioxide** and **methane**. **Human activities** have caused a rise in greenhouse gas concentrations, which **enhances** the greenhouse effect. So now **too much heat** is being trapped and the Earth is **getting warmer** — this is **global warming**.

4) Items about global warming on TV and in newspapers usually focus on cutting the levels of **carbon dioxide**, but the other greenhouse gases are important, too.

Visible and UV radiation from the Sun

Some infrared radiation emitted by the Earth is absorbed by greenhouse gases

Some infrared radiation emitted by the Earth escapes

5) When alkanes in **fossil fuels** are burned they also produce **water vapour**. People tend not to worry so much about water vapour in the atmosphere. There's always been lots of it, and unlike carbon dioxide, the levels have stayed pretty **constant** — and some of it gets removed every time it rains.

6) The other important greenhouse gas is **methane**. Methane's produced by rubbish rotting in **landfill sites**. Methane levels have also risen as we've had to grow more food for our rising population. **Cows** are responsible for large amounts of methane. From both ends.

Vegetarians can't feel entirely smug though. Paddy fields, in which rice is grown, kick out a fair amount of methane too.

Practice Questions

Q1 Which two compounds are produced when an alkane burns completely?

Q2 Why is the incomplete combustion of alkanes a problem?

Q3 Explain how burning fossil fuels may contribute to global warming.

Q4 Name three greenhouse gases.

Exam Questions

Q1 Heptane, C_7H_{16}, is an alkane present in some fuels.

   a) Write a balanced equation for the complete combustion of heptane.   [2 marks]

   b) Fuels often contain oxygenates such as methanol to ensure that the fuel burns completely.

      (i) What toxic compound can be produced by the incomplete combustion of alkanes such as heptane?   [1 mark]

      (ii) Apart from adding oxygenates, how else can this compound be removed from exhaust gases?   [1 mark]

Q2 Burning fossil fuels can cause a variety of environmental problems.

   a) Explain how oxides of nitrogen are produced in car engines.   [2 marks]

   b) Explain why burning fossil fuels in power stations can lead to acid rain and how this problem can be solved.   [3 marks]

Burn, baby, burn — so long as the combustion is complete...

Don't you just hate it when you come up with a great idea, then everyone picks holes in it? Well, just imagine if you were the one who thought of burning alkanes for fuel... it seemed like such a good idea at the time. Despite all the problems, we're still using them — and until we find some suitable alternatives, we all have to deal with the negative consequences.

Enthalpy Changes

A whole new section to enjoy — but don't forget, Big Brother is watching...

Chemical Reactions Usually Have Enthalpy Changes

When chemical reactions happen, there'll be a **change in energy**.
The souped-up chemistry term for this is **enthalpy change** —

> **Enthalpy change**, ΔH (delta H), is the heat energy transferred in a reaction at **constant pressure**. The units of ΔH are **kJ mol⁻¹**.

You write $\Delta H^{\ominus}$ to show that the elements were in their **standard states** and that the measurements were made under **standard conditions**. Standard conditions are **100 kPa (about 1 atm) pressure** and a stated temperature (e.g. ΔH_{298}). In this book, all the enthalpy changes are measured at 298 K (25 °C).

Reactions can be either Exothermic or Endothermic

> **Exothermic** reactions **give out** energy. ΔH is **negative**.

In exothermic reactions, the temperature often goes **up**.

Oxidation is exothermic. Here are two examples:

- The **combustion** of a fuel like methane $\longrightarrow$ $CH_{4(g)} + 2O_{2(g)} \longrightarrow CO_{2(g)} + 2H_2O_{(l)}$ $\Delta H^{\ominus}_{c,\,298} = -890$ kJ mol⁻¹ **exothermic**

- The oxidation of **carbohydrates**, such as glucose, $C_6H_{12}O_6$, in respiration.

> **Endothermic** reactions **absorb** energy. ΔH is **positive**.

In these reactions, the temperature often **falls**.

The **thermal decomposition** of calcium carbonate is endothermic.

$$CaCO_{3(s)} \longrightarrow CaO_{(s)} + CO_{2(g)} \quad \Delta H^{\ominus}_{r,\,298} = +178 \text{ kJ mol⁻¹ } \textbf{endothermic}$$

The main reactions of **photosynthesis** are also endothermic — sunlight supplies the energy.

Reactions are all about Breaking and Making Bonds

When reactions happen, **reactant bonds** are **broken** and **product bonds** are **formed**.

1) You **need** energy to break bonds, so bond breaking is **endothermic** (ΔH is **positive**). **Stronger** bonds take **more** energy to break.

2) Energy is **released** when bonds are formed, so this is **exothermic** (ΔH is **negative**). **Stronger** bonds release **more** energy when they form.

3) The **enthalpy change** for a reaction is the **overall effect** of these two changes. If you need **more** energy to **break** bonds than is released when bonds are made, ΔH is **positive**. If it's less, ΔH is negative.

You can only break bonds if you've got enough energy.

Mean Bond Enthalpies are not Exact

Water (H_2O) has got **two O–H bonds**. You'd think it'd take the same amount of energy to break them both... but it **doesn't**.

> The **first** bond, H–OH$_{(g)}$: E(H–OH) = +492 kJ mol⁻¹
> The **second** bond, H–O$_{(g)}$: E(H–O) = +428 kJ mol⁻¹
> (OH⁻ is a bit easier to break apart because of the extra electron repulsion.)
>
> So, the **mean** bond enthalpy is $\dfrac{492 + 428}{2}$ = **+460 kJ mol⁻¹**.

The data book says the bond enthalpy for O–H is +463 kJ mol⁻¹. It's a bit different because it's the average for a much bigger range of molecules, not just water. For example, it includes the O–H bonds in alcohols and carboxylic acids too.

Breaking bonds is always an endothermic process, so mean bond enthalpies are always **positive**.

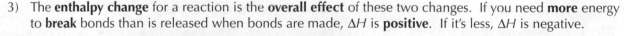

Enthalpy Changes

Enthalpy Changes Can Be Calculated using Average Bond Enthalpies

In any chemical reaction energy is **absorbed** to **break bonds** and **given out** during **bond formation**.
The difference between the energy absorbed and released is the overall **enthalpy change of reaction**:

Enthalpy Change of Reaction = Total Energy Absorbed − Total Energy Released

Example: Calculate the overall enthalpy change for this reaction:
$N_2 + 3H_2 \rightarrow 2NH_3$
Use the average bond enthalpy values in the table.

Bond	Average Bond Enthalpy
N≡N	945 kJ mol^{-1}
H–H	436 kJ mol^{-1}
N–H	391 kJ mol^{-1}

Bonds broken: 1 × N≡N bond broken = 1 × 945 = 945 kJ mol^{-1}
3 × H–H bonds broken = 3 × 436 = 1308 kJ mol^{-1}

Total Energy Absorbed = 945 + 1308 = **2253 kJ mol^{-1}**

Bonds formed: 6 × N–H bonds formed = 6 × 391 = 2346 kJ mol^{-1}

Total Energy Released = **2346 kJ mol^{-1}**

Now you just subtract 'total energy released' from 'total energy absorbed':

Enthalpy Change of Reaction = 2253 − 2346 = **−93 kJ mol^{-1}**

If you can't remember which value to subtract from which, just take the smaller number from the bigger one then add the sign at the end — positive if 'bonds broken' was the bigger number (endothermic), negative if 'bonds formed' was bigger (exothermic).

There are Different Types of ΔH

1) **Standard enthalpy change of reaction**, $\Delta H_r^{\ominus}$, is the enthalpy change when the reaction occurs in the **molar quantities** shown in the **chemical equation**, under standard conditions in their standard states.

2) **Standard enthalpy change of formation**, $\Delta H_f^{\ominus}$, is the enthalpy change when **1 mole** of a **compound** is formed from its **elements** in their standard states under standard conditions, e.g. $2C_{(s)} + 3H_{2(g)} + \frac{1}{2}O_{2(g)} \longrightarrow C_2H_5OH_{(l)}$

3) **Standard enthalpy change of combustion**, $\Delta H_c^{\ominus}$, is the enthalpy change when **1 mole** of a substance is completely **burned in oxygen** under standard conditions.

Practice Questions

Q1 Explain the terms exothermic and endothermic, giving an example in each case.
Q2 Is energy taken in or released when bonds are broken?
Q3 What is the mean bond enthalpy? How can you work it out?
Q4 Define standard enthalpy of formation and standard enthalpy of combustion.

Exam Questions

Q1 The table shows some average bond enthalpy values.

Bond	C–H	C=O	O=O	O–H
Average Bond Enthalpy (kJ mol^{-1})	435	805	498	464

The complete combustion of methane can be represented by the following equation:

$$CH_{4\,(g)} + 2O_{2\,(g)} \rightarrow CO_{2\,(g)} + 2H_2O_{\,(l)}$$

a) Use the table of bond enthalpies above to calculate the enthalpy change for the reaction. [4 marks]
b) Is the reaction endothermic or exothermic? Explain your answer. [1 mark]

Q2 Methanol, CH_3OH, when blended with petrol, can be used as a fuel. $\Delta H_{c,\,298}^{\ominus}$ [CH_3OH] = −726 kJ mol^{-1}.
a) Write an equation, including state symbols, for the standard enthalpy change of combustion of methanol. [2 marks]
b) Write an equation, including state symbols, for the standard enthalpy change of formation of methanol. [2 marks]
c) Liquid petroleum gas is a fuel that contains propane, C_3H_8.
Explain why the following equation does not represent a standard enthalpy change of combustion. [1 mark]

$$2C_3H_{8(g)} + 10O_{2(g)} \longrightarrow 8H_2O_{(g)} + 6CO_{2(g)} \quad \Delta H_{r,\,298} = -4113 \text{ kJ mol}^{-1}$$

I bonded with my friend — now we're waiting to be surgically separated...

What a lotta definitions. And you need to know them all. If you're going to bother learning them, you might as well do it properly and learn all the pernickety details. They probably seem about as useful as a dead fly in your custard right now, but all will be revealed over the next few pages. Learn them now, so you've got a bit of a head start.

Calculating Enthalpy Changes

Now you know what enthalpy changes are, here's how to calculate them...

You can find out Enthalpy Changes in the Lab

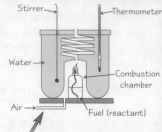

1) To measure the **enthalpy change** for a reaction, you only need to know **two things** —
   • The **number of moles** of the stuff that's reacting. • The change in **temperature**.

2) How you go about doing the experiment depends on what type of reaction it is.
   Some reactions will quite happily take place in a **container** and you can just stick a
   thermometer in to find out the temperature change. It's best to use a **polystyrene beaker**,
   so that you don't lose or gain much heat through the sides.

3) **Combustion reactions** are trickier because the reactant is burned in air. A **copper calorimeter** containing a **known mass of water** is often used. You burn a **known mass of the reactant** and record the **temperature change** of the water.

Calculate Enthalpy Changes Using the Equation q = mcΔT

It seems there's a snazzy equation for everything these days, and enthalpy change is no exception —

$q = mc\Delta T$ where, q = heat lost or gained (in joules). This is the same as the enthalpy change if the pressure is constant.

m = mass of water in the calorimeter, or solution in the polystyrene beaker (in grams)

c = specific heat capacity of water (4.18 J g⁻¹K⁻¹)

ΔT = the change in temperature of the water or solution

Example:

In a laboratory experiment, 1.16 g of an organic liquid fuel was completely burned in oxygen.

The heat formed during this combustion raised the temperature of 100 g of water from 295.3 K to 357.8 K.

Calculate the standard enthalpy of combustion, $\Delta H_c^\ominus$, of the fuel. Its M_r is 58.

1 First off, you need to calculate the **amount of heat** given out by the fuel using $q = mc\Delta T$.

$q = mc\Delta T$

$q = 100 \times 4.18 \times (357.8 - 295.3) = 26\,125$ J = 26.125 kJ ← *Change the amount of heat from J to kJ.*

Remember — m is the mass of water, NOT the mass of fuel.

2 Next you need to find out **how many moles** of fuel produced this heat. It's back to the old $n = \frac{mass}{M}$ equation.

$n = \frac{1.16}{58} = 0.02$ moles of fuel

3 The standard enthalpy of combustion involves 1 mole of fuel.

It's negative because combustion is an exothermic reaction.

So, the heat produced by 1 mole of fuel = $\frac{-26.125}{0.02}$

$\approx$ **-1306 kJ mol⁻¹**. This is the standard enthalpy change of combustion.

The actual $\Delta H_c^\ominus$ of this compound is -1615 kJ mol⁻¹ — loads of heat has been **lost** and not measured. E.g. it's likely a fair bit would escape through the **copper calorimeter** and also the fuel might not **combust completely**.

Hess's Law — the Total Enthalpy Change is Independent of the Route Taken

Hess's Law says that:

The **total enthalpy change** of a reaction is always **the same**, no matter **which route** is taken.

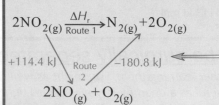

This law is handy for working out enthalpy changes that you **can't find directly** by doing an experiment.

Here's an example:
The **total enthalpy change** for route 1 is the **same as for route 2**.

So, $\Delta H_r = +114.4 + (-180.8) = -66.4$ kJ mol⁻¹.

Calculating Enthalpy Changes

Enthalpy Changes Can be *Worked Out Indirectly*

Enthalpy changes of formation are useful for calculating enthalpy changes you can't find directly.
You need to know $\Delta H_f^\ominus$ for **all** the reactants and products that are **compounds** — the value of $\Delta H_f^\ominus$ for elements is **zero**.

> The element's being formed from the element, so there's no change

Here's how to calculate $\Delta H_r^\ominus$ for this reaction: $SO_{2(g)} + 2H_2S_{(g)} \rightarrow 3S_{(s)} + 2H_2O_{(l)}$

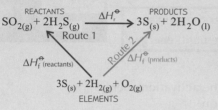

$$SO_{2(g)} + 2H_2S_{(g)} \xrightarrow{\;\Delta H_r^\ominus\;} 3S_{(s)} + 2H_2O_{(l)}$$

Route 1

$\Delta H_f^\ominus$ (reactants)   Route 2 $\Delta H_f^\ominus$ (products)

$$3S_{(s)} + 2H_{2(g)} + O_{2(g)}$$
ELEMENTS

$\Delta H_f^\ominus[SO_{2\,(g)}] = -297 \text{ kJ mol}^{-1}$

$\Delta H_f^\ominus[H_2S_{(g)}] = -20.2 \text{ kJ mol}^{-1}$

$\Delta H_f^\ominus[H_2O_{(l)}] = -286 \text{ kJ mol}^{-1}$

Using **Hess's Law**: Route 1 = Route 2

$\Delta H_r^\ominus$ + the sum of $\Delta H_f^\ominus$ (reactants) = the sum of $\Delta H_f^\ominus$ (products)

So, $\Delta H_r^\ominus$ = **the sum of** $\Delta H_f^\ominus$ **(products) – the sum of** $\Delta H_f^\ominus$ **(reactants)**

To find $\Delta H_r^\ominus$ of this reaction:

Just plug the numbers into the equation above:

$\Delta H_r^\ominus = [0 + (-286 \times 2)] - [-297 + (-20.2 \times 2)] = \mathbf{-234.6 \text{ kJ mol}^{-1}}$

$\Delta H_f^\ominus$ of sulfur is zero — it's an element.

There's 2 moles of H_2O and 2 moles of H_2S.

You can use a similar method to find an enthalpy change from **enthalpy changes of combustion**.

> The standard enthalpy changes are all measured at 298 K.

Here's how to calculate $\Delta H_f^\ominus$ of **ethanol**...

Using Hess's Law: Route 1 = Route 2

$\Delta H_f^\ominus$[ethanol] + $\Delta H_c^\ominus$[ethanol] = $2\Delta H_c^\ominus$[C] + $3\Delta H_c^\ominus$[H₂]

$\Delta H_f^\ominus$[ethanol] + (–1367) = (2 × –394) + (3 × –286)

$\Delta H_f^\ominus$[ethanol] = –788 + –858 – (–1367)

= **–279 kJ mol⁻¹**.

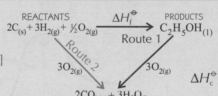

$$2C_{(s)} + 3H_{2(g)} + \tfrac{1}{2}O_{2(g)} \xrightarrow{\;\Delta H_f^\ominus\;} C_2H_5OH_{(l)}$$
Route 1

Route 2 $3O_{2(g)}$ $3O_{2(g)}$

$$2CO_{2(g)} + 3H_2O_{(l)}$$
COMBUSTION PRODUCTS

$\Delta H_c^\ominus[C_{(s)}] = -394 \text{ kJ mol}^{-1}$

$\Delta H_c^\ominus[H_{2\,(g)}] = -286 \text{ kJ mol}^{-1}$

$\Delta H_c^\ominus[\text{ethanol}_{(l)}] = -1367 \text{ kJ mol}^{-1}$

Practice Questions

Q1 Briefly describe an experiment that could be carried out to find the enthalpy change of a reaction.

Q2 What equation would you use to calculate the heat change in this experiment?

Q3 Why is the enthalpy change determined in a laboratory likely to be lower than the value shown in a data book?

Q4 What does Hess's Law state?

Q5 What is the standard enthalpy change of formation of any element?

Exam Questions

Q1 Using the facts that (at 298K) the standard enthalpy change of formation of $Al_2O_{3(s)}$ is –1676 kJ mol⁻¹ and the standard enthalpy change of formation of $MgO_{(s)}$ is –602 kJ mol⁻¹, calculate the enthalpy change of the following reaction.

$$Al_2O_{3(s)} + 3Mg_{(s)} \rightarrow 2Al_{(s)} + 3MgO_{(s)}$$

[3 marks]

Q2 A 50 cm³ sample of 0.200 M copper(II) sulfate solution placed in a polystyrene beaker gave a temperature increase of 2.6 K when excess zinc powder was added and stirred. Calculate the enthalpy change when 1 mole of zinc reacts. Assume that the specific heat capacity for the solution is 4.18 J g⁻¹K⁻¹. Ignore the increase in volume due to the zinc.

The equation for the reaction is: $Zn_{(s)} + CuSO_{4(aq)} \rightarrow Cu_{(s)} + ZnSO_{4(aq)}$

[8 marks]

To understand this lot, you're gonna need a bar of chocolate. Or two...

To get your head around those Hess diagrams, you're going to have to do more than skim them. You need to be able to use this stuff for any reaction they give you. It'll also help if you know the definitions for those standard enthalpy thingumabobs on page 45. If you didn't bother learning them, have a quick flick back and remind yourself about them.

Reaction Rates and Catalysts

The rate of a reaction is just how quickly it happens. Lots of things can make it go faster or slower.

Particles **Must** Collide to **React**

1) Particles in liquids and gases are **always moving** and **colliding** with **each other**. They **don't** react every time though — only when the **conditions** are right. A reaction **won't** take place between two particles **unless** —

> • They collide in the **right direction**. They need to be **facing** each other the right way.
>
> • They collide with at least a certain **minimum** amount of kinetic (movement) **energy**.

This stuff's called **Collision Theory**.

2) The **minimum amount of kinetic energy** particles need to react is called the **activation energy**. The particles need this much energy to **break the bonds** to start the reaction.

3) Reactions with **low activation energies** often happen **pretty easily**. But reactions with **high activation energies** don't. You need to give the particles extra energy by **heating** them.

To make this a bit clearer, here's an **enthalpy profile diagram**.

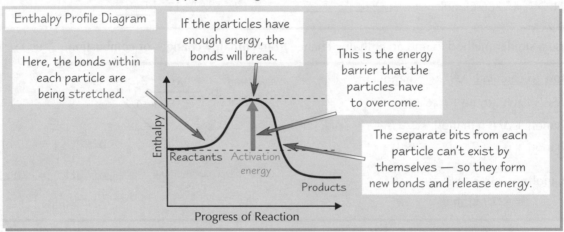

Molecules in a Gas **Don't** all have the **Same Amount of Energy**

Imagine looking down on Oxford Street when it's teeming with people. You'll see some people ambling along **slowly**, some hurrying **quickly**, but most of them will be walking with a **moderate speed**. It's the same with the **molecules** in a gas. Some **don't have much kinetic energy** and move **slowly**. Others have **loads of kinetic energy** and **whizz** along. But most molecules are somewhere **in between**.

If you plot a **graph** of the **numbers of molecules** in a gas with different **kinetic energies** you get a **Maxwell-Boltzmann distribution**. It looks like this —

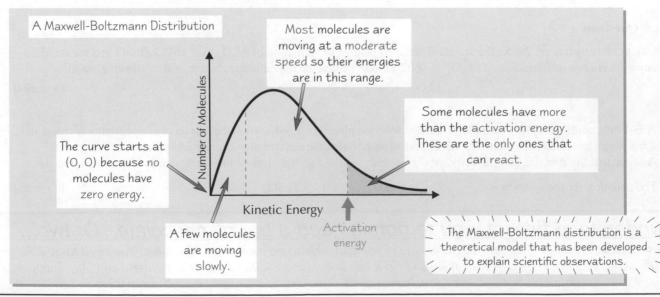

Reaction Rates and Catalysts

Increasing the Temperature makes Reactions Faster

1) If you increase the **temperature**, the particles will on average have more **kinetic energy** and will move **faster**.

2) So, a **greater proportion** of molecules will have at least the **activation energy** and be able to **react**. This changes the **shape** of the **Maxwell-Boltzmann distribution curve** — it pushes it over to the **right**.

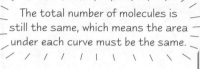
The total number of molecules is still the same, which means the area under each curve must be the same.

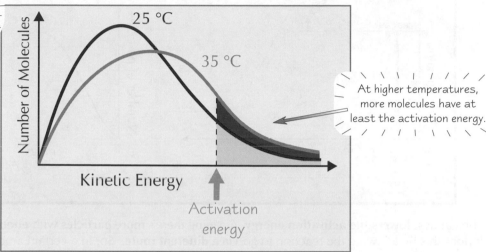

At higher temperatures, more molecules have at least the activation energy.

3) Because the molecules are flying about **faster**, they'll **collide more often**. This is **another reason** why increasing the temperature makes a reaction faster. So,

> **Small temperature increases** can lead to **large increases in reaction rate**.

Increasing Concentration also Increases the Rate of Reaction

1) If you increase the **concentration** of reactants in a **solution**, the particles will on average be **closer together**.

2) If they're closer, they'll **collide more often**. If there are **more collisions**, they'll have **more chances** to react.

3) If the reaction involves gases, increasing the **pressure** of the gases works in just the same way.

Catalysts Increase the Rate of Reactions Too

You can use **catalysts** to make chemical reactions happen **faster**. Learn this definition:

> A **catalyst** increases the **rate** of a reaction by providing an **alternative reaction pathway** with a **lower activation energy**. The catalyst is **chemically unchanged** at the end of the reaction.

There's more on this on the next page. Bet you can't wait.

1) Catalysts are **great**. They **don't** get used up in reactions, so you only need a **tiny bit** of catalyst to catalyse a **huge** amount of stuff. They **do** take part in reactions, but they're **remade** at the end.

2) Catalysts are **very fussy** about which reactions they catalyse. Many will usually **only** work on a single reaction.

3) Catalysts **save heaps of money** in industrial processes.

Reaction Rates and Catalysts

Enthalpy Profiles and Boltzmann Distributions Show Why Catalysts Work

If you look at an **enthalpy profile** together with a **Maxwell-Boltzmann Distribution**, you can see **why** catalysts work.

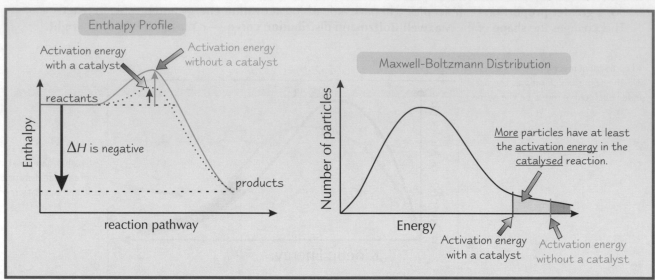

The catalyst **lowers the activation energy**, meaning there's **more particles** with **enough energy** to react when they collide. It does this by allowing the reaction to go **via a different route**. So, in a certain amount of time, **more particles react**.

Practice Questions

Q1 Does every particle collision result in a reaction? Explain your answer?

Q2 Explain the term 'activation energy'.

Q3 What does a Maxwell-Boltzmann distribution show?

Q4 Sketch a Maxwell-Boltzmann distribution for molecules at two different temperatures.

Q5 Explain what a catalyst is.

Q6 Draw a Maxwell-Boltzmann distribution diagram to show how a catalyst works.

Exam Questions

Q1 Nitrogen oxide (NO) and ozone (O_3) react to produce nitrogen dioxide (NO_2) and oxygen (O_2).
The collision between the two molecules does not always lead to a reaction, even if the molecules are
orientated correctly. Explain why this is.   [1 mark]

Q2 Use collision theory to explain why the reaction between
a solid and a liquid is generally faster than that between two solids.   [2 marks]

Q3 The decomposition of hydrogen peroxide, H_2O_2, into water and oxygen is catalysed by manganese(IV) oxide, MnO_2.
a) Write a balanced equation for the reaction.   [2 marks]
b) Sketch a Maxwell-Boltzmann distribution for the reaction.
Mark on the activation energy for the catalysed and uncatalysed process.   [3 marks]
c) Referring to your diagram from part b), explain how manganese(IV) oxide acts as a catalyst. [3 marks]
d) What would be the effect of raising the temperature of this reaction? Explain the effect.   [3 marks]

I'm a catalyst — I like to speed up arguments without getting too involved...

Whatever you do, don't confuse the effect of catalysts with the effect of a temperature change. They both mean more particles have the activation energy. Catalysts do this by lowering the activation energy, BUT increasing temperature does this by giving the particles more energy. Get these mixed up and you'll be the laughing stock of the Examiners' tea room.

Reversible Reactions

There's a lot of to-ing and fro-ing on this page. Mind your head doesn't start spinning.

Reversible Reactions Can Reach Dynamic Equilibrium

1) Lots of chemical reactions are **reversible** — they go **both ways**. To show a reaction's reversible, you stick in a $\rightleftharpoons$. Here's an example:

$$H_{2(g)} + I_{2(g)} \rightleftharpoons 2HI_{(g)}$$

This reaction can go in **either direction** —

forwards $H_{2(g)} + I_{2(g)} \rightarrow 2HI_{(g)}$ ....or backwards $2HI_{(g)} \rightarrow H_{2(g)} + I_{2(g)}$.

2) As the **reactants** get used up, the **forward** reaction **slows down** — and as more **product** is formed, the **reverse** reaction **speeds up**.

3) After a while, the forward reaction will be going at exactly the **same rate** as the backward reaction. The amounts of reactants and products **won't be changing** any more, so it'll seem like **nothing's happening**. It's a bit like you're **digging a hole**, while someone else is **filling it in** at exactly the **same speed**. This is called a **dynamic equilibrium**.

4) A **dynamic equilibrium** can only happen in a **closed system**. This just means nothing can get in or out.

Le Chatelier's Principle Predicts what will Happen if Conditions are Changed

If you **change** the **concentration**, **pressure** or **temperature** of a reversible reaction, you're going to **alter** the **position of equilibrium**. This just means you'll end up with **different amounts** of reactants and products at equilibrium.

If the position of equilibrium moves to the **left**, you'll get more **reactants**.

$$H_{2(g)} + I_{2(g)} \rightleftharpoons 2HI_{(g)}$$

Mr and Mrs Le Chatelier celebrate another successful year in the principle business

If the position of equilibrium moves to the **right**, you'll get more **products**.

$$H_{2(g)} + I_{2(g)} \rightleftharpoons 2HI_{(g)}$$

Le Chatelier's principle tells you how the **position of equilibrium** will change if a **condition changes**:

If there's a change in **concentration**, **pressure** or **temperature**, the equilibrium will move to help **counteract** the change.

So, basically, if you **raise the temperature**, the position of equilibrium will shift to try to **cool things down**. And, if you **raise the pressure or concentration**, the position of equilibrium will shift to try to **reduce it again**.

Catalysts Don't Affect The Position of Equilibrium

Catalysts have **NO EFFECT** on the **position of equilibrium**. They **can't** increase **yield** — but they **do** mean equilibrium is reached **faster**.

Reversible Reactions

Here's Some **Handy Rules** for Using **Le Chatelier's Principle**

CONCENTRATION $2SO_{2(g)} + O_{2(g)} \rightleftharpoons 2SO_{3(g)}$

1) If you **increase the concentration** of a **reactant** (SO_2 or O_2), the equilibrium tries to **get rid** of the extra reactant. It does this by making **more product** (SO_3). So the equilibrium's shifted to the **right**.

2) If you **increase the concentration** of the **product** (SO_3), the equilibrium tries to remove the extra product. This makes the **reverse reaction** go faster. So the equilibrium shifts to the **left**.

3) **Decreasing** the concentrations has the **opposite effect**.

PRESSURE   (changing this only affects **equilibria involving gases**)

1) **Increasing** the pressure shifts the equilibrium to the side with **fewer** gas molecules. This **reduces** the pressure.

2) **Decreasing** the pressure shifts the equilibrium to the side with **more** gas molecules. This **raises** the pressure again.

> There's 3 moles on the left, but only 2 on the right.
> So, an increase in pressure shifts the equilibrium to the right. $\Longrightarrow$ $2SO_{2(g)} + O_{2(g)} \rightleftharpoons 2SO_{3(g)}$

TEMPERATURE

1) **Increasing** the temperature means **adding heat**.
The equilibrium shifts in the **endothermic (positive ΔH) direction** to absorb this heat.

2) **Decreasing** the temperature **removes heat**.
The equilibrium shifts in the **exothermic (negative ΔH) direction** to try to replace the heat.

3) If the forward reaction's **endothermic**, the reverse reaction will be **exothermic**, and vice versa.

> This reaction's exothermic in the forward direction.
> If you increase the temperature, the equilibrium shifts to the left to absorb the extra heat.
>
> Exothermic $\Longrightarrow$
> $2SO_{2(g)} + O_{2(g)} \rightleftharpoons 2SO_{3(g)}$ $\Delta H = -197$ kJ mol^{-1}
> $\Longleftarrow$ Endothermic

Right, you've got to be able to **apply** this Le Chatelier's Principle stuff to industrial processes — like the production of **ethanol** and **methanol**.

Ethanol can be formed from Ethene and Steam

1) **Ethanol** is produced via a **reversible exothermic reaction** between **ethene** and **steam**:

$$C_2H_{4(g)} + H_2O_{(g)} \rightleftharpoons C_2H_5OH_{(g)}   \Delta H = -46 \text{ kJ mol}^{-1}$$

2) The reaction is carried out at a pressure of **60-70 atmospheres** and a temperature of **300 °C**, with a catalyst of **phosphoric acid**.

The Conditions Chosen are a Compromise

1) Because it's an **exothermic reaction**, **lower** temperatures favour the forward reaction.
This means that at lower temperatures **more** ethane and steam is converted to ethanol — you get a better **yield**.

2) But **lower temperatures** mean a **slower rate of reaction**. You'd be **daft** to try to get a **really high yield** of ethanol if it's going to take you 10 years. So the 300 °C is a **compromise** between **maximum yield** and a **faster reaction**.

3) **Higher pressures** favour the **forward reaction**, so a pressure of **60-70 atmospheres** is used — **high pressure** moves the reaction to the side with **fewer molecules of gas**. **Increasing the pressure** also increases the **rate** of reaction.

4) Cranking up the pressure as high as you can sounds like a great idea so far. But **high pressures** are **expensive** to produce. You need **stronger pipes** and **containers** to withstand high pressure. And, in this process, increasing the pressure can also cause **side reactions** to occur.

5) So the **60-70 atmospheres** is a **compromise** between **maximum yield** and **expense**.
In the end, it all comes down to **minimising costs**.

Recycling Unreacted Ethene Also Saves Money

1) Only a **small proportion** of the ethene reacts each time the gases pass through the catalyst.

2) To save money and raw materials, the **unreacted ethene** is separated from the liquid ethanol and **recycled** back into the reactor. Thanks to this around **95%** of the ethene is eventually converted to ethanol.

Reversible Reactions

Methanol can be Produced from Hydrogen and Carbon Monoxide

1) **Methanol** is also made industrially in a **reversible reaction**. It's made from **hydrogen** and **carbon monoxide**:

$$2H_{2(g)} + CO_{(g)} \rightleftharpoons CH_3OH_{(g)} \qquad \Delta H = \text{-90 kJ mol}^{-1}$$

Industrial conditions — **pressure:** 50-100 atmospheres, **temperature:** 250 °C, **catalyst:** mixture of copper, zinc oxide and aluminium oxide

2) Just like with the production of **ethanol**, the conditions used are a **compromise** between keeping **costs** low and **yield** high.

Methanol and ethanol are used as fuels in some forms of motor racing.

Methanol and Ethanol are Important Liquid Fuels

1) Methanol is mainly used to make other chemicals, but both **methanol** and **ethanol** can also be used as **fuels for cars** — either on their own, or added to petrol.

2) Ethanol and methanol are thought of as **greener** than petrol — they can be made from **renewable resources** and they produce **fewer pollutants** (like NO_x and CO).

3) Methanol and ethanol can both be **carbon neutral fuels** (pretty much). See page 76 for more on why ethanol is thought of as carbon neutral.

Something is carbon neutral if it has no net annual carbon (greenhouse gas) emissions to the atmosphere.

Practice Questions

Q1 Using an example, explain the terms 'reversible' and 'dynamic equilibrium'.

Q2 If the equilibrium moves to the right, do you get more products or reactants?

Q3 A reaction at equilibrium is endothermic in the forward direction. What happens to the position of equilibrium as the temperature is increased?

Q4 Write down the equation for making ethanol from ethene and steam.

Q5 What does 'carbon neutral' mean?

Exam Questions

Q1 Nitrogen and oxygen gases were reacted together in a closed flask and allowed to reach equilibrium with the nitrogen monoxide formed. The forward reaction is endothermic.

$$N_{2(g)} + O_{2(g)} \rightleftharpoons 2NO_{(g)}$$

a) State Le Chatelier's principle. [1 mark]

b) Explain how the following changes would affect the position of equilibrium of the above reaction:
   (i)   Pressure is **increased**. [2 marks]
   (ii) Temperature is **reduced**. [2 marks]
   (iii) Nitrogen monoxide is removed. [1 mark]

c) What would be the effect of a catalyst on the composition of the equilibrium mixture? [1 mark]

Q2 The manufacture of ethanol can be represented by the reaction: $C_2H_{4(g)} + H_2O_{(g)} \rightleftharpoons C_2H_5OH_{(g)}$ $\Delta H = \text{-46 kJ mol}^{-1}$
Typical conditions are 300 °C and 60-70 atmospheres.

a) Explain, in molecular terms, why a temperature lower than the one quoted is not used. [3 marks]

b) Explain why a pressure higher than the one quoted is not often used. [2 marks]

Only going forward cos we can't find reverse...

*Equilibria never do what you want them to do. They always **oppose** you. Be sure you know what happens to an equilibrium if you change the conditions. A word about pressure — if there's the same number of gas moles on each side of the equation, then you can raise the pressure as high as you like and it won't make a blind bit of difference to the position of equilibrium.*

Redox Reactions

This double page has more occurences of "oxidation" than the Beatles' "All You Need is Love" features the word "love".

If Electrons are Transferred, it's a **Redox Reaction**

1) A **loss** of electrons is called **oxidation**. A **gain** in electrons is called **reduction**.

2) Reduction and oxidation happen **simultaneously** — hence the term "**redox**" reaction.

3) An **oxidising agent accepts** electrons and gets reduced.

4) A **reducing agent donates** electrons and gets oxidised.

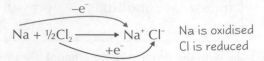

$$Na + \tfrac{1}{2}Cl_2 \xrightarrow[+e^-]{-e^-} Na^+ Cl^-$$

Na is oxidised
Cl is reduced

Sometimes it's easier to talk about **Oxidation States**

(It's also called oxidation <u>number</u>.)

There are lots of rules. Take a deep breath...

1) All atoms are treated as **ions** for this, even if they're covalently bonded.

2) Uncombined **elements** have an oxidation state of **0**.

3) Elements just bonded to **identical atoms**, like O_2 and H_2, also have an oxidation state of **0**.

4) The oxidation state of a simple **monatomic ion**, e.g. Na^+, is the same as its **charge**.

5) In **compounds** or **compound ions**, the **overall oxidation state** is just the ion charge.

SO_4^{2-} — **overall oxidation state = –2**,
oxidation state of **O = –2** (total = –8),
so oxidation state of **S = +6**

> Within an ion, the most electronegative element has a negative oxidation state (equal to its ionic charge). Other elements have more positive oxidation states.

6) The sum of the oxidation states for a **neutral compound** is 0.

Fe_2O_3 — **overall oxidation state = 0**, oxidation state of **O = –2** (total = –6), so oxidation state of **Fe = +3**

7) Combined **oxygen** is nearly always –2, except in peroxides, where it's –1, (and in the fluorides OF_2, where it's +2, and O_2F_2, where it's +1 (and O_2 where it's 0).

In H_2O, oxidation state of O = –2, but in H_2O_2, oxidation state of H has to be +1 (an H atom can only lose one electron), so oxidation state of **O = –1**

8) Combined **hydrogen** is +1, except in metal hydrides where it is –1 (and H_2 where it's 0).

In **HF**, oxidation state of H = +1, but in **NaH**, oxidation state of H = –1

Roman Numerals Give Oxidation States

Sometimes, oxidation states aren't clear from the formula of a compound.

If you see **Roman numerals** in a chemical name, it's an **oxidation number**.

E.g. copper has oxidation state **+2** in **copper(II) sulfate** and manganese has oxidation state **+7** in a **manganate(VII) ion** (MnO_4^-)

Hands up if you like Roman numerals...

Redox Reactions

You can Write **Half-Equations** and Combine them into **Redox Equations**

1) **Ionic half-equations** show oxidation or reduction.

2) You can **combine** half-equations for different oxidising or reducing agents together to make **full equations** for redox reactions.

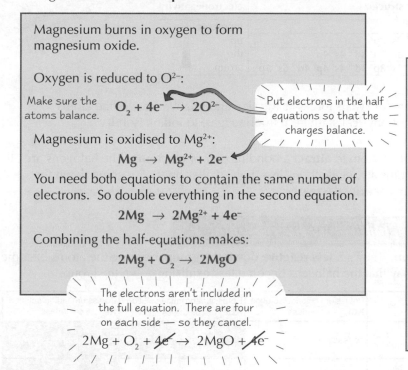

Magnesium burns in oxygen to form magnesium oxide.

Oxygen is reduced to O^{2-}:

Make sure the atoms balance.

$$O_2 + 4e^- \rightarrow 2O^{2-}$$

Put electrons in the half equations so that the charges balance.

Magnesium is oxidised to Mg^{2+}:

$$Mg \rightarrow Mg^{2+} + 2e^-$$

You need both equations to contain the same number of electrons. So double everything in the second equation.

$$2Mg \rightarrow 2Mg^{2+} + 4e^-$$

Combining the half-equations makes:

$$2Mg + O_2 \rightarrow 2MgO$$

The electrons aren't included in the full equation. There are four on each side — so they cancel.

$$2Mg + O_2 + \cancel{4e^-} \rightarrow 2MgO + \cancel{4e^-}$$

Aluminium reacts with chlorine to form aluminium chloride.

Aluminium is oxidised to Al^{3+}:

$$Al \rightarrow Al^{3+} + 3e^-$$

Chlorine is reduced to Cl^-:

$$Cl_2 + 2e^- \rightarrow 2Cl^-$$

Make sure the atoms and charges balance.

Now make sure the equations each contain the same number of electrons.

$$Al \rightarrow Al^{3+} + 3e^- \xrightarrow{\times 2} 2Al \rightarrow 2Al^{3+} + 6e^-$$

$$Cl_2 + 2e^- \rightarrow 2Cl^- \xrightarrow{\times 3} 3Cl_2 + 6e^- \rightarrow 6Cl^-$$

Combining the half-equations makes:

$$2Al + 3Cl_2 \rightarrow 2AlCl_3$$

Practice Questions

Q1 What is a reducing agent?

Q2 What is the usual oxidation state of oxygen combined with another element?

Q3 What is the oxidation state of hydrogen in H_2 gas?

Q4 What always appears in a half-equation that does not appear in a full equation?

Exam Questions

Q1 Lithium oxide forms when lithium is burned in air.
The equation for the combustion of lithium is: $4Li_{(s)} + O_{2(g)} \rightarrow 2Li_2O_{(s)}$

a) Define oxidation and reduction in terms of the movement of electrons. [1 mark]

b) What is the oxidation state of lithium in:
(i) Li   (ii) Li_2O [2 marks]

c) Write half-equations for the reaction of lithium with oxygen.
State which reactant is oxidised and which is reduced. [5 marks]

Q2 Halogens are powerful oxidising agents.
The half-equation for chlorine acting as an oxidising agent is: $Cl_2 + 2e^- \rightarrow 2Cl^-$

a) Define the term oxidising agent in terms of electron movement. [1 mark]

b) Write a balanced half-equation for the oxidation of indium metal to form In^{3+} ions. [2 marks]

c) Use your answer to b) and the equation above to form a balanced equation
for the reaction of indium with chlorine by combining half-equations. [2 marks]

Redox — relax in a lovely warm bubble bath...

Ionic equations are so evil even Satan wouldn't mess with them. But they're on the syllabus, so you can't ignore them. Have a flick back to p16 if they're freaking you out.

And while we're on the oxidation page, I suppose you ought to learn the most famous memory aid thingy in the world...

OIL RIG
- **Oxidation Is Loss**
- **Reduction Is Gain**
(of electrons)

Group 7 — The Halogens

Here comes a page jam-packed with golden nuggets of halogen fun. Oh yes, I kid you not.
This page is the Alton Towers of AS Chemistry... white-knuckle excitement all the way...

The word halogen should be used when describing the atom (X) or molecule (X_2), but the word halide is used to describe the negative ion (X^-).

Halogens are the Highly Reactive Non-Metals of Group 7

The table below gives some of the main properties of the first 4 halogens.

halogen	formula	colour	physical state	electronic structure	electronegativity
fluorine	F_2	pale yellow	gas	$1s^2\ 2s^2\ 2p^5$	increases
chlorine	Cl_2	green	gas	$1s^2\ 2s^2\ 2p^6\ 3s^2\ 3p^5$	up
bromine	Br_2	red-brown	liquid	$1s^2\ 2s^2\ 2p^6\ 3s^2\ 3p^6\ 3d^{10}\ 4s^2\ 4p^5$	the
iodine	I_2	grey	solid	$1s^2\ 2s^2\ 2p^6\ 3s^2\ 3p^6\ 3d^{10}\ 4s^2\ 4p^6\ 4d^{10}\ 5s^2\ 5p^5$	group

1) **Their boiling points increase down the group**
This is due to the increasing strength of the **Van der Waals forces** as the size and relative mass of the atoms increases. This trend is shown in the changes of **physical state** from fluorine (gas) to iodine (solid).

2) **Electronegativity decreases down the group**.
Electronegativity, remember, is the tendency of an atom to **attract** a bonding pair of **electrons**. The halogens are all highly electronegative elements. But larger atoms attract electrons **less** than smaller ones. So, going down the group, as the atoms become **larger**, the electronegativity **decreases**.

Halogens Displace Less Reactive Halide Ions from Solution

1) When the halogens react, they **gain an electron**. They get **less reactive down the group**, because the atoms become larger (and less electronegative). So you can say that the halogens become **less oxidising** down the group.

2) The **relative oxidising strengths** of the halogens can be seen in their **displacement reactions** with the halide ions:

A **halogen** will **displace a halide** from solution if the halide is **below it** in the periodic table.

	Potassium chloride solution $KCl_{(aq)}$ - colourless	Potassium bromide solution $KBr_{(aq)}$ - colourless	Potassium iodide solution $KI_{(aq)}$ - colourless
Chlorine water $Cl_{2(aq)}$ - colourless	no reaction	orange solution (Br_2) formed	brown solution (I_2) formed
Bromine water $Br_{2(aq)}$ - orange	no reaction	no reaction	brown solution (I_2) formed
Iodine solution $I_{2(aq)}$ - brown	no reaction	no reaction	no reaction

3) These displacement reactions can be used to help **identify** which halogen (or halide) is present in a solution.

Halogen	Displacement reaction	Ionic equation
Cl	chlorine (Cl_2) will displace bromide (Br^-) and iodide (I^-)	$Cl_{2(aq)} + 2Br^-_{(aq)} \rightarrow 2Cl^-_{(aq)} + Br_{2(aq)}$ $Cl_{2(aq)} + 2I^-_{(aq)} \rightarrow 2Cl^-_{(aq)} + I_{2(aq)}$
Br	bromine (Br_2) will displace iodide (I^-)	$Br_{2(aq)} + 2I^-_{(aq)} \rightarrow 2Br^-_{(aq)} + I_{2(aq)}$
I	no reaction with F^-, Cl^-, Br^-	

Chlorine and Sodium Hydroxide make Bleach

If you mix chlorine gas with dilute sodium hydroxide at **room temperature**, you get **sodium chlorate(I) solution**, $NaClO_{(aq)}$, which just happens to be common household **bleach**.

$$2NaOH_{(aq)} + Cl_{2(g)} \rightarrow NaClO_{(aq)} + NaCl_{(aq)} + H_2O_{(l)}$$

Ox. state: $\quad 0 \qquad +1 \qquad -1$

The oxidation state of Cl goes up **and** down. This is **disproportionation**.

ClO^- is the chlorate(I) ion. Chloride's oxidation state is +1 in this ion.

The sodium chlorate(I) solution (bleach) has loads of uses — it's used in **water treatment**, to bleach **paper** and **textiles**... and it's good for **cleaning toilets**, too. Handy...

Group 7 — The Halogens

Chlorine is used to Kill Bacteria in Water

When you mix chlorine with water,
it undergoes disproportionation.

You end up with a mixture of
hydrochloric acid and **chloric(I) acid**
(also called hypochlorous acid).

> Aqueous chloric(I) acid **ionises**
> to make **chlorate(I) ions**
> (also called hypochlorite ions).

Chlorate(I) ions **kill bacteria**.

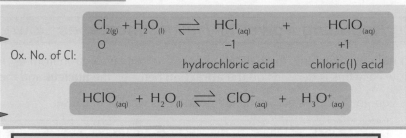

Ox. No. of Cl:

$$Cl_{2(g)} + H_2O_{(l)} \rightleftharpoons HCl_{(aq)} + HClO_{(aq)}$$
$$0 \qquad\qquad\qquad -1 \qquad\qquad +1$$

hydrochloric acid   chloric(I) acid

$$HClO_{(aq)} + H_2O_{(l)} \rightleftharpoons ClO^-_{(aq)} + H_3O^+_{(aq)}$$

So, **adding chlorine** (or a compound containing
chlorate(I) ions) to water can make it safe to **drink** or **swim** in.

1) In the UK our drinking water is **treated** to make it safe. **Chlorine** is an important part of water treatment:

   - It **kills disease-causing microorganisms**.
   - Some chlorine persists in the water and **prevents reinfection** further down the supply.
   - It prevents the growth of algae, eliminating **bad tastes** and **smells**, and **removes discolouration** caused by organic compounds.

2) However, there are risks from using chlorine to treat water:

   - **Chlorine gas** is **very harmful** if it's breathed in — it irritates the **respiratory system**. **Liquid chlorine** on the skin or eyes causes severe **chemical burns**. Accidents involving chlorine could be really serious, or fatal.
   - Water contains a variety of organic compounds, e.g. from the decomposition of plants. Chlorine reacts with these compounds to form **chlorinated hydrocarbons**, e.g. chloromethane (CH_3Cl) — and many of these chlorinated hydrocarbons are carcinogenic (cancer-causing). However, this increased cancer risk is small compared to the risks from untreated water — a cholera epidemic, say, could kill thousands of people.

3) There are ethical considerations too. We don't get a **choice** about having our water chlorinated — some people object to this as forced 'mass medication'.

Practice Questions

Q1 Place the halogens F, Cl, Br and I in order of increasing: (a) boiling point, (b) electronegativity.

Q2 What would be seen when chlorine water is added to potassium iodide solution?

Q3 How is common household bleach formed?

Q4 Write the equation for the reaction of chlorine with water. State underneath the oxidation numbers of chlorine.

Exam Questions

Q1 a) Write an ionic equation for the reaction between iodine solution and sodium astatide (NaAt). [1 mark]

    b) For the equation in (a), deduce which substance is oxidised. [1 mark]

Q2 a) Describe and explain the trends in:
       (i) the boiling points of Group 7 elements as you move down the Periodic Table. [3 marks]
       (ii) electronegativity of Group 7 elements as you move down the Periodic Table. [3 marks]

    b) Which halogen is the most powerful oxidising agent? [1 mark]

Q3 Chlorine is added to the water in public swimming baths in carefully controlled quantities.

    a) Write an equation for the reaction of chlorine with water [2 marks]

    b) Why is chlorine added to the water in swimming baths,
and why must the quantity added be carefully controlled? [2 marks]

Don't skip this page — it could cost you £31 000...

Let me explain... the other night I was
watching Who Wants to Be a Millionaire,
and this question was on for £32 000:

Which of the these elements is a halogen?
A Argon   B Nitrogen
C Fluorine   D Sodium

Bet Mr Redmond from Wiltshire
wishes he paid more attention in
Chemistry now, eh. Ha sucker...

Halide Ions

OK, a quick reminder of the basics first. Halides are compounds with the −1 halogen ion (e.g. Cl⁻, Br⁻, I⁻) like KI, HCl, NaBr. They all end in "-ide" — chloride, bromide, iodide. Got that? Good. Now, you're ready to go in...

The **Reducing Power** of Halides **Increases** Down the Group...

To reduce something, the halide ion needs to lose an electron from its outer shell. How easy this is depends on the **attraction** between the **nucleus** and the outer **electrons**.

As you go down the group, the attraction gets **weaker** because:

> 1) the ions get bigger, so the electrons are **further** away from the positive nucleus
>
> 2) there are extra inner electron shells, so there's a greater **shielding** effect.

An example of them doing this is the good old halogen / halide displacement reaction (the one you learned on p56... yes, that one). And here comes some more examples to learn...

...which Explains their Reactions with **Sulfuric Acid**

All the halides react with concentrated sulfuric acid to give a **hydrogen halide** as a product to start with. But what happens next depends on which halide you've got...

Reaction of NaF or NaCl with H_2SO_4

$$NaF_{(s)} + H_2SO_{4(aq)} \rightarrow NaHSO_{4(s)} + HF_{(g)}$$

$$NaCl_{(s)} + H_2SO_{4(aq)} \rightarrow NaHSO_{4(s)} + HCl_{(g)}$$

1) Hydrogen fluoride (HF) or hydrogen chloride gas (HCl) is formed. You'll see misty fumes as the gas comes into contact with moisture in the air.

2) But HF and HCl aren't strong enough reducing agents to reduce the sulfuric acid, so the reaction stops there.

3) It's not a redox reaction — the oxidation states of the halide and sulfur stay the same (−1 and +6).

Reaction of NaBr with H_2SO_4

$$NaBr_{(s)} + H_2SO_{4(aq)} \rightarrow NaHSO_{4(s)} + HBr_{(g)}$$

$$2HBr_{(aq)} + H_2SO_{4(aq)} \rightarrow Br_{2(g)} + SO_{2(g)} + 2H_2O_{(l)}$$

| ox. state of S: | | +6 | → | +4 | | reduction |
| ox. state of Br: | −1 | | | → 0 | | oxidation |

1) The first reaction gives misty fumes of hydrogen bromide gas (HBr).

2) But the HBr is a stronger reducing agent than HCl and reacts with the H_2SO_4 in a redox reaction.

3) The reaction produces choking fumes of SO_2 and orange fumes of Br_2.

Reaction of NaI with H_2SO_4

$$NaI_{(s)} + H_2SO_{4(aq)} \rightarrow NaHSO_{4(s)} + HI_{(g)}$$

$$2HI_{(g)} + H_2SO_{4(aq)} \rightarrow I_{2(s)} + SO_{2(g)} + 2H_2O_{(l)}$$

| ox. state of S: | | +6 | → | +4 | | reduction |
| ox. state of I: | −1 | | → 0 | | | oxidation |

1) Same initial reaction giving HI gas.

2) The HI then reduces H_2SO_4 like above.

3) But HI (being well 'ard as far as reducing agents go) keeps going and reduces the SO_2 to H_2S.

$$6HI_{(g)} + SO_{2(g)} \rightarrow H_2S_{(g)} + 3I_{2(s)} + 2H_2O_{(l)}$$

| ox. state of S: | | +4 | → | −2 | | reduction |
| ox. state of I: | −1 | | → | 0 | | oxidation |

H_2S gas is toxic and smells of bad eggs. A bit like my mate Andy at times...

Halide Ions

Silver Nitrate Solution is used to Test for Halides

The test for halides is dead easy. First you add **dilute nitric acid** to remove ions which might interfere with the test. Then you just add **silver nitrate solution** ($AgNO_{3\,(aq)}$). A **precipitate** is formed (of the silver halide).

$$Ag^+_{(aq)} + X^-_{(aq)} \rightarrow AgX_{(s)} \quad \text{...where X is F, Cl, Br or I}$$

1) The **colour** of the precipitate identifies the halide.

SILVER NITRATE TEST FOR HALIDE IONS...

Fluoride F^-: no precipitate

Chloride Cl^-: white precipitate

Bromide Br^-: cream precipitate

Iodide I^-: yellow precipitate

2) Then to be extra sure, you can test your results by adding **ammonia solution**. Each silver halide has a different solubility in ammonia.

SOLUBILITY OF SILVER HALIDE PRECIPITATES IN AMMONIA...

Chloride Cl^-: white precipitate, dissolves in dilute $NH_{3(aq)}$

Bromide Br^-: cream precipitate, dissolves in conc. $NH_{3(aq)}$

Iodide I^-: yellow precipitate, insoluble in conc. $NH_{3(aq)}$

Practice Questions

Q1 Give two reasons why a bromide ion is a more powerful reducing agent than a chloride ion.

Q2 Name the gaseous products formed when sodium bromide reacts with concentrated sulfuric acid.

Q3 What is produced when potassium iodide reacts with concentrated sulfuric acid?

Q4 How would you test whether an aqueous solution contained chloride ions?

Exam Questions

Q1 Describe the tests you would carry out in order to distinguish between solid samples of sodium chloride and sodium bromide using:

a) silver nitrate solution and aqueous ammonia,

b) concentrated sulfuric acid.

For each test, state your observations and write equations for the reactions which occur. [11 marks]

Q2 The halogen below iodine in Group 7 is astatine (At). Predict, giving an explanation, whether or not:

a) hydrogen sulfide gas would be evolved when concentrated sulfuric acid is added to a solid sample of sodium astatide. [4 marks]

b) silver astatide will dissolve in concentrated ammonia solution. [3 marks]

[Sing along with me] "Why won't this section end... Why won't this section end..."

AS Chemistry. What a bummer, eh... No one ever said it was going to be easy. Not even your teacher would be that cruel. There are plenty more equations on this page to learn. As well as that, make sure you really understand everything... the trend in the reducing power of halides... and the reactions with sulfuric acid... And no, you can't swap to English. Sorry.

Group 2 — The Alkaline Earth Metals

Group 2, AKA the alkaline earth metals, are in the "s block" of the periodic table. There's three pages about these jolly fellas and their compounds, so we've got a lot to do — best get on...

1) Group 2 Elements **Lose Two Electrons** when they React

Element	Atom	Ion
Be	$1s^2 2s^2$	$1s^2$
Mg	$1s^2 2s^2 2p^6 3s^2$	$1s^2 2s^2 2p^6$
Ca	$1s^2 2s^2 2p^6 3s^2 3p^6 4s^2$	$1s^2 2s^1 2p^6 3s^2 3p^6$

Group 2 elements all have two electrons in their outer shell (s^2).

They lose their two outer electrons to form **2+ ions**. Their ions then have every atom's dream electronic structure — that of a **noble gas**.

2) Atomic Radius **Increases** Down the Group

This is because of the extra **electron shells** as you go down the group.

3) Ionisation Energy **Decreases** Down the Group

1) Each element down Group 2 has an **extra electron shell** compared to the one above.

2) The extra inner shells **shield** the outer electrons from the attraction of the nucleus.

3) Also, the extra shell means that the outer electrons are **further away** from the nucleus, which greatly reduces the nucleus's attraction.

Both of these factors make it **easier** to remove outer electrons, resulting in a **lower ionisation energy**.

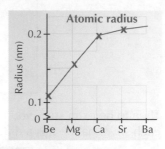

Mr Kelly has one final attempt at explaining electron shielding to his students...

The positive charge of the nucleus does increase as you go down a group (due to the extra protons), but this effect is overridden by the effect of the extra shells.

4) Reactivity **Increases** Down the Group

1) As you go down the group, the **ionisation energies** decrease. This is due to the increasing atomic radius and shielding effect (see above).

2) When Group 2 elements react they **lose electrons**, forming positive ions. The easier it is to lose electrons (i.e. the lower the first and second ionisation energies), the more reactive the element, so **reactivity increases** down the group.

5) Melting Points Generally **Decrease** Down the Group

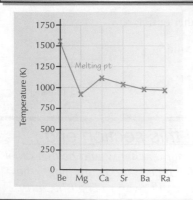

1) The Group 2 elements have typical **metallic structures**, with the electrons of their outer shells being **delocalised**.

2) Going down the group the metallic ions get **bigger** — so they have a smaller **charge/volume ratio**. But the number of delocalised electrons per atom doesn't change (it's always 2) — so the delocalised electrons get more **spread out**.

3) These two factors mean there's reduced attraction of the positive ions to the 'sea' of delocalised electrons. So it takes **less energy** to break the bonds, which means lower melting points generally down the group. However, there's a big 'blip' at magnesium, because the crystal structure (the arrangement of the metallic ions) changes.

Group 2 — The Alkaline Earth Metals

Group 2 Elements React With Water

When Group 2 elements react, they are **oxidised** from a state of **0** to **+2**, forming M^{2+} ions.

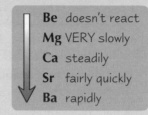

$$M \rightarrow M^{2+} + 2e^-$$

Oxidation state: **0**   **+2**   E.g.

$$Ca \rightarrow Ca^{2+} + 2e^-$$

0   **+2**

The Group 2 metals react with water to give a **metal hydroxide and hydrogen**.

Oxidation state:

$$M_{(s)} + 2H_2O_{(l)} \rightarrow M(OH)_{2\,(aq)} + H_{2\,(g)}$$
$$\phantom{M_{(s)} + 2H_2O_{(l)} \rightarrow} \mathbf{0} \mathbf{+2}$$

e.g.

$$Ca_{(s)} + 2H_2O_{(l)} \rightarrow Ca(OH)_{2\,(aq)} + H_{2\,(g)}$$

They react **more readily** down the group because the **ionisation energies** decrease.

Be	doesn't react
Mg	VERY slowly
Ca	steadily
Sr	fairly quickly
Ba	rapidly

Right, that's enough about the Group 2 elements. From here on, we're looking at their cuddly compounds...

(Sorry, just trying to liven things up a bit.)

Solubility Trends Depend on the Compound Anion

Generally, compounds of Group 2 elements that contain **singly charged** negative ions (e.g. OH^-) **increase** in solubility down the group, whereas compounds that contain **doubly charged** negative ions (e.g. SO_4^{2-}) **decrease** in solubility down the group.

Group 2 element	hydroxide (OH^-)	sulfate (SO_4^{2-})
magnesium	least soluble	most soluble
calcium		
strontium		
barium	most soluble	least soluble

Compounds like magnesium hydroxide which have **very low** solubilities are said to be **sparingly soluble**.

Most sulfates are soluble in water, but **barium sulfate** is **insoluble**.

The test for sulfate ions makes use of this property...

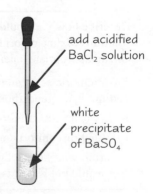

add acidified $BaCl_2$ solution

white precipitate of $BaSO_4$

Test for sulfate ions

If acidified barium chloride ($BaCl_2$) is added to a solution containing sulfate ions then a white precipitate of barium sulfate is formed.

$$Ba^{2+}_{(aq)} + SO_4^{2-}_{(aq)} \rightarrow BaSO_{4\,(s)}$$

You need to acidify the barium chloride (with, say, hydrochloric acid) to get rid of any lurking sulfites or carbonates.

Group 2 — The Alkaline Earth Metals

Group 2 Compounds are used to Neutralise Acidity

Group 2 elements are known as the **alkaline earth metals**, and many of their common compounds are used for neutralising acids. Here are a couple of common examples:

1) Calcium hydroxide (slaked lime, $Ca(OH)_2$) is used in **agriculture** to neutralise acid soils.

2) Magnesium hydroxide ($Mg(OH)_2$) is used in some indigestion tablets as an **antacid**.

In both cases, the ionic equation for the neutralisation is
$$H^+_{(aq)} + OH^-_{(aq)} \rightarrow H_2O_{(l)}$$

Daisy the cow *

Barium Sulfate is Used in 'Barium Meals'

CHRIS PRIEST / SCIENCE PHOTO LIBRARY

X-rays are great for finding broken bones, but they pass straight through soft tissue — so soft tissues, like the digestive system, don't show up on conventional X-ray pictures.

1) Barium sulfate is **opaque** to X-rays — they won't pass through it. It's used in 'barium meals' to help diagnose problems with the oesophagus, stomach or intestines.

2) A patient swallows the barium meal, which is a suspension of **barium sulfate**. The barium sulfate **coats** the tissues, making them show up on the X-rays, showing the structure of the organs.

Practice Questions

Q1 Which is the least reactive metal in Group 2? Why does reactivity with water increase down Group 2?

Q2 Which of the following increases in size down Group 2? **atomic radius, first ionisation energy, boiling point**

Q3 Which is less soluble, barium sulfate or magnesium sulfate?

Q4 How is the solubility of magnesium hydroxide often described?

Q5 Give a use of magnesium hydroxide.

Exam Questions

Q1 Use the electron configurations of magnesium and calcium to help explain the difference between their first ionisation energies. [5 marks]

Q2 The table shows the atomic radii of three elements from Group 2.

Element	Atomic radius/nm
X	0.089
Y	0.198
Z	0.176

a) Predict which element would react most rapidly with water. [1 mark]

b) Explain your answer. [2 marks]

Q2 Describe how you could use barium chloride solution to distinguish between solutions of zinc chloride and zinc sulfate. Give the expected observations and an appropriate balanced equation including state symbols. [4 marks]

Q4 Hydrochloric acid can be produced in excess quantities in the stomach, causing indigestion. Antacid tablets often contain sodium hydrogencarbonate ($NaHCO_3$), which reacts with the acid to form a salt, carbon dioxide and water.

a) Write an equation for the neutralisation of hydrochloric acid with sodium hydrogencarbonate. [1 mark]

b) What discomfort could be caused by the carbon dioxide produced? [1 mark]

c) From your knowledge of Group 2 compounds, choose an alternative antacid that would not give this problem and write an equation for its reaction with hydrochloric acid. [2 marks]

Bored of Group 2 trends? Me too. Let's play noughts and crosses...

x	0	
x	0	x
0	0	

Noughts and crosses is pretty rubbish really, isn't it? It's always a draw. Ho hum. Back to Chemistry then, I guess...

* She wanted to be in the book. I said OK

Extraction of Metals

Metals are handy for making metal things. Sadly, you don't just find big lumps of pure metal lying about, ready to use...

Sulfide Ores are Usually Converted to Oxides First

1) An **ore** is a natural substance that a **metal** can be economically extracted from. In other words, a rock you can get quite a bit of metal out of.

2) Metals are often found in ores as **sulfides** (such as lead sulfide and zinc sulfide), or **oxides** (like titanium dioxide and iron(III) oxide). The metal **element** needs to be removed from these compounds — that's where the chemistry comes in.

3) The first step to extract a metal from a **sulfide ore** is to turn it into an **oxide**. This is done by **roasting** the sulfide in air.

When extracting metals, Jimmy liked to use his ores and cart.

E.g.

| zinc sulfide | + | oxygen | → | zinc oxide | + | sulfur dioxide |
| $2ZnS_{(s)}$ | + | $3O_{2(g)}$ | → | $2ZnO_{(s)}$ | + | $2SO_{2(g)}$ |

4) Here's the bad news: **sulfur dioxide** gas causes **acid rain**. Acid rain can cause harm to plants and aquatic life, and damage limestone buildings, so the sulfur dioxide can't be **released** into the atmosphere.

5) But here's the good news: by converting the sulfur dioxide to **sulfuric acid** a pollutant is avoided, and a **valuable product** is made — sulfuric acid's in demand because it's used in many chemical and manufacturing processes.

Oxides are Reduced to the Metal

1) The method for **reducing** the oxide depends on the metal you're trying to extract.

2) **Carbon** (as coke — a solid fuel made from coal) and **carbon monoxide** are used as **reducing agents** for quite a few metals — usually the ones that are **less reactive** than carbon.

You need to know these three **examples** of the extraction of metals with carbon and carbon monoxide:

REDUCTION OF IRON(III) OXIDE

Iron(III) oxide is reduced by **carbon** or **carbon monoxide** to iron and carbon dioxide.

$$2Fe_2O_3 \; + \; 3C \; \rightarrow \; 4Fe \; + \; 3CO_2$$
$$Fe_2O_3 \; + \; 3CO \; \rightarrow \; 2Fe \; + \; 3CO_2$$

This happens in a blast furnace at temperatures greater than 700 °C.

REDUCTION OF MANGANESE(IV) OXIDE (MANGANESE DIOXIDE)

Manganese(IV) oxide is reduced with **carbon** (as coke) or **carbon monoxide** in a blast furnace.

$$MnO_2 \; + \; C \; \rightarrow \; Mn \; + \; CO_2$$
$$MnO_2 \; + \; 2CO \; \rightarrow \; Mn \; + \; 2CO_2$$

This needs higher temperatures than iron(III) oxide — about 1200 °C.

REDUCTION OF COPPER CARBONATE

Copper can be extracted using **carbon**.

One ore of copper is **malachite**, containing $CuCO_3$. This can be heated directly with carbon.

$$2CuCO_3 \; + \; C \; \rightarrow \; 2Cu \; + \; 3CO_2$$

Another method involves heating the carbonate until it decomposes, then reducing the oxide with carbon.

$$CuCO_3 \; \rightarrow \; CuO \; + \; CO_2$$
$$2CuO \; + \; C \; \rightarrow \; 2Cu \; + \; CO_2$$

Carbon and **carbon monoxide** are the first choice for extracting metals because they're **cheap**.
But they're not always suitable — some metals have to be extracted by other methods.
You'll see three examples on the next page...

Extraction of Metals

Tungsten is Extracted Using Hydrogen

1) **Tungsten** can be extracted from its oxide with carbon, but that can leave **impurities** which make the metal more **brittle**. If pure tungsten is needed, the ore is reduced using **hydrogen** instead.

$$WO_{3(s)} + 3H_{2(g)} \rightarrow W_{(s)} + 3H_2O_{(g)}$$

This happens in a furnace at temperatures above 700 °C.

2) Tungsten is the **only metal** reduced on a large scale using **hydrogen**.
Hydrogen is more **expensive** but it's worth the extra cost to get pure tungsten, which is much easier to work with.

3) Hydrogen is **highly explosive** when mixed with air though, which is a bit of a hazard.

Aluminium is Extracted by Electrolysis

1) **Aluminium** is **too reactive** to extract using reduction by carbon. A very **high temperature** is needed, so extracting aluminium by reduction is **too expensive** to make it worthwhile.

2) Aluminium's ore is called **bauxite** — it's aluminium oxide, Al_2O_3, with various impurities.
First of all, these impurities are removed. Next, it's dissolved in **molten cryolite** (sodium aluminium fluoride, Na_3AlF_6), which lowers its **melting point** from a scorching 2050 °C, to a cool **970 °C**.
This reduces the operating costs.

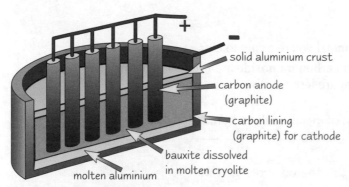

solid aluminium crust

carbon anode (graphite)

carbon lining (graphite) for cathode

bauxite dissolved in molten cryolite

molten aluminium

The current used in electrolysis is high (200 000 A), so the process is carried out where cheap electricity is available, often near hydroelectric power stations.

ELECTROLYSIS OF ALUMINIUM

1) Aluminium is produced at the **cathode** and collects as the molten liquid at the bottom of the cell.

$$Al^{3+} + 3e^- \rightarrow Al$$

2) Oxygen is produced at the **anode**.

$$2O^{2-} \rightarrow O_2 + 4e^-$$

Titanium is used in the bodies of modern planes.

Titanium is Great but a bit too Expensive

1) Titanium is a pretty **abundant** metal in the Earth's crust. In its pure form, titanium is a **strong**, **light** metal that is highly resistant to **corrosion**. Pretty much perfect really, so how come it's not used more...
Well basically, it's just a bit too **difficult** and **expensive** to produce.

2) The main ore is rutile (titanium(IV) oxide, TiO_2). You can't extract titanium from it by carbon reduction because you get titanium carbide which ruins it... $TiO_{2\,(s)} + 3C_{\,(s)} \rightarrow TiC_{\,(s)} + 2CO_{\,(g)}$

THE EXTRACTION OF TITANIUM

...is a **batch** process with several stages.

1) The ore is converted to **titanium(IV) chloride** by heating it to about 900 °C with carbon in a stream of chlorine gas.

$$TiO_{2\,(s)} + 2Cl_{2\,(g)} + 2C_{\,(s)} \rightarrow TiCl_{4\,(g)} + 2CO_{\,(g)}$$

2) The titanium chloride is purified by **fractional distillation** under an inert atmosphere of argon or nitrogen.

3) Then the chloride gets reduced in a **furnace** at almost 1000 °C. It's heated with a **more reactive** metal such as sodium or magnesium. An inert atmosphere is used to prevent side reactions.

$$TiCl_{4\,(g)} + 4Na_{\,(l)} \rightarrow Ti_{\,(s)} + 4NaCl_{\,(l)}$$
$$TiCl_{4\,(g)} + 2Mg_{\,(l)} \rightarrow Ti_{\,(s)} + 2MgCl_{2\,(l)}$$

Na and Mg are reducing agents.

Extraction of Metals

Recycling can be Good for the Environment and Save Money

Once you've got the metal out of the ore, you can keep **recycling** it again and again.
As usual, there are pros and cons:

Advantages of recycling metals:

- Saves raw materials — ores are a finite resource.
- Saves energy — recycling metals takes less energy than extracting metal. This saves money too.
- Reduces waste sent to landfill.
- Mining damages the landscape and spoil heaps are ugly. Recycling metals reduces this.

Disadvantages of recycling metals:

- Collecting and sorting metals from other waste can be difficult and expensive.
- The purity of recycled metal varies — there's usually other metals and other impurities mixed in.
- Recycling metals may not produce a consistent supply to meet demand.

Scrap Iron can be used in Copper Extraction

1) Some scrap metal can be put to other uses. For example, **scrap iron** can be used to extract **copper** from solution. This method is mainly used with **low grade** ore — ore that only contains a **small percentage** of copper.

2) Acidified water **dissolves** the copper compounds in the ore. The solution is collected and **scrap iron** is then added. The iron dissolves and **reduces** the copper(II) ions. The copper precipitates out of the solution. $Cu^{2+}_{(aq)} + Fe_{(s)} \rightarrow Cu_{(s)} + Fe^{2+}_{(aq)}$

3) This process produces copper **more slowly** than carbon reduction and has a **lower yield**, which is why it's not used with ores that have a high copper content. It's **cheaper** than carbon reduction though, because you don't need **high temperatures**, and better for the environment because there's no **CO_2** produced.

Practice Questions

Q1 What is used to reduce manganese(IV) oxide to manganese?

Q2 Write the equation for the conversion of zinc sulfide to zinc oxide.

Q3 Write the equation for the displacement of titanium from titanium chloride using sodium.

Q4 Why is sodium (or magnesium) chosen to reduce titanium chloride?

Q5 Give one environmental and one economic reason for recycling metals.

Exam Questions

Q1 The iron ore in a blast furnace contains a mixture of oxides, one of which is Fe_3O_4.
When Fe_3O_4 is reduced, both carbon and carbon monoxide act as reducing agents.
Write equations to show
a) how carbon monoxide reduces Fe_3O_4. [1 mark]
b) how carbon reduces Fe_3O_4. [1 mark]

Q2 Hydrogen is used to extract tungsten from its ore.
Give an advantage and a disadvantage of using hydrogen in place of carbon. [2 marks]

Q3 Aluminium is extracted from its purified ore by electrolysis.
a) What important step is taken to reduce the cost of extracting aluminium? [2 marks]
b) Write equations for the reactions occurring at each electrode. [2 marks]
c) Explain why aluminium is more expensive to extract than iron. [1 mark]

Extraction can be heavy going — in fact, it's like pulling teeth...

It might look like there are loads of equations to learn here — in fact, come to think of it, there are quite a few. But at least most of those oxide reduction ones are pretty similar, e.g. oxide + carbon monoxide → metal + carbon dioxide. There's actually nothing too hard on these pages at all. But there is plenty to get stuck into with your revision shovel.

Synthesis of Chloroalkanes

Wow, you've reached the last section of the book already — time flies when you're having fun...

Alkanes **Don't React** with Most Chemicals

1) The C–C bonds and C–H bonds in alkanes are pretty **non-polar**.
   But most chemicals are **polar** — like water, haloalkanes, acids and alkalis.

2) Polar chemicals are attracted to the **polar groups** on molecules they attack.
   Alkanes don't have any polar groups, so they **don't** react with polar chemicals.

3) Alkanes **will** react with some **non-polar** things though — such as oxygen or the halogens.
   But they'll **only** bother if you give them enough **energy**.

If all this polar talk means nothing to you, flick back to p28. It's all explained there.

Halogens React with **Alkanes**, Forming **Haloalkanes**   (P68 has more detail on haloalkanes.)

1) Halogens react with alkanes in **photochemical** reactions. Photochemical reactions are started by **ultraviolet** light.

2) A hydrogen atom is **substituted** (replaced) by chlorine or bromine. This is a **free-radical substitution reaction**.

   Free radicals are particles with an unpaired electron, written like this — $Cl\cdot$ or $CH_3\cdot$
   You get them when bonds split equally, and they're highly reactive.

Chlorine and **methane** react with a bit of a bang to form **chloromethane**:

$$CH_4 + Cl_2 \xrightarrow{U.V.} CH_3Cl + HCl$$

The **reaction mechanism** has three stages:

Initiation reactions — free radicals are produced.

1) Sunlight provides enough energy to break the Cl-Cl bond — this is **photodissociation**.

$$Cl_2 \xrightarrow{U.V.} 2Cl\cdot$$

2) The bond splits **equally** and each atom gets to keep one electron.
   The atom becomes a highly reactive **free radical**, $Cl\cdot$, because of its **unpaired electron**.

Propagation reactions — free radicals are used up and created in a chain reaction.

1) $Cl\cdot$ attacks a **methane** molecule: $Cl\cdot + CH_4 \rightarrow CH_3\cdot + HCl$

2) The new **methyl free radical**, $CH_3\cdot$, can attack another Cl_2 molecule: $CH_3\cdot + Cl_2 \rightarrow CH_3Cl + Cl\cdot$

3) The new $Cl\cdot$ can attack **another** CH_4 molecule, and so on, until all the Cl_2 or CH_4 molecules are wiped out.

Termination reactions — free radicals are mopped up.

1) If two free radicals join together, they make a **stable molecule**.
2) There are **heaps** of possible termination reactions.
   Here's a couple of them to give you the idea: $Cl\cdot + CH_3\cdot \rightarrow CH_3Cl$
   $$CH_3\cdot + CH_3\cdot \rightarrow C_2H_6$$

Some products formed will be trace impurities in the final sample.

More substitutions
What happens now **depends** on whether there's too much **chlorine** or too much **methane**:

1) If the **chlorine's** in excess, $Cl\cdot$ free radicals will start attacking chloromethane, producing **dichloromethane** CH_2Cl_2, **trichloromethane** $CHCl_3$, and **tetrachloromethane** CCl_4.

2) **But if the methane's** in excess, then the product will mostly be **chloromethane**.

Chloroalkanes and **Chlorofluoroalkanes** are Used as **Solvents**

1) **Chlorofluorocarbons** (**CFCs**) are haloalkane
   molecules where all of the hydrogen atoms have
   been replaced by **chlorine** and **fluorine** atoms. E.g.

2) Both **CFCs** and **chloroalkanes** can be used as **solvents** —
   they both used to be used in dry cleaning and degreasing.

trichlorofluoromethane   chlorotrifluoromethane

Synthesis of Chloroalkanes

Chlorine Atoms are Destroying The Ozone Layer

1) Ozone in the upper atmosphere acts as a **chemical sunscreen**. It absorbs a lot of the **ultraviolet radiation** which can cause sunburn or even skin cancer.

2) Ozone's **formed naturally** when an **oxygen molecule** is **broken down** into **two free radicals** by **ultraviolet radiation**. The free radicals **attack** other oxygen molecules forming **ozone**.
Just like this:

$$O_2 + h\nu \rightarrow O\bullet + O\bullet \implies O_2 + O\bullet \rightarrow O_3$$

a quantum of UV radiation

You've heard of how the **ozone layer's** being destroyed by **CFCs**, right. Well, here's what's happening.

1) **Chlorine free radicals**, $Cl\bullet$, are formed when the C–Cl bonds in **CFCs** are broken down by **ultraviolet radiation**.

E.g. $CCl_3F_{(g)} \rightarrow CCl_2F\bullet_{(g)} + Cl\bullet_{(g)}$

2) These free radicals are **catalysts**. They react with **ozone** to form an **intermediate** ($ClO\bullet$), and an oxygen molecule.

These are all gases, so it's homogeneous catalysis.

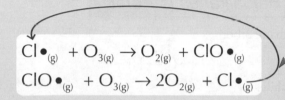
$$Cl\bullet_{(g)} + O_{3(g)} \rightarrow O_{2(g)} + ClO\bullet_{(g)}$$
$$ClO\bullet_{(g)} + O_{3(g)} \rightarrow 2O_{2(g)} + Cl\bullet_{(g)}$$

The chlorine free radical is regenerated. It goes straight on to attack another ozone molecule. It only takes one little chlorine free radical to destroy loads of ozone molecules.

3) So the **overall reaction** is...
$$2O_{3(g)} \rightarrow 3O_{2(g)}$$
... and $Cl\bullet$ is the catalyst.

CFCs Are Now Banned

1) CFCs are pretty **unreactive**, **non-flammable** and **non-toxic**. They used to be used in fire extinguishers, as propellants in aerosols, as the coolant gas in fridges and to foam plastics to make insulation and packaging materials.

2) In the 1970s scientists discovered that CFCs were causing **damage** to the **ozone layer**.
The **advantages** of CFCs couldn't outweigh the **environmental problems** they were causing, so they were **banned**.

3) Chemists have developed **alternatives** to CFCs. **HCFCs (hydrochlorofluorocarbons)** and **HFCs (hydrofluorocarbons)** are less dangerous than CFCs, so they're being used as temporary alternatives until safer products are developed.

4) Most aerosols now have been replaced by **pump spray systems** or use **nitrogen** as the propellant. Many industrial fridges use **ammonia** or **hydrocarbons** as the coolant gas, and **carbon dioxide** is used to make foamed polymers.

Practice Questions

Q1 What's a free radical?

Q2 What's photodissociation?

Q3 Give two uses of CFCs.

Q4 Describe how ozone is beneficial.

Q5 Write an equation to show what happens when UV radiation breaks down CFCs.

Q6 What is the formula for ozone?

Q7 What products are formed when a chlorine free radical reacts with an ozone molecule?

Exam Question

Q1 The alkane ethane is a saturated hydrocarbon. It is mostly unreactive, but will react with chlorine in a photochemical reaction.

   (a) What is a saturated hydrocarbon?   [2 marks]

   (b) Why is ethane unreactive with most reagents?   [2 marks]

   (c) Write an equation and outline the mechanism for the photochemical reaction of chlorine with ethane. Assume ethane is in excess. What type of mechanism is it?   [8 marks]

This stuff is like...totally radical, man...

Mechanisms are an absolute pain in the bum to learn, but unfortunately reactions are what Chemistry's all about. If you don't like it, you should have taken art — no mechanisms in that, just pretty pictures. Ah well, there's no going back now. You've just got to sit down and learn the stuff. Keep hacking away at it, till you know it all off by heart.

Haloalkanes

Don't worry if you see haloalkanes called halogenoalkanes. It's a government conspiracy to confuse you.

Haloalkanes are Alkanes with Halogen Atoms

A **haloalkane** is an alkane with at least one **halogen atom** in place of a hydrogen atom.

E.g.

Cl—C—Cl (with Cl below, H above)	H—C—C—C—H structure	F—C—C—H structure
trichloromethane	2-iodopropane	2-bromo-2-chloro-1, 1, 1-trifluoroethane

The Carbon–Halogen Bond in Haloalkanes is Polar

1) Halogens are much more **electronegative** than carbon. So, the **carbon-halogen bond** is **polar**.

2) The **δ+ carbon** doesn't have enough electrons. This means it can be attacked by a **nucleophile**. A nucleophile's an **electron-pair donor**. It donates an electron pair to somewhere without enough electrons.

3) **OH⁻**, **CN⁻** and **NH₃** are all **nucleophiles** which react with haloalkanes.

$-\overset{|}{\underset{|}{C}}{}^{\delta+}-Br^{\delta-}$

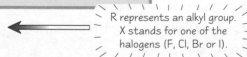

Haloalkanes are special amongst alkanes...

Haloalkanes can be Hydrolysed to make Alcohols

For example, bromoethane can be hydrolysed to ethanol. You have to use **warm aqueous sodium** or **potassium hydroxide** or it won't work. It's a **nucleophilic substitution reaction**.

Here's how it happens:

Curly arrows (and that's an official term) show the movement of an electron pair.

The OH⁻ ion acts as a nucleophile, attacking the slightly positive carbon atom.

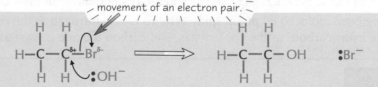

The C-Br bond is polar. The C$^{\delta+}$ attracts a lone pair of electrons from the OH⁻ ion.

The C-Br bond breaks and both the electrons are taken by the Br. A new bond forms between the C and the OH⁻ ion.

Here's the general equation for this reaction: **R–X + OH⁻ → ROH + X⁻**

R represents an alkyl group. X stands for one of the halogens (F, Cl, Br or I).

Iodoalkanes are Hydrolysed the Fastest

1) The **carbon-halogen bond strength** (or enthalpy) decides **reactivity**. For any reaction to occur the carbon-halogen bond needs to **break**.

2) The **C-F bond** is the **strongest** — it has the highest **bond enthalpy**. So **fluoroalkanes** are hydrolysed **more slowly** than other haloalkanes.

3) The **C-I bond** has the **lowest bond enthalpy**, so it's easier to break. This means that **iodoalkanes** are hydrolysed more **quickly**.

bond	bond enthalpy kJ mol⁻¹
C–F	467
C–Cl	346
C–Br	290
C–I	228

Faster hydrolysis as bond enthalpy decreases (the bonds are getting weaker).

Haloalkanes

Haloalkanes React With Ammonia to Form Amines

If you **warm** a haloalkane with excess **ethanolic** ammonia, the **ammonia** swaps places with the **halogen** — yes, it's another one of those **nucleophilic substitution reactions**.

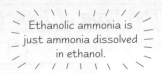

Ethanolic ammonia is just ammonia dissolved in ethanol.

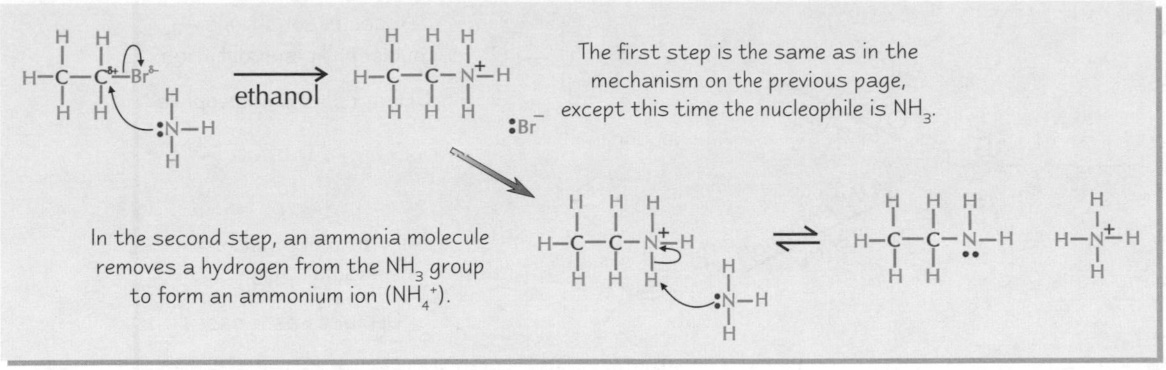

The first step is the same as in the mechanism on the previous page, except this time the nucleophile is NH_3.

In the second step, an ammonia molecule removes a hydrogen from the NH_3 group to form an ammonium ion (NH_4^+).

The ammonium ion can react with the bromine ion to form ammonium bromide. So the overall reaction is this:

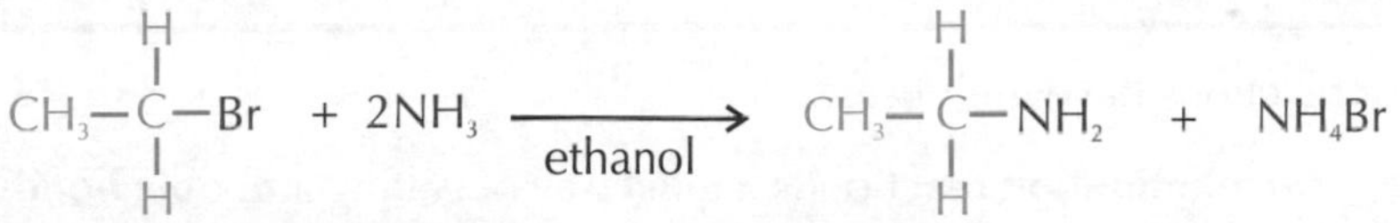

$$CH_3-\overset{\displaystyle H}{\underset{\displaystyle H}{C}}-Br \; + \; 2NH_3 \; \xrightarrow{\text{ethanol}} \; CH_3-\overset{\displaystyle H}{\underset{\displaystyle H}{C}}-NH_2 \; + \; NH_4Br$$

You can use Haloalkanes to form Nitriles

If you **warm** a haloalkane with **ethanolic potassium cyanide** you get a **nitrile**.
It's yet another **nucleophilic substitution reaction** — the **cyanide ion**, CN^-, is the **nucleophile**.

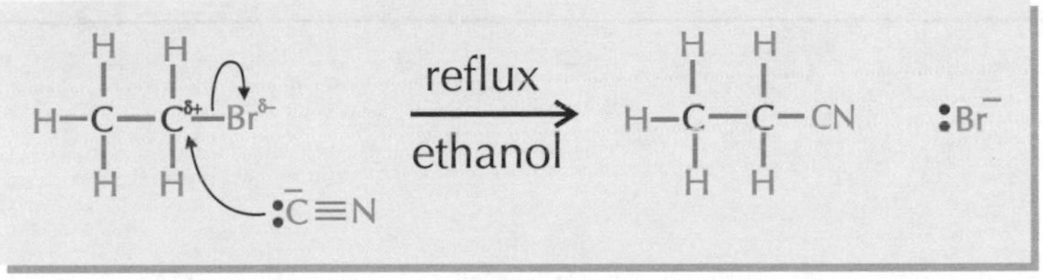

Nitriles have $-C\equiv N$ groups.

Haloalkanes also Undergo Elimination Reactions

If you warm a haloalkane with hydroxide ions dissolved in **ethanol** instead of water, an **elimination reaction** happens, and you end up with an **alkene**. This is how you do it:

1) Heat the mixture **under reflux** or you'll lose volatile stuff.

$$CH_3CHBrCH_3 + KOH \xrightarrow[\text{reflux}]{\text{ethanol}} CH_2=CH_2CH_3 + H_2O + KBr$$

These conditions are anhydrous (there's no water).

2) Here's how the reaction works:

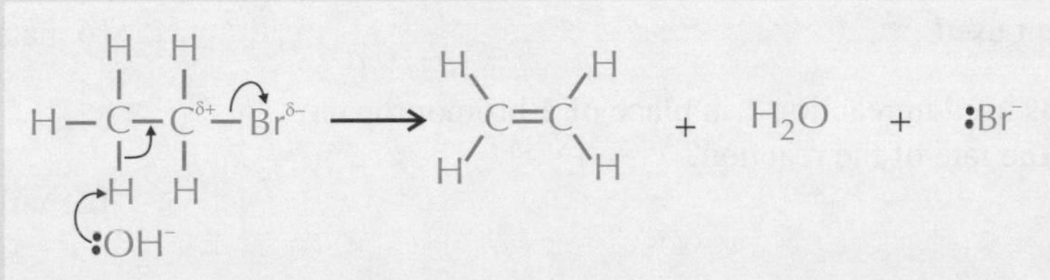

OH^- acts as a base and takes a proton, H^+, from the carbon on the left. This makes water. The left carbon now has a spare electron, so it forms a double bond with the other carbon. To form the double bond, the right carbon has to let go of the Br, which drops off as a Br^- ion.

3) This is an example of an **elimination reaction**. In an elimination **reaction**, a **small group** of atoms breaks away from a larger molecule. This **small group** is **not replaced** by anything else (whereas it would be in a substitution reaction).

In the reaction above, H and Br have been eliminated from CH_3CH_2Br to leave $CH_2=CH_2$

Haloalkanes

The Type of Reaction Depends on the Conditions

You can control what **type of reaction** happens by **changing the conditions**.

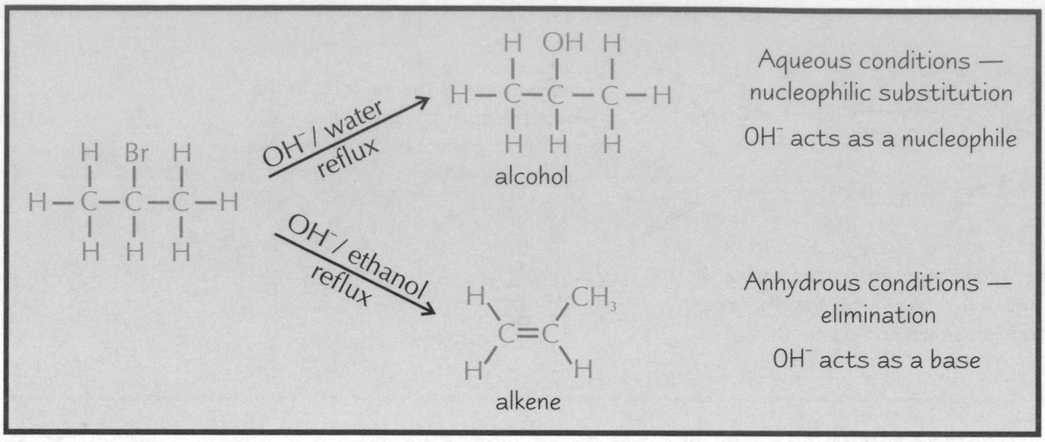

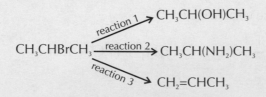

Aqueous conditions — nucleophilic substitution

OH⁻ acts as a nucleophile

alcohol

Anhydrous conditions — elimination

OH⁻ acts as a base

alkene

Both of these reactions have their uses...

1) The **elimination** reaction is a **good way** of getting a **double bond** into a molecule. Loads of other organic synthesis reactions use **alkenes**, so the elimination reaction is a good starting point for making lots of different organic chemicals.

2) The **substitution** reaction allows you to produce any **alcohol** molecule that you need. And alcohols can be the starting point for synthesis reactions that produce **aldehydes**, **ketones**, **esters**, and **carboxylic acids**.

So haloalkanes are very useful as a **starting material** for making other organic compounds.

Practice Questions

Q1 What is a haloalkane?

Q2 What is a nucleophile?

Q3 Why is the carbon-halogen bond polar?

Q4 Why does iodoethane react faster than chloro- or bromoethane with warm, aqueous sodium hydroxide?

Exam Question

Q1 Three reactions of 2-bromopropane, $CH_3CHBrCH_3$, are shown below.

$$CH_3CHBrCH_3 \xrightarrow{\text{reaction 1}} CH_3CH(OH)CH_3$$
$$\xrightarrow{\text{reaction 2}} CH_3CH(NH_2)CH_3$$
$$\xrightarrow{\text{reaction 3}} CH_2=CHCH_3$$

a) For each reaction, name the reagent and solvent used.   [6 marks]

b) Under the same conditions, 2-iodopropane was used in reaction 1 in place of 2-bromopropane. What difference (if any) would you expect in the rate of the reaction? Explain your answer.   [2 marks]

If you don't learn this — you will be eliminated. Resistance is nitrile...

Polar bonds get in just about every area of Chemistry. If you still think they're something to do with either bears or mints, flick back to page 28 and have a good read. Make sure you learn these reactions, and the mechanisms, as well as which bonds are hydrolysed fastest. This stuff's always coming up in exams. Ruin the examiner's day and get it right.

Reactions of Alkenes

I'll warn you now — some of this stuff gets a bit heavy — but stick with it, as it's pretty important.

Alkenes are **Unsaturated Hydrocarbons**

1) Alkenes have the **general formula** C_nH_{2n}. They're just made of carbon and hydrogen atoms, so they're **hydrocarbons**.

2) Alkene molecules **all** have at least one **C=C double covalent bond**. Molecules with C=C double bonds are **unsaturated** because they can make more bonds with extra atoms in **addition** reactions.

3) Because there's two pairs of electrons in the C=C double bond, it has a really **high electron density**. This makes alkenes pretty reactive.

Here are a few pretty diagrams of **alkenes**:

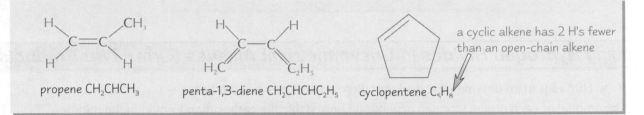

propene CH₂CHCH₃   penta-1,3-diene CH₂CHCHC₂H₅   cyclopentene C₅H₈

a cyclic alkene has 2 H's fewer than an open-chain alkene

Electrophilic Addition Reactions *Happen to Alkenes*

Electrophilic addition reactions aren't too complicated...

1) The **double bonds** open up and atoms are **added** to the carbon atoms.

2) Electrophilic addition reactions happen because the double bond has got plenty of **electrons** and is easily attacked by **electrophiles**.

> **Electrophiles** are **electron-pair acceptors** — they're usually a bit short of electrons, so they're <u>attracted</u> to areas where there's lots of them about.
>
> Here's a few examples:
> - **Positively charged ions**, like H^+, NO_2^+.
> - **Polar molecules** — the δ+ atom is attracted to places with lots of electrons

See page 28 for a reminder about polar molecules.

3) The double bond is also **nucleophilic** — it's attracted to places that don't have enough **electrons**.

Use **Bromine Water** *to Test for C=C Double Bonds*

When you shake an alkene with **orange bromine water**, the solution quickly **decolourises**. Bromine is added across the double bond to form a colourless **dibromoalkane** — this happens by **electrophilic addition**. Here's the mechanism...

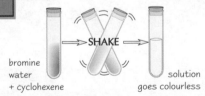

bromine water + cyclohexene → SHAKE → solution goes colourless

$$H_2C=CH_2 + Br_2 \rightarrow CH_2BrCH_2Br$$

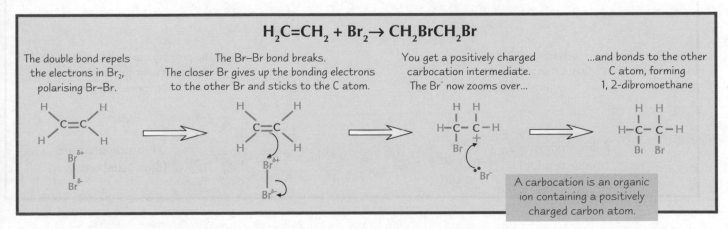

The double bond repels the electrons in Br₂, polarising Br–Br.

The Br–Br bond breaks. The closer Br gives up the bonding electrons to the other Br and sticks to the C atom.

You get a positively charged carbocation intermediate. The Br⁻ now zooms over...

...and bonds to the other C atom, forming 1, 2-dibromoethane

A carbocation is an organic ion containing a positively charged carbon atom.

Reactions of Alkenes

Alkenes also Undergo *Addition* with *Hydrogen Halides*

Alkenes also undergo **addition** reactions with hydrogen bromide — to form **bromoalkanes**.
This is the reaction between **ethene** and HBr:

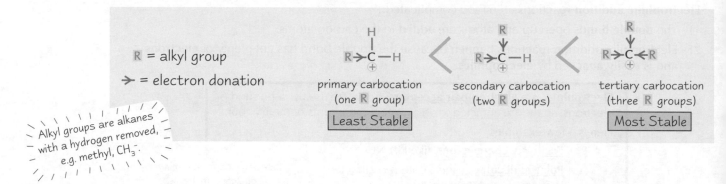

$$C_2H_4 + HBr \rightarrow C_2H_5Br$$

Other alkenes react in a similar way.

Adding *Hydrogen Halides* to *Unsymmetrical Alkenes* Forms *Two Products*

1) If the HBr adds to an **unsymmetrical** alkene, there are two possible products.

2) The amount of each product formed depends on how **stable** the **carbocation** formed in the middle of the reaction is.

3) Carbocations with more **alkyl groups** are more stable because the alkyl groups feed **electrons** towards the positive charge. The **more stable carbocation** is much more likely to form.

R = alkyl group

➤ = electron donation

primary carbocation
(one R group)

Least Stable

secondary carbocation
(two R groups)

tertiary carbocation
(three R groups)

Most Stable

Alkyl groups are alkanes with a hydrogen removed, e.g. methyl, CH₃.

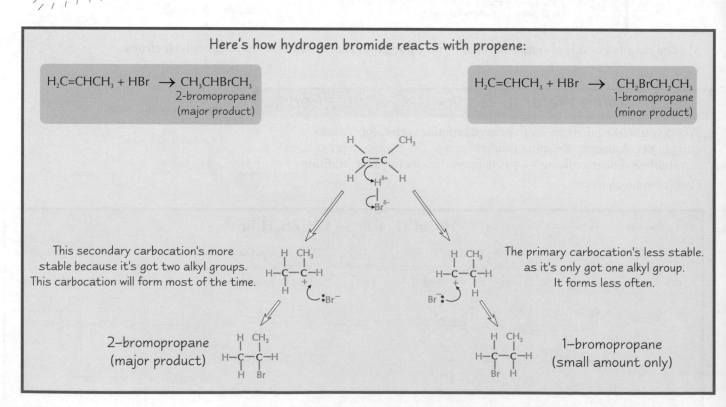

Here's how hydrogen bromide reacts with propene:

$H_2C=CHCH_3 + HBr \rightarrow CH_3CHBrCH_3$
2-bromopropane
(major product)

$H_2C=CHCH_3 + HBr \rightarrow CH_2BrCH_2CH_3$
1-bromopropane
(minor product)

This secondary carbocation's more stable because it's got two alkyl groups. This carbocation will form most of the time.

The primary carbocation's less stable. as it's only got one alkyl group. It forms less often.

2–bromopropane
(major product)

1–bromopropane
(small amount only)

Reactions of Alkenes

Reacting Alkenes with *Water* and an H_2SO_4 Catalyst Makes *Alcohols*

Alcohols are produced industrially by **hydrating alkenes** in the presence of an **acid catalyst**, such as sulfuric acid:

1) Cold concentrated **sulfuric acid** reacts with an alkene in an **electrophilic addition** reaction.

$$H_2C = CH_2 \quad + \quad H_2SO_4 \quad \longrightarrow \quad CH_3CH_2OSO_2OH$$
ethene $\qquad$ sulfuric acid $\qquad$ ethyl hydrogen sulfate

2) If you then add cold **water** and warm the product, it's **hydrolysed** to form an alcohol.

$$CH_3CH_2OSO_2OH \quad + \quad H_2O \longrightarrow \quad CH_3CH_2OH \quad + \quad H_2SO_4$$
ethyl hydrogen sulfate $\qquad\qquad\qquad$ ethanol

3) The **sulfuric acid** isn't used up — it acts as a **catalyst**.

> Hydrolysis is the breaking of covalent bonds by reaction with water.

So the overall reaction is:

$$H_2C = CH_2 \quad + \quad H_2O \quad \xrightarrow{H_2SO_4} \quad C_2H_5OH$$

Ethanol is Manufactured by *Steam Hydration*

1) Ethene can be **hydrated** by **steam** at 300 °C and a pressure of 60 atm. It needs a solid **phosphoric(V) acid catalyst**.

2) The reaction's **reversible** and the reaction yield is low — only about 5%. This sounds rubbish, but you can **recycle** the unreacted ethene gas, making the overall yield a much more profitable **95%**.

$$H_2C=CH_{2(g)} + H_2O_{(g)} \underset{\substack{300\,^\circ C \\ 60\ atm}}{\overset{H_3PO_4}{\rightleftharpoons}} CH_3CH_2OH_{(g)}$$

Practice Questions

Q1 What's the general formula for an alkene?

Q2 What is an electrophile?

Q3 Why do alkenes react with electrophiles?

Q4 What is a carbocation?

Q5 What conditions are needed to produce ethanol from ethene?

Exam Question

Q1 But-1-ene is an alkene. Alkenes contain at least one C=C double bond.
a) Describe how bromine water can be used to test for C=C double bonds. [2 marks]

b) Name the reaction mechanism involved in the above test. [2 marks]

c) Hydrogen bromide will react with but-1-ene by this mechanism, producing two isomeric products.
(i) Write a mechanism for the reaction of HBr with $CH_2=CHCH_2CH_3$, showing the formation of the major product only. Name the product. [3 marks]
(ii) Explain why it is the major product for this reaction. [2 marks]

This section is free from all GM ingredients...

Wow... these pages really are jam-packed. There's not one, not two, but three mechanisms to learn. And learn them you must. They mightn't be as handy in real life as a tin opener, but you won't need a tin opener in the exam. Get the book shut and scribble them out. Make sure your arrows start at the electron pair and finish exactly where the electrons are going.

E/Z Isomers and Polymers

The chemistry on these pages isn't so bad. And don't be too worried when I tell you that a good working knowledge of German would be useful. It's not absolutely essential... and you'll be fine without.

Double Bonds Can't Rotate

1) Carbon atoms in a C=C double bond and the atoms bonded to these carbons all lie in the **same plane** (they're **planar**).
Because of the way they're arranged, they're actually said to be **trigonal planar** — the atoms attached to each double-bond carbon are at the corners of an imaginary equilateral triangle.

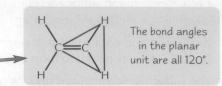
The bond angles in the planar unit are all 120°.

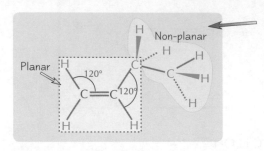

2) Ethene, **C_2H_4** (like in the diagram above) is completely planar, but in larger alkenes, only the >C=C< unit is planar.

3) Another important thing about C=C double bonds is that atoms **can't rotate** around them like they can around single bonds. In fact, double bonds are fairly **rigid** — they don't bend much either.

4) Even though atoms can't rotate about the **double bond**, things can still rotate about any **single bonds** in the molecule — like in this molecule of pent-2-ene.

5) The **restricted rotation** around the C=C double bond is what causes **E/Z isomerism**.

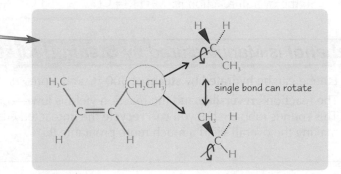
single bond can rotate

E/Z isomerism is a Type of Stereoisomerism

1) **Stereoisomers** have the same structural formula but a **different arrangement** in space.
(Just bear with me for a moment... that will become clearer, I promise.)

2) Because of the **lack of rotation** around the double bond, some **alkenes** can have stereoisomers.

3) Stereoisomers happen when the two double-bonded carbon atoms each have **different atoms** or **groups** attached to them. Then you get an 'E-isomer' and a 'Z-isomer'.

For example, the double-bonded carbon atoms in but-2-ene each have an **H** and a **CH_3** group attached.

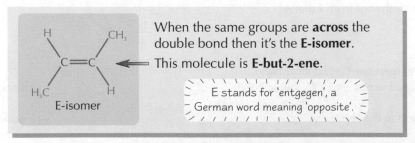
When the same groups are **across** the double bond then it's the **E-isomer**.
This molecule is **E-but-2-ene**.
E stands for 'entgegen', a German word meaning 'opposite'.

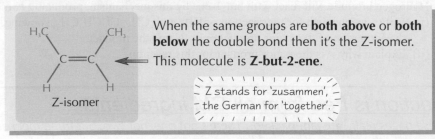
When the same groups are **both above** or **both below** the double bond then it's the Z-isomer.
This molecule is **Z-but-2-ene**.
Z stands for 'zusammen', the German for 'together'.

E/Z Isomers and Polymers

Alkenes *Join Up* to form *Addition Polymers*

1) The **double bonds** in alkenes can open up and join together to make long chains called **polymers**. It's kind of like they're holding hands in a big line. The individual, small alkenes are called **monomers**.

2) This is called **addition polymerisation**. For example, **poly(ethene)** is made by the **addition polymerisation** of ethene.

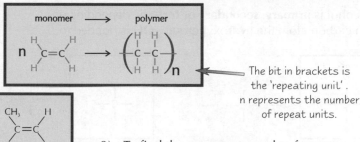

The bit in brackets is the 'repeating unit'. n represents the number of repeat units.

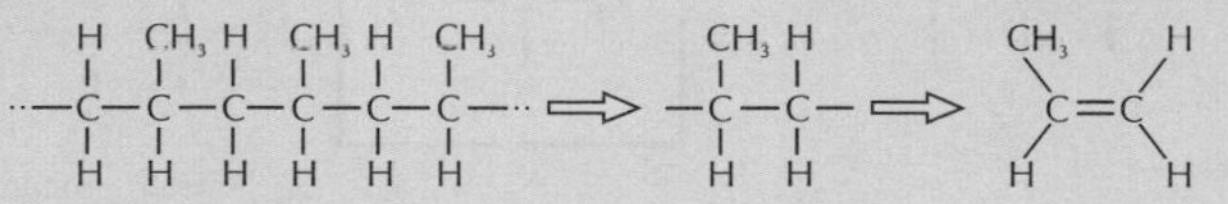

3) To find the **monomer** used to form an addition polymer, take the **repeating unit** and add a **double bond**.

polymer
poly(propene)

repeating unit

monomer propene

4) Because of the loss of the double bond, poly(alkenes), like alkanes, are **unreactive**.

5) Different polymer **structures** have different **properties**, which means they're suited to different **uses**. Some typical uses of **poly(ethene)** and **poly(propene)** are shown in the table.

6) Many polymers are difficult to dispose of and are made from **non-renewable** oil fractions, so it makes sense to **recycle** them. For example, poly(propene) is recycled — it can be **melted** and **remoulded**.

	Properties	Uses
Low density poly(ethene)	Soft Flexible	Plastic bags Squeezy bottles
Poly(propene)	Tough Strong	Bottle crates Rope

Practice Questions

Q1 Why is an ethene molecule said to be planar?

Q2 What feature of alkene molecules gives rise to E/Z isomerism?

Q3 Which of the following is the Z-isomer of but-2-ene?

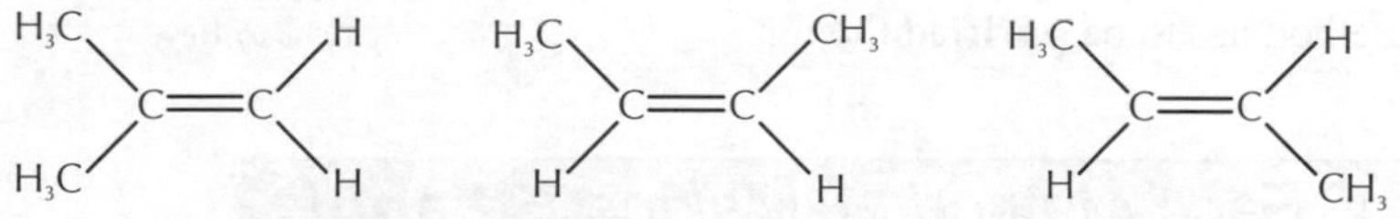

Q4 Define the term 'stereoisomers'.

Q5 Give a typical use for poly(ethene), and one for poly(propene).

Exam Questions

Q1 a) Draw and name the E/Z isomers of pent-2-ene. [4 marks]
   b) Explain why alkenes can have E/Z isomers but alkanes cannot. [2 marks]

Q2 Part of the structure of a polymer is shown on the right. 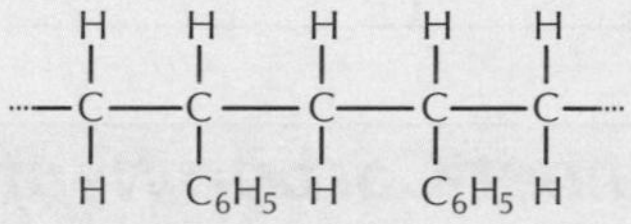
   a) (i) Draw the repeating unit of the polymer. [1 mark]
      (ii) Draw the monomer from which the polymer was formed. [1 mark]

   b) Poly(tetrafluoroethene) is made from the monomer shown on the right.
      Draw part of the polymer consisting of three of the repeating units. [1 mark]

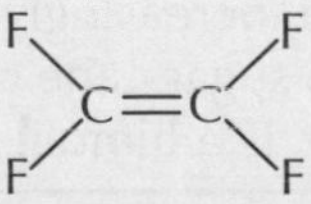

And there you have it, folks — two E/Z pages in an AS Chemistry book...

To remember which is which out of E and Z isomers, remember that Z-isomers are the ones with the groups on 'ze zame zide'. Or if you prefer, you could learn to speak German... It's the double bond in alkenes that gives them their special powers — things can't rotate around it, so there are stereoisomers. And it can open up, so they can form addition polymers.

Alcohols

Alcohol — evil stuff, it is. I could start preaching, but I won't, because this page is enough to put you off alcohol for life...

Alcohols can be **Primary**, **Secondary** or **Tertiary**

1) The alcohol homologous series has the **general formula $C_nH_{2n+1}OH$**.

2) An alcohol is **primary**, **secondary** or **tertiary**, depending on which carbon atom the hydroxyl group **–OH** is bonded to...

> The naming rules on page 36-37 apply to alcohols too. The suffix for alcohols is –ol, and you have to say which carbon the –OH group is attached to.

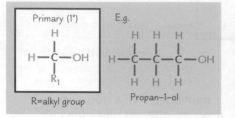

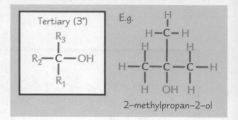

Ethanol can be Produced Industrially by **Fermentation**

At the moment most industrial ethanol is produced by **steam hydration of ethene** with a **phosphoric acid catalyst** (see page 73). The ethene comes from cracking heavy fractions of crude oil. But in the future, when crude oil supplies start **running out**, petrochemicals like ethene will be expensive — so producing ethanol by **fermentation** will become much more important...

Industrial Production of Ethanol by Fermentation

1) Fermentation is an **exothermic** process, carried out by **yeast** in **anaerobic conditions** (without oxygen).

2) Yeast produces an **enzyme** which converts sugars, such as glucose, into **ethanol** and **carbon dioxide**.

3) The enzyme works at an **optimum** (ideal) temperature of **30-40 °C**. If it's too cold, the reaction is **slow** — if it's too hot, the enzyme is **denatured** (damaged).

4) When the solution reaches about **15% ethanol**, the yeast dies. **Fractional distillation** is used to increase the concentration of ethanol.

5) Fermentation is **low-tech** — it uses cheap equipment and **renewable resources**. The ethanol produced by this method has to be **purified** though.

$$C_6H_{12}O_{6(aq)} \xrightarrow[\text{yeast}]{30\text{-}40°C} 2C_2H_5OH_{(aq)} + 2CO_{2(g)}$$
glucose

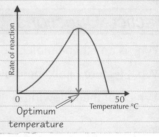

Learn the **Pros and Cons** of Each Method of Making **Ethanol**

Here's a quick summary of the **advantages** and **disadvantages** of the two methods of making ethanol.

Method	Rate of reaction	Quality of product	Raw material	Process/Costs
Hydration of ethene	Very fast	Pure	Ethene from oil — a finite resource	Continuous process, so expensive equipment needed, but low labour costs.
Fermentation	Very slow	Very impure — needs further processing	Sugars — a renewable resource	Batch process, so cheap equipment needed, but high labour costs.

Ethanol is an **Almost Carbon Neutral** Fuel

1) Ethanol is also being used increasingly as a **fuel**, particularly in countries with few oil reserves. E.g. in Brazil, **sugars** from sugar cane are **fermented** to produce alcohol, which is added to petrol. Ethanol, made in this way, is a **biofuel**.

> **A biofuel is a fuel that's made from biological material that's recently died.**

2) Ethanol is thought of as a **carbon neutral** fuel, because all the CO_2 released when the fuel is burned was removed by the crop as it grew. **BUT** — there are still carbon emissions if you consider the **whole** process. E.g. making the fertilisers and powering agricultural machinery will probably involve burning fossil fuels.

Alcohols

Alcohols can be **Dehydrated** to Form **Alkenes**

1) You can make ethene by **eliminating** water from **ethanol** in a **dehydration reaction** (i.e. elimination of **water**).

$$C_2H_5OH \longrightarrow CH_2=CH_2 + H_2O$$

In an elimination reaction, a small group of atoms breaks away from a larger molecule. It's not replaced by anything else.

Here's how you can go about it:

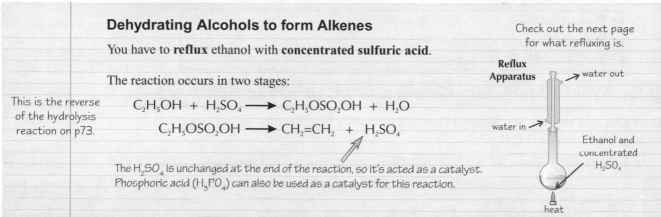

Dehydrating Alcohols to form Alkenes

You have to **reflux** ethanol with **concentrated sulfuric acid**.

Check out the next page for what refluxing is.

The reaction occurs in two stages:

This is the reverse of the hydrolysis reaction on p73.

$$C_2H_5OH + H_2SO_4 \longrightarrow C_2H_5OSO_2OH + H_2O$$
$$C_2H_5OSO_2OH \longrightarrow CH_2=CH_2 + H_2SO_4$$

The H_2SO_4 is unchanged at the end of the reaction, so it's acted as a catalyst. Phosphoric acid (H_3PO_4) can also be used as a catalyst for this reaction.

Reflux Apparatus → water out
water in → Ethanol and concentrated H_2SO_4
heat

2) This reaction allows you to produce alkenes from **renewable** resources.

Because you can produce ethanol by fermentation of glucose, which you can get from plants.

3) This is important, because it means that you can produce **polymers** (poly(ethene), for example) **without** needing **oil**.

Practice Questions

Q1 What is the general formula for an alcohol?

Q2 Write down two advantages and two disadvantages of the fermentation method of producing ethanol.

Q3 What is a biofuel? What does carbon-neutral mean? Is ethanol a carbon-neutral biofuel?

Q4 Describe how to form an alkene from an alcohol using an acid catalysed elimination reaction.

Exam Questions

Q1 Butanol C_4H_9OH has four chain and positional isomers.
Name each isomer and class it as a primary, secondary or tertiary alcohol. [8 marks]

Q2 Ethanol is a useful alcohol.
a) State whether ethanol is a primary, secondary or tertiary alcohol, and explain why. [2 marks]
b) Industrially, ethanol can be produced by fermentation of glucose, $C_6H_{12}O_6$.
   (i) Write a balanced equation for this reaction. [1 mark]
   (ii) State the optimum conditions for fermentation. [3 marks]
c) At present most ethanol is produced by the acid-catalysed hydration of ethene.
   Why is this? Why might this change in the future? [3 marks]

Euuurghh, what a page... I think I need a drink...

Not much to learn here — a few basic definitions, an industrial process, the advantages and disadvantages of it compared to another industrial process, a bit about biofuels, and a lovely two-stage reaction... Like I said, not much here at all. Think I'm going to faint. [THWACK]

Oxidising Alcohols

Another page of alcohol reactions. Probably not what you wanted for Christmas... But at least you are almost at the end of the section... and the book for that matter... and your wits, probably.

How Much an Alcohol can be **Oxidised** Depends on its **Structure**

The simple way to oxidise alcohols is to burn them. But you don't get the most exciting products by doing this. If you want to end up with something more interesting at the end, you need a more sophisticated way of oxidising...

You can use the **oxidising agent acidified potassium dichromate(VI)** to **mildly** oxidise alcohols.

> • **Primary** alcohols are oxidised to **aldehydes** and then to **carboxylic acids**.
> • **Secondary** alcohols are oxidised to **ketones** only.
> • **Tertiary** alcohols aren't oxidised.

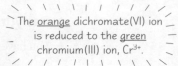
The orange dichromate(VI) ion is reduced to the green chromium(III) ion, Cr^{3+}.

Learn What **Aldehydes**, **Ketones** and **Carboxylic Acids** are

Aldehydes and **ketones** are **carbonyl** compounds — they have the functional group C=O. Their general formula is $C_nH_{2n}O$.

Aldehydes have a **hydrogen** and **one alkyl group** attached to the carbonyl carbon atom. E.g. Their suffix is **-al**. You don't have to say which carbon the functional group is on — it's always on carbon-1.

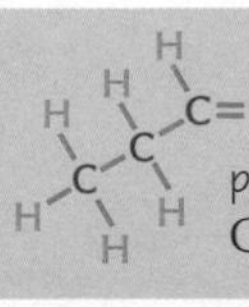

propanal CH_3CH_2CHO

Ketones have **two alkyl groups** attached to the carbonyl carbon atom.
Their suffix is **-one**. For ketones with five or more carbons, you always have to say which carbon the functional group is on. (If there are other groups attached, such as methyl groups, you have to say it for four-carbon ketones too.)

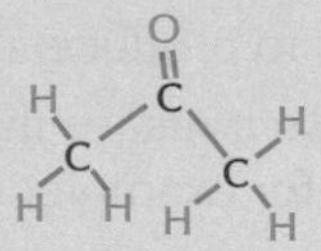

propanone CH_3COCH_3

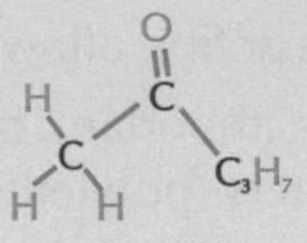

pentan-2-one $CH_3COC_3H_7$

Carboxylic acids have a different functional group...

Carboxylic acids have a **COOH group** at the end of their carbon chain. Their suffix is **-oic**.

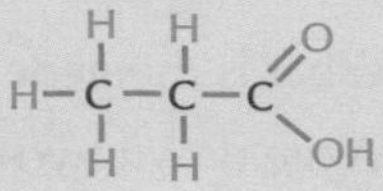

propanoic acid CH_3CH_2COOH

Primary Alcohols will Oxidise to **Aldehydes** and **Carboxylic Acids**

$$R-CH_2-OH + [O] \longrightarrow R-C{\overset{O}{\underset{H}{}}} + [O] \xrightarrow{reflux} R-C{\overset{O}{\underset{OH}{}}}$$
$$+ H_2O$$
primary alcohol   aldehyde   carboxylic acid

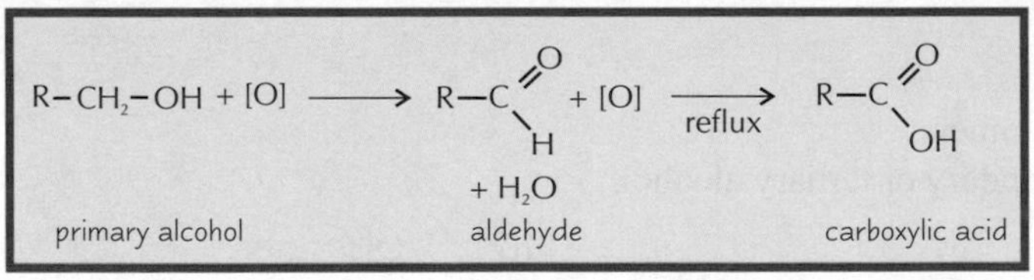

[O] = oxidising agent

You can control how **far** the alcohol is oxidised by controlling the **reaction conditions**:

Oxidising Primary Alcohols

1) Gently heating ethanol with potassium dichromate(VI) solution and sulfuric acid in a test tube should produce "apple" smelling **ethanal** (an aldehyde). However, it's **really tricky** to control the amount of heat and the aldehyde is usually oxidised to form "vinegar" smelling **ethanoic acid**.

2) To get just the **aldehyde**, you need to get it out of the oxidising solution **as soon** as it's formed. You can do this by gently heating excess alcohol with a **controlled** amount of oxidising agent in **distillation apparatus**, so the aldehyde (which boils at a lower temperature than the alcohol) is distilled off **immediately**.

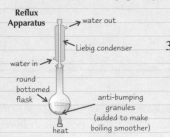
Reflux Apparatus
water out
Liebig condenser
water in
round bottomed flask
anti-bumping granules (added to make boiling smoother)
heat

3) To produce the **carboxylic acid**, the alcohol has to be **vigorously oxidised**. The alcohol is mixed with excess oxidising agent and heated under **reflux**. Heating under reflux means you can increase the **temperature** of an organic reaction to boiling without losing **volatile** solvents, reactants or products. Any vaporised compounds are cooled, condense and drip back into the reaction mixture. Handy, hey.

Oxidising Alcohols

Secondary Alcohols will Oxidise to Ketones

1) Refluxing a secondary alcohol, e.g. propan-2-ol, with acidified dichromate(VI) will produce a **ketone**.

2) Ketones can't be oxidised easily, so even prolonged refluxing won't produce anything more.

Tertiary Alcohols can't be Oxidised Easily

Tertiary alcohols don't react with potassium dichromate(VI) at all — the solution stays orange. The only way to oxidise tertiary alcohols is by **burning** them.

Use Oxidising Agents to Distinguish Between Aldehydes and Ketones

Aldehydes and ketones can be distinguished using **oxidising agents** — aldehydes are easily oxidised but ketones aren't.

1) **Fehling's solution** and **Benedict's solution** are both deep blue Cu^{2+} complexes, which reduce to brick-red Cu_2O when warmed with an aldehyde, but stay blue with a ketone.

2) **Tollen's reagent** is $[Ag(NH_3)_2]^+$ — it's reduced to **silver** when warmed with an aldehyde, but not with a ketone. The silver will coat the inside of the apparatus to form a **silver mirror**.

Practice Questions

Q1 What's the difference between the structure of an aldehyde and a ketone?

Q2 What will acidified potassium dichromate(VI) oxidise secondary alcohols to?

Q3 What is the colour change when potassium dichromate(VI) is reduced?

Q4 Describe two tests you can use to distinguish between a sample of an aldehyde and a sample of a ketone.

Exam Question

Q1 A student wanted to produce the aldehyde propanal from propanol, and set up a reflux apparatus using a suitable oxidising agent.

a) (i) Suggest an oxidising agent that the student could use. [1 mark]
   (ii) Draw the structural formula of propanal. [1 mark]

b) The student tested his product and found that he had not produced propanal.
   (i) Describe a test for an aldehyde. [2 marks]
   (ii) What is the student's product? [1 mark]
   (iii) Write equations to show the two-stage reaction. Use [O] to represent the oxidising agent. [2 marks]
   (iv) What technique should the student have used and why? [2 marks]

c) The student also tried to oxidise 2-methylpropan-2-ol, unsuccessfully.
   (i) Draw the full structural formula for 2-methylpropan-2-ol. [1 mark]
   (ii) Why is it not possible to oxidise 2-methylpropan-2-ol with an oxidising agent? [1 mark]

I.... I .... just can't do it, R2...

Don't give up now. Only as a fully-trained Chemistry Jedi, with the force as your ally, can you take on the Examiner. If you quit now, if you choose the easy path as Wader did, all the marks you've fought for will be lost. Be strong. Don't give in to hate — that leads to the dark side... (Only two more double pages to go before you're done with this section...)

Analytical Techniques

Get ready for the thrilling climax of the book — and watch out for the twist at the end...

Mass Spectrometry Can Help to Identify Compounds

1) You saw on pages 8-9 how **mass spectrometry** can be used to find **relative isotopic masses**, the **abundance** of different isotopes, and the **relative molecular mass**, M_r, of a compound.

2) Remember — to find the relative molecular mass of a compound you look at the **molecular ion peak** (the **M peak**) on the spectrum. Molecular ions are formed when molecules have **electrons** knocked off. The mass/charge value of the molecular ion peak is the molecular mass.

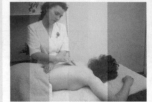
Assuming the ion has 1+ charge, which it normally will have.

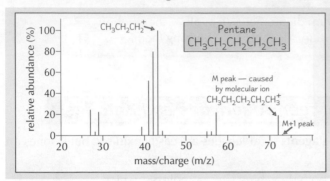

Pentane
$CH_3CH_2CH_2CH_2CH_3$

M peak — caused by molecular ion $CH_3CH_2CH_2CH_2CH_3^+$

M+1 peak

Here's the mass spectrum of pentane. Its M peak is at 72 — so the compound's M_r is 72.

For most <u>organic compounds</u> the M peak is the one with the second highest mass/charge ratio. The smaller peak to the right of the M peak is called the M+1 peak — it's caused by the presence of the carbon isotope ^{13}C (you don't need to worry about this at AS).

A massage spectrum

The Molecular Ion can be Broken into Smaller Fragments

The bombarding electrons make some of the molecular ions break up into **fragments**. The fragments that are ions show up on the mass spectrum, making a **fragmentation pattern**. Fragmentation patterns are actually pretty cool because you can use them to identify **molecules** and even their **structure**.

For propane, the molecular ion is $CH_3CH_2CH_3^+$, and the fragments it breaks into include CH_3^+ ($M_r = 15$) and $CH_3CH_2^+$ ($M_r = 29$).

Only the **ions** show up on the mass spectrum — the **free radicals** are 'lost'.

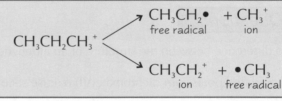

$CH_3CH_2\bullet$ free radical $+ CH_3^+$ ion

$CH_3CH_2CH_3^+$

$CH_3CH_2^+$ ion $+ \bullet CH_3$ free radical

To work out the structural formula, you've got to work out what **ion** could have made each peak from its **m/z value**. (You assume that the m/z value of a peak matches the **mass** of the ion that made it.)

Example: Use this mass spectrum to work out the structure of the molecule:

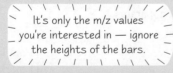
It's only the m/z values you're interested in — ignore the heights of the bars.

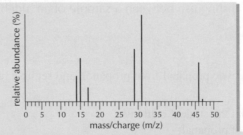

Fragment	Molecular Mass
CH_3	15
C_2H_5	29
C_3H_7	43
OH	17

1. Identify the fragments

This molecule's got a peak at 15 m/z, so it's likely to have a **CH₃ group**.

It's also got a peak at 17 m/z, so it's likely to have an **OH group**.

Other ions are matched to the peaks here:

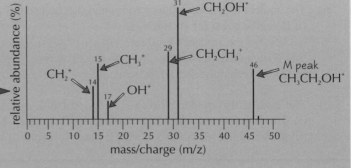

CH_2^+ · 14 · 15 CH_3^+ · 17 OH^+ · 29 $CH_2CH_3^+$ · 31 CH_2OH^+ · 46 M peak $CH_3CH_2OH^+$

2. Piece them together to form a molecule with the correct M_r

Ethanol has all the fragments on this spectrum.

```
    H   H
    |   |
H—C—C—O—H
    |   |
    H   H
```

Ethanol's **molecular mass** is 46.
This should be the same as the m/z value of the M peak — it is.

Analytical Techniques

If you've got some stuff and don't know what it is, don't taste it. Stick it in an infrared spectrometer instead.
Infrared spectroscopy produces scary looking graphs. But just learn the basics, and you'll be fine.

Infrared Spectroscopy Helps You Identify Organic Molecules

1) In infrared (IR) spectroscopy, a beam of **IR radiation** is passed through a sample of a chemical.

2) The IR radiation is absorbed by the **covalent bonds** in the molecules, increasing their **vibrational** energy.

3) **Bonds between different atoms** absorb **different frequencies** of IR radiation. Bonds in different **places** in a molecule absorb different frequencies too — so the O–H group in an **alcohol** and the O–H in a **carboxylic acid** absorb different frequencies. This table shows what **frequencies** different bonds absorb:

Functional group	Where it's found	Frequency/ Wavenumber (cm^{-1})	Type of absorption
C–H	most organic molecules	2800 - 3100	strong, sharp
O–H	alcohols	3200 - 3550	strong, broad
O–H	carboxylic acids	2500 - 3300	medium, broad
N–H	amines (e.g. methylamine, CH_3NH_2)	3200 - 3500	strong, sharp
C=O	aldehydes, ketones, carboxylic acids	1680 - 1750	strong, sharp
C–X	haloalkanes	500 - 1000	strong, sharp

This tells you what the peak on the graph will look like.

You don't need to learn this data, but you do need to understand how to use it.

4) An infrared spectrometer produces a **graph** that shows you what frequencies of radiation the molecules are absorbing. So you can use it to identify the **functional groups** in a molecule:

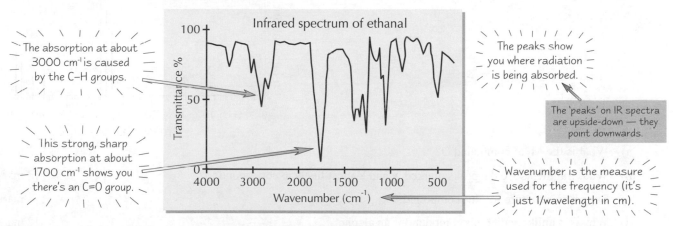

The absorption at about 3000 cm⁻¹ is caused by the C–H groups.

This strong, sharp absorption at about 1700 cm⁻¹ shows you there's an C=O group.

The peaks show you where radiation is being absorbed.

The 'peaks' on IR spectra are upside-down — they point downwards.

Wavenumber is the measure used for the frequency (it's just 1/wavelength in cm).

The Fingerprint Region Identifies a Molecule

1) The region between **1000 cm⁻¹ and 1550 cm⁻¹** on the spectrum is called the **fingerprint** region. It's **unique** to a **particular compound**. You can check this region of an unknown compound's IR spectrum against those of known compounds. If it **matches up** with one of them, hey presto — you know what the molecule is.

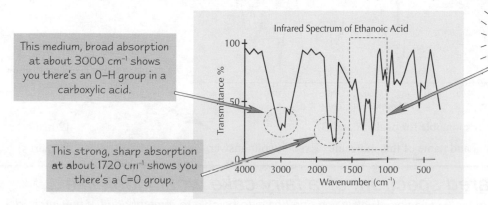

This medium, broad absorption at about 3000 cm⁻¹ shows you there's an O–H group in a carboxylic acid.

This strong, sharp absorption at about 1720 cm⁻¹ shows you there's a C=O group.

This is the fingerprint region. If you see an infrared spectrum of an unknown molecule that has the same pattern in this area, you can be sure that it's ethanoic acid.

Clark began to regret having an infrared mechanism installed in his glasses.

2) Infrared spectroscopy can also be used to find out how **pure** a compound is, and identify any impurities. Impurities produce **extra peaks** in the fingerprint region.

Analytical Techniques

Infrared Energy **Absorption** is Linked to **Global Warming**

1) Some of the electromagnetic radiation emitted by the **Sun** is in the form of **infrared radiation**.

2) Molecules of **greenhouse gases**, like **carbon dioxide**, **methane** and **water vapour**, are really good at absorbing infrared energy — so if the amounts of them in the atmosphere increase, it leads to **global warming**. There's lots more about how this happens on pages 42-43.

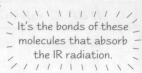
It's the bonds of these molecules that absorb the IR radiation.

Practice Questions

Q1 What is meant by the molecular ion?

Q2 What is the M peak?

Q3 How do fragments get formed?

Q4 Which parts of a molecule absorb infrared energy?

Q5 Why do most infrared spectra of organic molecules have a strong, sharp peak at around 3000 cm^{-1}?

Q6 On an infrared spectrum, what is meant by the 'fingerprint region'?

Exam Question

Q1 On the right is the mass spectrum of an organic compound, Q.

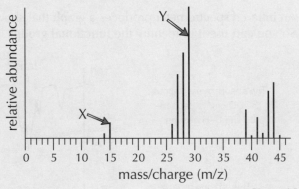

a) What is the M_r of compound Q? [1 mark]

b) What fragments are the peaks marked X and Y most likely to correspond to? [2 marks]

c) Suggest a structure for this compound. [1 mark]

d) Why is it unlikely that this compound is an alcohol? [2 marks]

Q2 A molecule with a molecular mass of 74 produces the following IR spectrum.

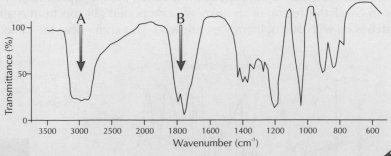

Use the infrared absorption data on the previous page.

a) Which functional groups are responsible for peaks A and B? [2 marks]

b) Suggest the molecular formula and name of this molecule. Explain your answer. [3 marks]

I wonder what the infrared spectrum of a fairy cake would look like...

I don't suppose I'll ever know. Very squiggly I imagine. Luckily you don't have to be able to remember what any infrared spectrum graphs look like. But you definitely need to know how to interpret them. And don't worry, I haven't forgotten I said there was twist at the end... erm... hydrogen was my sister all along... and all the elements went to live in Jamaica. The End.

Practical and Investigative Skills

You're going to have to do some practical work too — and once you've done it, you have to make sense of your results...

Make it a **Fair Test** — Control your **Variables**

You probably know this all off by heart but it's easy to get mixed up sometimes. So here's a quick recap:

> **Variable** — A variable is a **quantity** that has the **potential to change**, e.g. mass.
> There are two types of variable commonly referred to in experiments:
> * **Independent variable** — the thing that you **change** in an experiment.
> * **Dependent variable** — the thing that you **measure** in an experiment.

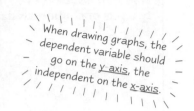

When drawing graphs, the dependent variable should go on the y-axis, the independent on the x-axis.

So, if you're investigating the effect of **temperature** on rate of reaction
using the apparatus on the right, the variables will be:

Independent variable	Temperature
Dependent variable	Amount of oxygen produced — you can measure this by collecting it in a gas syringe
Other variables — you MUST keep these the same	Concentration and volume of solutions, mass of solids, pressure, the presence of a catalyst and the surface area of any solid reactants

Know Your Different Sorts of **Data**

Experiments always involve some sort of measurement to provide **data**.
There are different types of data — and you need to know what they are.

> **Discrete** — you get discrete data by **counting**. E.g. the number of bubbles produced in a reaction would be discrete. You can't have 1.25 bubbles. That'd be daft. Shoe size is another good example of a discrete variable.

> **Continuous** — a continuous variable can have **any value** on a scale. For example, the volume of gas produced or the mass of products from a reaction. You can never measure the exact value of a continuous variable.

> **Categoric** — a categoric variable has values that can be sorted into **categories**. For example, the colours of solutions might be blue, red and green. Or types of material might be wood, steel, glass.

> **Ordered (ordinal)** — Ordered data is similar to categoric, but the categories can be **put in order**. For example, if you classify reactions as 'slow', 'fairly fast' and 'very fast' you'd have ordered data.

Organise Your Results in a **Table** — And Watch Out For **Anomalous** Ones

Before you start your experiment, make a **table** to write your results in.
You'll need to repeat each test at least three times to check your results are reliable.

This is the sort of table you might end up with when you investigate the effect of **temperature** on **reaction rate**.
(You'd then have to do the same for **different temperatures**.)

Temperature	Time (s)	Volume of gas evolved (cm³) Run 1	Volume of gas evolved (cm³) Run 2	Volume of gas evolved (cm³) Run 3	Average volume of gas evolved (cm³)
	10	8	7	8	**7.7**
20 °C	**20**	17	19	20	**18.7**
	30	28	(20)	30	**29**

Find the average of each set of repeated values.

You need to add them all up and divide by how many there are.

E.g.: (8 + 7 + 8) ÷ 3 = 7.7 cm³

Watch out for **anomalous results**. These are ones that don't fit in with the other values and are likely
to be wrong. They're likely to be due to random errors — here the syringe plunger may have got stuck.
Ignore anomalous results when you calculate the average.

Practical and Investigative Skills

Graphs: *Line, Bar or Scatter* — Use the *Best Type*

You'll usually be expected to make a **graph** of your results. Not only are graphs **pretty**, they make your data **easier to understand** — so long as you choose the right type.

Line graphs are best when you have **two sets of continuous data**. For example:

Use simple scales — this'll make it easier to plot points.

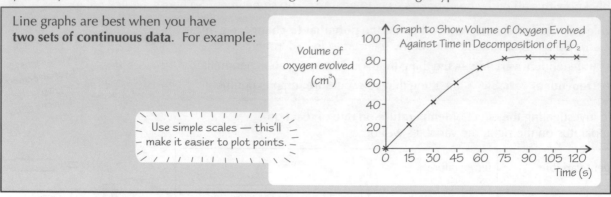

Graph to Show Volume of Oxygen Evolved Against Time in Decomposition of H_2O_2

You should use a bar chart when one of your data sets is **categoric or ordered data**. For example:

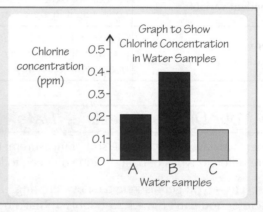

Graph to Show Chlorine Concentration in Water Samples

Scatter plots are great for showing how two sets of data are related (or **correlated**).

Don't try to join all the points — draw a **line of best fit** to show the **trend**.

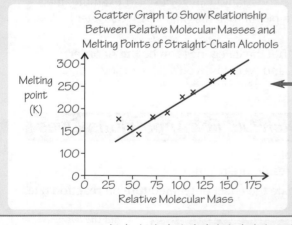

Scatter Graph to Show Relationship Between Relative Molecular Masses and Melting Points of Straight-Chain Alcohols

Scatter Graphs Show the Relationship Between Variables

Correlation describes the **relationship** between two variables — the independent one and the dependent one.

Data can show:

1) **Positive correlation** — as one variable **increases** the other **increases**. The graph on the left shows positive correlation.

2) **Negative correlation** — as one variable **increases** the other **decreases**.

3) **No correlation** — there is **no relationship** between the two variables.

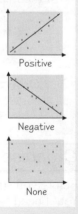

— There are also pie charts. These are normally used to display categoric data. —

Remember These *Important Points* When *Drawing Graphs*

Whatever type of graph you make, you'll ONLY get full marks if you:

• Choose a sensible scale — don't do a tiny graph in the corner of the paper.

• Label both axes — including units.

• Plot your points accurately — using a sharp pencil.

Practical and Investigative Skills

Correlation *Doesn't Necessarily* Mean **Cause** — Don't Jump to Conclusions

1) Ideally, only **two** quantities would **ever** change in any experiment — everything else would remain **constant**.

2) But in experiments or studies outside the lab, you **can't** usually control all the variables.
So even if two variables are correlated, the change in one may **not** be causing the change in the other. Both changes might be caused by a **third variable**.

Watch out for bias too — for instance, a bottled water company might point these studies out to people without mentioning any of the doubts.

Example

Some studies have found a correlation between **drinking chlorinated tap water** and the risk of developing certain cancers. So some people argue that this means water shouldn't have chlorine added.

BUT it's hard to control all the variables between people who drink tap water and people who don't. It could be many lifestyle factors.

Or, the cancer risk could be affected by something else in tap water — or by whatever the non-tap water drinkers drink instead...

Don't Get **Carried Away** When Drawing Conclusions

The **data** should always **support** the conclusion. This may sound obvious but it's easy to **jump** to conclusions. Conclusions have to be **specific** — not make sweeping generalisations.

Example

The rate of an enzyme-controlled reaction was measured at **10 °C, 20 °C, 30 °C, 40 °C, 50 °C and 60 °C**. All other variables were kept constant, and the results are shown in this graph.

A science magazine **concluded** from this data that enzyme X works best at **40 °C**. The data **doesn't** support this.

The enzyme **could** work best at 42 °C or 47 °C but you can't tell from the data because **increases** of **10 °C** at a time were used.

The rate of reaction at in-between temperatures **wasn't** measured.

All you know is that it's faster at **40 °C** than at any of the other temperatures tested.

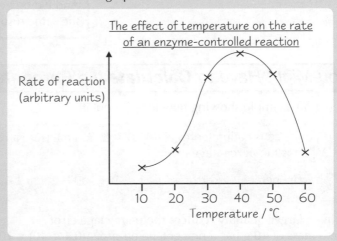

The effect of temperature on the rate of an enzyme-controlled reaction

Rate of reaction (arbitrary units)

Temperature / °C

Example

The experiment above **ONLY** gives information about this particular enzyme-controlled reaction. You can't conclude that **all** enzyme-controlled reactions happen faster at a particular temperature — only this one. And you can't say for sure that doing the experiment at, say, a different constant pressure, wouldn't give a different optimum temperature.

Practical and Investigative Skills

You Need to Look **Critically** at Your Results

There are a few bits of lingo that you need to understand.
They'll be useful when you're evaluating how convincing your results are.

VALID RESULTS

Valid results answer the original question. For example, if you haven't **controlled all the variables** your results won't be valid, because you won't be testing just the thing you wanted to.

ACCURATE RESULTS

Accurate results are those that are **really close** to the **true** answer.

PRECISE RESULTS

These are results taken using **sensitive instruments** that measure in **small increments**, e.g. pH measured with a meter (pH 7.692) will be **more precise** than pH measured with paper (pH 7).

It's possible for results to be precise **but not** accurate, e.g. a balance that weighs to 1/1000 th of a gram will give precise results but if it's not **calibrated** properly the results won't be accurate.

RELIABLE RESULTS

Reliable means the results can be **consistently reproduced** in independent experiments. And if the results are reproducible they're more likely to be **true**. If the data isn't reliable for whatever reason you **can't draw** a valid **conclusion**.

For experiments, the **more repeats** you do, the **more reliable** the data. If you get the **same result** twice, it could be the correct answer. But if you get the same result **20 times**, it'd be much more reliable. And it'd be even more reliable if everyone in the class got about the same results using different apparatus.

You Might Have to **Calculate** the **Percentage Error** of a Measurement

Here's an example showing how to go about it:

A balance is calibrated to within 0.1 g. You measure a mass as **4 g**. What is the **percentage error**?

The percentage error is: $(0.1 \div 4) \times 100 = \textbf{2.5\%}$

Just work out what percentage of your measurement the possible error is.

Using a **larger** quantity **reduces** the percentage error.
E.g. a mass of **40 g** has a percentage error of: $(0.1 \div 40) \times 100 = \textbf{0.25\%}$

Work **Safely** and **Ethically** — Don't Blow Up the Lab or Harm Small Animals

In any experiment you'll be expected to show that you've thought about the **risks and hazards**. It's generally a good thing to wear a lab coat and goggles, but you may need to take additional safety measures, depending on the experiment. For example, anything involving nasty gases will need to be done in a fume cupboard.

You need to make sure you're working **ethically** too. This is most important if there are other people or animals involved. You have to put their welfare first.

A2-Level
Chemistry

Exam Board: AQA

Rate Equations

Sorry — this first section's a bit mathsy. Just take a deep breath, dive in, and don't bash your head on the bottom.

The **Reaction Rate** tells you How Fast **Reactants** are Converted to **Products**

The **reaction rate** is the **change in the amount** of reactants or products **per unit time** (normally per second). If the reactants are in **solution**, the rate will be **change in concentration per second** and the units will be **moldm^{-3}s^{-1}**.

If you draw a graph of the **amount of reactant or product against time** for a reaction, the rate at any point is given by the **gradient** at that point on the graph. If the graph's a curve, you have to draw a **tangent** to the curve and find the gradient of that.

A tangent is a line that just touches a curve and has the same gradient as the curve does at that point.

A graph of the **concentration of a reactant against time** might look something like this:

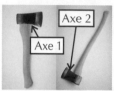

Duncan always took care to label his axes.

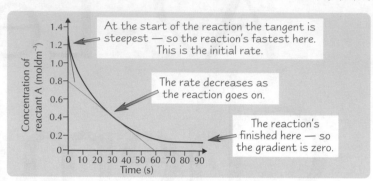

At the start of the reaction the tangent is steepest — so the reaction's fastest here. This is the initial rate.

The rate decreases as the reaction goes on.

The reaction's finished here — so the gradient is zero.

The **Rate Equation** links **Reaction Rate** to **Reactant Concentrations**

Rate equations look ghastly, but all they're really telling you is how the **rate** is affected by the **concentrations of reactants**. For a general reaction: **A + B → C + D**, the **rate equation** is:

The units of rate are moldm^{-3}s^{-1}.

$$\text{Rate} = k[\text{A}]^m[\text{B}]^n$$

Remember — square brackets mean the concentration of whatever's inside them.

1) **m** and **n** are the **orders of the reaction** with respect to reactant A and reactant B. **m** tells you how the **concentration of reactant A** affects the **rate** and **n** tells you the same for **reactant B**.

> If [A] changes and the rate **stays the same**, the order of reaction with respect to A is **0**.
> *So if [A] doubles, the rate will stay the same. If [A] triples, the rate will stay the same.*
>
> If the rate is **proportional to [A]**, then the order of reaction with respect to A is **1**.
> *So if [A] doubles, the rate will double. If [A] triples, the rate will triple.*
>
> If the rate is **proportional to [A]^2**, then the order of reaction with respect to A is **2**.
> *So if [A] doubles, the rate will be $2^2 = 4$ times faster. If [A] triples, the rate will be $3^2 = 9$ times faster.*

2) The **overall order of the reaction** is **m + n**.

3) You can only find **orders of reaction** from **experiments**. You **can't** work them out from chemical equations.

4) **k** is the **rate constant** — the bigger it is, the **faster** the reaction. The rate constant is **always the same** for a certain reaction at a **particular temperature** — but if you **increase** the temperature, the rate constant rises too.

> When you **increase** the **temperature** of a reaction, the **rate of reaction increases** — you're increasing the **number of collisions** between reactant molecules, and also the **energy** of each collision. But the **concentrations** of the reactants and the **orders of reaction** stay the same. So the value of **k** must **increase** for the rate equation to balance.

> **Example:** The chemical equation below shows the acid-catalysed reaction between propanone and iodine.
>
> $$CH_3COCH_{3(aq)} + I_{2(aq)} \xrightarrow{H^+_{(aq)}} CH_3COCH_2I_{(aq)} + H^+_{(aq)} + I^-_{(aq)}$$
>
> This reaction is **first order** with respect to propanone and H$^+_{(aq)}$ and **zero order** with respect to iodine. Write down the rate equation for this reaction.
>
> *Even though H$^+_{(aq)}$ is a catalyst, rather than a reactant, it can still be in the rate equation.*
>
> The **rate equation** is: rate = $k[CH_3COCH_3]^1[H^+]^1[I_2]^0$
> But [X]^1 is usually written as **[X]**, and [X]^0 equals **1** so is usually **left out** of the rate equation.
> So you can **simplify** the rate equation to: **rate = $k[CH_3COCH_3][H^+]$**
>
> *Think about the indices laws from maths.*

Rate Equations

The *Initial Rates Method* can be used to work out *Rate Equations* too

The **initial rate of a reaction** is the rate right at the **start** of the reaction. You can find this from a **concentration-time** graph by calculating the **gradient** of the **tangent** at **time = 0**.

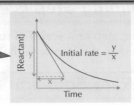

But you might get questions where they change more than one concentration, like the example below.

Here's a quick explanation of how to use the **initial rates method**:

1) Repeat the experiment several times using **different initial concentrations** of reactants. You should usually only change **one** of the concentrations at a time, keeping the rest constant.

2) Calculate the **initial rate** for each experiment using the method above.

3) Finally, see how the **initial concentrations** affect the **initial rates** and figure out the **order** for each reactant. The example below shows you how to do this. Once you know the **orders**, you can work out the rate equation.

Example:

The table on the right shows the results of a series of initial rate experiments for the reaction:

$$NO_{(g)} + CO_{(g)} + O_{2(g)} \rightarrow NO_{2(g)} + CO_{2(g)}$$

The experiments were carried out at a constant temperature.

Write down the rate equation for the reaction.

Experiment number	[NO] (moldm^{-3})	[CO] (moldm^{-3})	[O$_2$] (moldm^{-3})	Initial rate (moldm^{-3}s^{-1})
1	2.0×10^{-2}	1.0×10^{-2}	1.0×10^{-2}	0.17
2	6.0×10^{-2}	1.0×10^{-2}	1.0×10^{-2}	1.53
3	2.0×10^{-2}	2.0×10^{-2}	1.0×10^{-2}	0.17
4	4.0×10^{-2}	1.0×10^{-2}	2.0×10^{-2}	0.68

1) Look at experiments 1 and 2 — when **[NO] triples** (and all the other concentrations stay constant) the rate is **nine times** faster, and $9 = 3^2$. So the reaction is **second order** with respect to NO.

2) Look at experiments 1 and 3 — when **[CO]** doubles (but all the other concentrations stay constant), the rate **stays the same**. So the reaction is **zero order** with respect to CO.

3) Look at experiments 1 and 4 — the rate of experiment 4 is **four times faster** than experiment 1. The reaction is **second order** with respect to **[NO]**, so the rate will **quadruple** when you **double** [NO]. But in experiment 4, **[O$_2$]** has also been **doubled**. As doubling [O$_2$] hasn't had any additional effect on the rate, the reaction must be **zero order** with respect to O$_2$.

4) Now that you know the order with respect to each reactant you can write the rate equation: **rate = k[NO]^2** .

You can calculate k at this temperature by putting the concentrations and the initial rate from one of the experiments into the rate equation (see page 91).

Practice Questions

Q1 Explain what the term 'zero order' means.

Q2 How would the value of k change if you decreased the temperature of a chemical reaction?

Q3 Describe how you could find the initial rate of a reaction.

Exam Question

1 This table shows the results of a series of initial rate experiments for the reaction between substances D and E at a constant temperature:

Experiment	[D] (moldm^{-3})	[E] (moldm^{-3})	Initial rate × 10^{-3} (moldm^{-3}s^{-1})
1	0.2	0.2	1.30
2	0.2	0.4	2.60
3	0.1	0.1	0.65

a) Find the orders of reaction with respect to reactants D and E. Explain your reasoning. [5 marks]

b) Write the rate equation for the reaction in terms of k, the rate constant. [1 mark]

c) The experiments were conducted at a temperature of 300 °C.
State what will happen to the value of the rate constant, k, if the experiments are repeated at 500 °C. [1 mark]

Rate of A2 Chemistry Revision = k [Student] [Tea] ...

These rate equations might look a bit odd, but knowing how to use them can be really handy. Learn what all the different bits of the equation mean, and how to use it to find the overall order of reaction. And make sure that you understand the initial rates method, and that you can use it to work out orders of reaction — it's something examiners love to ask about.

More on Rates and Rate Equations

If you want to work out the mechanism of a reaction then you're going to have to know about the rate-determining step...

The **Rate-Determining Step** is the **Slowest Step** in a Multi-Step Reaction

Mechanisms can have **one step** or a **series of steps**. In a series of steps, each step can have a **different rate**.
The **overall rate** is decided by the step with the **slowest** rate — the **rate-determining step**. ⟵ *Otherwise known as the rate-limiting step.*

Reactants in the **Rate Equation** Affect the **Rate**

The rate equation is handy for working out the **mechanism** of a chemical reaction.
You need to be able to pick out which reactants from the chemical equation are involved in the **rate-determining step**:

> If a reactant appears in the **rate equation**, it must affect the **rate**.
> So this reactant, or something derived from it, must be in the **rate-determining step**.
> If a reactant **doesn't** appear in the **rate equation**, then it **won't** be involved
> in the **rate-determining step** (and neither will anything derived from it).

Catalysts can appear in rate equations, so they can be in rate-determining steps too.

Some **important points** to remember about rate-determining steps and mechanisms are:
1) The rate-determining step **doesn't** have to be the first step in a mechanism.
2) The reaction mechanism **can't** usually be predicted from **just** the chemical equation.

Orders of Reaction Provide More Information About the **Rate-Determining Step**

The **order of a reaction** with respect to a reactant shows the **number of molecules** of that reactant that are involved in the **rate-determining step**.

So, if a reaction's second order with respect to X, there'll be two molecules of X in the rate-determining step.

For example, the mechanism for the reaction between **chlorine free radicals** and **ozone**, O_3, consists of **two steps**:

$$Cl\bullet_{(g)} + O_{3(g)} \rightarrow ClO\bullet_{(g)} + O_{2(g)} \text{ — slow (rate-determining step)}$$
$$ClO\bullet_{(g)} + O\bullet_{(g)} \rightarrow Cl\bullet_{(g)} + O_{2(g)} \text{ — fast}$$

$Cl\bullet$ and O_3 must both be in the rate equation, so the rate equation will be: **rate = $k[Cl\bullet]^m[O_3]^n$**.
There's only **one** $Cl\bullet$ molecule and **one** O_3 molecule in the rate-determining step, so the **orders**, m and n, are both **1**.
So the rate equation is **rate = $k[Cl\bullet][O_3]$**.

The **Rate-Determining Step** Can Help You Work Out the **Reaction Mechanism**

If you know which reactants are in the **rate-determining step**, you can work out the reaction **mechanism**.

There are two possible mechanisms for the nucleophile **OH⁻** substituting for **Br** in 2-bromo-2-methylpropane:

$$CH_3{-}\underset{\underset{CH_3}{|}}{\overset{\overset{CH_3}{|}}{C}}{-}Br + OH^- \rightarrow CH_3{-}\underset{\underset{CH_3}{|}}{\overset{\overset{CH_3}{|}}{C}}{-}OH + Br^-$$

or

$$CH_3{-}\underset{\underset{CH_3}{|}}{\overset{\overset{CH_3}{|}}{C}}{-}Br \rightarrow CH_3{-}\underset{\underset{CH_3}{|}}{\overset{\overset{CH_3}{|}}{C}}{^+} + Br^- \text{ — slow (rate-determining step)}$$

$$CH_3{-}\underset{\underset{CH_3}{|}}{\overset{\overset{CH_3}{|}}{C}}{^+} + OH^- \rightarrow CH_3{-}\underset{\underset{CH_3}{|}}{\overset{\overset{CH_3}{|}}{C}}{-}OH \text{ — fast}$$

The actual **rate equation** was worked out using rate experiments. It is: **rate = $k[(CH_3)_3CBr]$**
OH⁻ isn't in the **rate equation**, so it **can't** be involved in the rate-determining step.
The **second mechanism** is correct because OH⁻ **isn't** in the rate-determining step.

Nitrogen monoxide can react with **oxygen** to produce **nitrogen dioxide**: $2NO + O_2 \rightarrow 2NO_2$
The rate equation for this reaction is: **rate = $k[NO]^2[O_2]$**.
So you know that the **rate-determining step** must involve **2 molecules of NO** and **1 molecule of O_2**.

The reaction **mechanism** for this reaction is made up of **two steps**:

Step 1 — $NO + NO \rightarrow N_2O_2$
Step 2 — $N_2O_2 + O_2 \rightarrow 2NO_2$

Neither step 1 nor step 2 contains all the molecules you'd expect from the rate equation. But in step 2 there's an **intermediate** molecule, N_2O_2, that's **derived from** 2 molecules of NO. So step 2 is the rate-determining step.

More on Rates and Rate Equations

Leaving reaction mechanisms behind for a moment, and getting back to rate equations...

You can Calculate the **Rate Constant** from the **Orders** and **Rate of Reaction**

Once the rate and the orders of the reaction have been found by experiment, you can work out the **rate constant**, *k*. The **units of *k* vary**, so you'll need to work them out too.

> **Example:** The reaction below is second order with respect to NO and zero order with respect to CO and O_2.
> At a certain temperature, the rate is 1.76×10^{-3} moldm^{-3}s^{-1}, when $[NO_{(g)}] = [CO_{(g)}] = [O_{2(g)}] = 2.00 \times 10^{-3}$ moldm^{-3}.
> Find the value of the rate constant, *k*, at this temperature.
>
> $$NO_{(g)} + CO_{(g)} + O_{2(g)} \rightarrow NO_{2(g)} + CO_{2(g)}$$
>
> First write out the **rate equation**:   Rate $= k[NO]^2[CO]^0[O_2]^0 = k[NO]^2$
>
> Next insert the **concentration** and the **rate**. **Rearrange** the equation and calculate the value of *k*:
>
> Rate $= k[NO]^2$, so, $1.76 \times 10^{-3} = k \times (2.00 \times 10^{-3})^2 \Rightarrow k = \dfrac{1.76 \times 10^{-3}}{(2.00 \times 10^{-3})^2} = 440$
>
> Find the **units of *k*** by putting the other units in the rate equation:
>
> Rate $= k[NO]^2$, so moldm^{-3}s$^{-1} = k \times$ (moldm^{-3})$^2 \Rightarrow k = \dfrac{\text{moldm}^{-3}\text{s}^{-1}}{(\text{moldm}^{-3})^2} = \dfrac{\text{s}^{-1}}{\text{moldm}^{-3}} = $ dm^{3}mol^{-1}s^{-1}
>
> So the answer is: ***k* = 440 dm^{3}mol^{-1}s^{-1}**

Practice Questions

Q1 What is meant by the 'rate-determining step' of a reaction?

Q2 Explain how knowing the rate equation for a reaction can help you to predict a reaction mechanism.

Q3 For a reaction between three reactants, X, Y and Z, the rate-determining step is X + 2Y → M + N.
What will the order of reaction be with respect to :   a) reactant X?   b) reactant Y?   c) reactant Z?

Exam Questions

1 For the reaction $CH_3COOH_{(aq)} + C_2H_5OH_{(aq)} \rightarrow CH_3COOC_2H_{5(aq)} + H_2O_{(l)}$, the rate equation is Rate $= k[CH_3COOH][H^+]$.
   What can you deduce about the role that H^+ plays in the reaction? Explain your answer. [3 marks]

2 The following reaction is first order with respect to H_2 and first order with respect to ICl.
   $$H_{2(g)} + 2ICl_{(g)} \rightarrow I_{2(g)} + 2HCl_{(g)}$$
   a) Write the rate equation for this reaction in terms of *k*, the rate constant. [1 mark]
   b) The mechanism for this reaction consists of two steps.
      (i) Identify the molecules that are in the rate-determining step. Justify your answer. [3 marks]
      (ii) A chemist suggested the following mechanism for the reaction:
      $$2ICl_{(g)} \rightarrow I_{2(g)} + Cl_{2(g)} \qquad \text{slow}$$
      $$H_{2(g)} + Cl_{2(g)} \rightarrow 2HCl_{(g)} \qquad \text{fast}$$
      Suggest, with reasons, whether this mechanism is likely to be correct. [2 marks]

3 The following reaction is second order with respect to NO and first order with respect to H_2.
   $$2NO_{(g)} + 2H_{2(g)} \rightarrow 2H_2O_{(g)} + N_{2(g)}$$
   a) Write a rate equation for the reaction in terms of *k*, the rate constant. [1 mark]
   b) The rate of the reaction at 800 °C was determined to be 0.00267 moldm^{-3}s^{-1}
      when $[H_2] = 0.0020$ moldm^{-3} and $[NO] = 0.0040$ moldm^{-3}.
      (i) Calculate a value for the rate constant at 800 °C, including units. [3 marks]
      (ii) Predict the effect on the rate constant of decreasing the temperature of the reaction to 600 °C. [1 mark]

I found rate-determining step aerobics a bit on the slow side...

These pages show you how rate equations, orders of reaction, and reaction mechanisms all tie together and how each actually means something in the grand scheme of A2 Chemistry. It's all very profound. So get it all learnt and answer the questions and then you'll have plenty of time to practise the quickstep for your Strictly Come Dancing routine.

The Equilibrium Constant

There's a lot of toing and froing coming up. You won't know whether you're coming or going.

At **Equilibrium** the Amounts of Reactants and Products **Stay the Same**

1) Lots of changes are **reversible** — they can go **both ways**. To show a change is reversible, you stick in a ⇌.

2) As the **reactants** get used up, the **forward** reaction **slows down** — and as more **product** is formed, the **reverse** reaction **speeds up**. After a while, the forward reaction will be going at exactly the **same rate** as the backward reaction.
The amounts of reactants and products **won't be changing** any more, so it'll seem like **nothing's happening**. It's a bit like you're **digging a hole** while someone else is **filling it in** at exactly the **same speed**. This is called a **dynamic equilibrium**.

3) Equilibria can be set up in **physical** systems, e.g.:

   When **liquid bromine** is shaken in a closed flask, some of it changes to orange **bromine gas**. After a while, **equilibrium** is reached — bromine liquid is **still** changing to bromine gas and bromine gas is still changing to bromine liquid, but they are changing at the **same rate**.

   $$Br_{2(l)} \rightleftharpoons Br_{2(g)}$$

   ...and **chemical** systems, e.g.:

   If **hydrogen gas** and **iodine gas** are mixed together in a closed flask, **hydrogen iodide** is formed.

   $$H_{2(g)} + I_{2(g)} \rightleftharpoons 2HI_{(g)}$$

   Imagine that **1.0 mole** of hydrogen gas is mixed with **1.0 mole** of iodine gas at a constant temperature of **640 K**. When this mixture reaches equilibrium, there will be **1.6 moles** of hydrogen iodide and **0.2 moles** of both hydrogen gas and iodine gas. No matter how long you leave them at this temperature, the **equilibrium** amounts **never change**. As with the physical system, it's all a matter of the forward and backward rates **being equal**.

4) A **dynamic equilibrium** can only happen in a **closed system** at a **constant temperature**.

   A closed system just means nothing can get in or out.

K_c is the **Equilibrium Constant**

If you know the **molar concentration** of each substance at equilibrium, you can work out the **equilibrium constant**, K_c. Your value of K_c will only be true for that particular **temperature**.

Before you can calculate K_c, you have to write an **expression** for it. Here's how:

For the general reaction **aA + bB ⇌ dD + eE**, $K_c = \dfrac{[D]^d[E]^e}{[A]^a[B]^b}$

The products go on the top line. The square brackets, [ ], mean concentration in moldm⁻³.

The lower-case letters a, b, d and e are the number of moles of each substance.

So for the reaction $H_{2(g)} + I_{2(g)} \rightleftharpoons 2HI_{(g)}$, $K_c = \dfrac{[HI]^2}{[H_2]^1[I_2]^1} = \dfrac{[HI]^2}{[H_2][I_2]}$.

These little numbers look like the orders of reaction that you saw in rate equations. But they're not — they're the number of moles.

Calculate K_c by **Sticking Numbers** into the Expression

If you know the **equilibrium concentrations**, just bung them in your expression. Then with a bit of help from the old calculator, you can work out the **value** for K_c. The **units** are a bit trickier though — they **vary**, so you have to work them out after each calculation.

Example: If the volume of the closed flask in the hydrogen iodide example above is 2.0 dm³, what is the equilibrium constant for the reaction at 640 K? The equilibrium concentrations are:
[HI] = 0.8 moldm⁻³, [H₂] = 0.1 moldm⁻³, and [I₂] = 0.1 moldm⁻³ .

Just stick the concentrations into the **expression** for K_c: $K_c = \dfrac{[HI]^2}{[H_2][I_2]} = \dfrac{0.8^2}{0.1 \times 0.1} = 64$ ← This is the value of K_c.

To work out the **units** of K_c put the units in the expression instead of the numbers:

$K_c = \dfrac{(\text{moldm}^{-3})^2}{(\text{moldm}^{-3})(\text{moldm}^{-3})}$ — the concentration units cancel, so there are **no units** for K_c .

So K_c is just **64**.

The Equilibrium Constant

You Might Need to **Work Out** the **Equilibrium Concentrations**

You might have to figure out some of the **equilibrium concentrations** before you can find K_c:

Example: 0.20 moles of phosphorus(V) chloride decomposes at 600 K in a vessel of 5.00 dm³. The equilibrium mixture is found to contain 0.08 moles of chlorine. Write the expression for K_c and calculate its value, including units.

$$PCl_{5(g)} \rightleftharpoons PCl_{3(g)} + Cl_{2(g)}$$

First find out how many moles of PCl_5 and PCl_3 there are at equilibrium:

The **equation** tells you that when **1 mole of PCl_5** decomposes, **1 mole of PCl_3** and **1 mole of Cl_2** are formed. So if 0.08 moles of chlorine are produced at equilibrium, then there will be **0.08 moles** of PCl_3 as well. 0.08 mol of PCl_5 must have decomposed, so there will be **0.12 moles** left (0.2 – 0.08).

Divide each number of moles by the volume of the flask to give the molar concentrations:

$$[PCl_3] = [Cl_2] = 0.08 \div 5.00 = \textbf{0.016 moldm}^{-3} \qquad [PCl_5] = 0.12 \div 5.00 = \textbf{0.024 moldm}^{-3}$$

Put the concentrations in the expression for K_c and calculate it:
$$K_c = \frac{[PCl_3][Cl_2]}{[PCl_5]} = \frac{[0.016][0.016]}{[0.024]} = \textbf{0.011}$$

Now find the units of K_c:
$$K_c = \frac{(moldm^{-3})(moldm^{-3})}{moldm^{-3}} = \textbf{moldm}^{-3} \qquad \text{So } \textbf{K}_c = \textbf{0.011 moldm}^{-3}$$

K_c can be used to Find **Concentrations** in an **Equilibrium Mixture**

Example: When ethanoic acid was allowed to reach equilibrium with ethanol at 25 °C, it was found that the equilibrium mixture contained 2.0 moldm⁻³ ethanoic acid and 3.5 moldm⁻³ ethanol. The K_c of the equilibrium is 4.0 at 25 °C. What are the concentrations of the other components?

$$CH_3COOH_{(l)} + C_2H_5OH_{(l)} \rightleftharpoons CH_3COOC_2H_{5(l)} + H_2O_{(l)}$$

Put all the values you know in the K_c expression:
$$K_c = \frac{[CH_3COOC_2H_5][H_2O]}{[CH_3COOH][C_2H_5OH]} \Rightarrow 4.0 = \frac{[CH_3COOC_2H_5][H_2O]}{2.0 \times 3.5}$$

Rearranging this gives: $[CH_3COOC_2H_5][H_2O] = 4.0 \times 2.0 \times 3.5 = 28.0$

From the equation, you know that $[CH_3COOC_2H_5] = [H_2O]$, so: $[CH_3COOC_2H_5] = [H_2O] = \sqrt{28} = 5.3$ moldm⁻³

The concentration of $CH_3COOC_2H_5$ and H_2O is 5.3 moldm⁻³

If **Conditions Change** the **Position of Equilibrium** Will Move

If you **change** the **concentration**, **pressure** or **temperature** of a reversible reaction, you're going to **alter** the **position of equilibrium**. This just means you'll end up with **different amounts** of reactants and products at equilibrium.

If the position of equilibrium moves to the **left**, you'll get more **reactants**. $\mathbf{H_{2(g)} + I_{2(g)}} \rightleftharpoons 2HI_{(g)}$

If the position of equilibrium moves to the **right**, you'll get more **products**. $H_{2(g)} + I_{2(g)} \rightleftharpoons \mathbf{2HI_{(g)}}$

There's a rule that lets you predict how the **position of equilibrium** will change if a **condition changes**. This rule is known as **Le Chatelier's Principle**. Here it is:

If there's a change in **concentration**, **pressure** or **temperature**, the equilibrium will move to help **counteract** the change.

So, basically, if you **raise the temperature**, the position of equilibrium will shift to try to **cool things down**. And if you **raise the pressure or concentration**, the position of equilibrium will shift to try to **reduce it**.

The Equilibrium Constant

Temperature Changes Alter K_c — Concentration Changes Don't

TEMPERATURE

1) If you **increase** the temperature, you **add heat**. The equilibrium shifts in the **endothermic (positive ΔH) direction** to absorb the heat.

2) **Decreasing** the temperature **removes heat**. The equilibrium shifts in the **exothermic (negative ΔH) direction** to try to replace the heat.

3) If the forward reaction's **endothermic**, the reverse reaction will be **exothermic**, and vice versa.

4) If the change means **more product** is formed, K_c **will rise**. If it means **less product** is formed, then K_c **will decrease**.

The reaction below is exothermic in the forward direction. If you increase the temperature, the equilibrium shifts to the left to absorb the extra heat. This means that less product's formed.

Exothermic $\Longrightarrow$
$$2SO_{2(g)} + O_{2(g)} \rightleftharpoons 2SO_{3(g)} \quad \Delta H = -197 \text{ kJmol}^{-1}$$
$\Longleftarrow$ Endothermic

$$K_c = \frac{[SO_3]^2}{[SO_2]^2[O_2]}$$

Since there's less product formed, K_c decreases.

CONCENTRATION

The value of the **equilibrium constant**, K_c, is **fixed** at a given temperature. So if the concentration of one thing in the equilibrium mixture **changes** then the concentrations of the others must change to keep the value of K_c the same.

$$CH_3COOH_{(l)} + C_2H_5OH_{(l)} \rightleftharpoons CH_3COOC_2H_{5(l)} + H_2O_{(l)}$$

If you increase the concentration of CH_3COOH then the equilibrium will move to the right to get rid of the extra CH_3COOH — so more $CH_3COOC_2H_5$ and H_2O are produced. This keeps the equilibrium constant the same.

Catalysts have **NO EFFECT** on the **position of equilibrium** or on the value of K_c. They **can't** increase yield — but they **do** mean equilibrium is approached **faster**.

Practice Questions

Q1 Write an expression for the equilibrium constant, K_c, for the reaction $N_{2(g)} + 3H_{2(g)} \rightleftharpoons 2NH_{3(g)}$.

Q2 If a reversible reaction has an exothermic forward reaction, how will increasing the temperature affect the equilibrium position?

Q3 How does adding a catalyst to a reaction affect the equilibrium position and the value of K_c?

Exam Questions

1 A sample of pure hydrogen iodide is placed in a sealed flask, and heated to 443 °C.
The following equilibrium is established: $2HI_{(g)} \rightleftharpoons H_{2(g)} + I_{2(g)}$ ($K_c = 0.02$)
At equilibrium the concentration of I_2 is found to be 0.77 moldm^{-3}. Find the equilibrium concentration of HI. [4 marks]

2 Nitrogen dioxide dissociates according to the equation $2NO_{2(g)} \rightleftharpoons 2NO_{(g)} + O_{2(g)}$.
When 42.5 g of nitrogen dioxide were heated in a vessel of volume 22.8 dm^3 at 500 °C, 14.1 g of oxygen were found in the equilibrium mixture.
a) Calculate (i) the number of moles of nitrogen dioxide originally. [1 mark]
(ii) the number of moles of each gas in the equilibrium mixture. [3 marks]
b) Write an expression for K_c for this reaction. Find the value of K_c at this temperature, and give its units. [5 marks]

3 This equation shows the equilibrium that is set up when ethanoic acid is added to water:
$$CH_3COOH_{(aq)} \rightleftharpoons CH_3COO^-_{(aq)} + H^+_{(aq)}$$
a) Write an expression for K_c for this reaction. [1 mark]
b) State the effect on the equilibrium constant, K_c, of increasing the concentration of $CH_3COOH_{(aq)}$. [1 mark]
The forward reaction is endothermic.
c) Describe how the equilibrium position would change if the temperature was increased. Explain your answer. [2 marks]
d) State the effect on the equilibrium constant, K_c, of increasing the reaction temperature. [1 mark]

A big K_c means heaps of product...

Most organic reactions and plenty of inorganic reactions are reversible. Sometimes the backwards reaction's about as speedy as a dead snail though, so some reactions might be thought of as only going one way. It's like if you're walking forwards, continental drift could be moving you backwards at the same time, just reeeeally slowly.

Acids, Bases and pH

Remember this stuff? Well, it's all down to Brønsted and Lowry — they've got a lot to answer for.

An Acid **Releases** Protons — a Base **Accepts** Protons

Brønsted-Lowry acids are **proton donors** — they release **hydrogen ions** (H^+) when they're mixed with water. You never get H^+ ions by themselves in water though — they're always combined with H_2O to form **hydroxonium ions**, H_3O^+.

HA is just any old acid. ⟹ $HA_{(aq)} + H_2O_{(l)} \rightarrow H_3O^+_{(aq)} + A^-_{(aq)}$

Brønsted-Lowry bases do the opposite — they're **proton acceptors**. When they're in solution, they grab **hydrogen ions** from water molecules.

B is just a random base. ⟹ $B_{(aq)} + H_2O_{(l)} \rightarrow BH^+_{(aq)} + OH^-_{(aq)}$

Acids and Bases can be **Strong** or **Weak**

These are really all reversible reactions, but the equilibrium lies extremely far to the right.

1) **Strong acids dissociate** (or ionise) **almost completely** in water — **nearly all** the H^+ ions will be released. **Hydrochloric acid** is a strong acid — $HCl_{(g)} +$ water $\rightarrow H^+_{(aq)} + Cl^-_{(aq)}$.
   Strong bases (like sodium hydroxide) **ionise almost completely** in water too. E.g. $NaOH_{(s)} +$ water $\rightarrow Na^+_{(aq)} + OH^-_{(aq)}$.

2) **Weak acids** (e.g. ethanoic or citric) dissociate only very **slightly** in water — so only small numbers of H^+ ions are formed. An **equilibrium** is set up which lies well over to the **left**. E.g. $CH_3COOH_{(aq)} \rightleftharpoons CH_3COO^-_{(aq)} + H^+_{(aq)}$.
   Weak bases (such as ammonia) **only slightly dissociate** in water too. E.g. $NH_{3(aq)} + H_2O_{(l)} \rightleftharpoons NH_4^+_{(aq)} + OH^-_{(aq)}$.
   Just like with weak acids, the equilibrium lies well over to the **left**.

Protons are **Transferred** when **Acids** and **Bases** React

Acids can't just throw away their protons — they can only get rid of them if there's a **base** to accept them. In this reaction the **acid**, HA, **transfers** a proton to the **base**, B:

$HA_{(aq)} + B_{(aq)} \rightleftharpoons BH^+_{(aq)} + A^-_{(aq)}$

It's an **equilibrium**, so if you add more **HA** or **B**, the position of equilibrium moves to the **right**. But if you add more **BH⁺** or **A⁻**, the equilibrium will move to the **left**. See page 93 for the rule about this.

When an acid is added to **water**, water acts as the **base** and accepts the proton:

$HA_{(aq)} + H_2O_{(l)} \rightleftharpoons H_3O^+_{(aq)} + A^-_{(aq)}$ ⟵ *The equilibrium's far to the left for weak acids, and far to the right for strong acids.*

Water Dissociates **Slightly**

Water dissociates into **hydroxonium ions** and **hydroxide ions**.

So this equilibrium exists in water:

$H_2O_{(l)} + H_2O_{(l)} \rightleftharpoons H_3O^+_{(aq)} + OH^-_{(aq)}$ or more simply $H_2O_{(l)} \rightleftharpoons H^+_{(aq)} + OH^-_{(aq)}$

And, just like for any other equilibrium reaction, you can apply the equilibrium law and write an expression for the **equilibrium constant**: $K_c = \dfrac{[H^+][OH^-]}{[H_2O]}$

Water only dissociates a **tiny amount**, so the equilibrium lies well over to the **left**. There's so much water compared to the amounts of H^+ and OH^- ions that the concentration of water is considered to have a **constant** value.

So if you multiply the expression you wrote for K_c (which is a constant) by $[H_2O]$ (another constant), you get a **constant**. This new constant is called the **ionic product of water** and it is given the symbol K_w.

$K_w = K_c \times [H_2O] = [H^+][OH^-] \Rightarrow$ $K_w = [H^+][OH^-]$ ⟵ *The units of K_w are always mol^2dm^{-6}.*

K_w always has the **same value** for an aqueous solution at a **given temperature**.
For example, at 298 K (25 °C), K_w has a value of $1.00 \times 10^{-14}\ mol^2dm^{-6}$.

> In **pure water**, there is always **one H^+ ion** for **each OH⁻ ion**. So $[H^+] = [OH^-]$.
> That means if you are dealing with **pure water**, then you can say that $K_w = [H^+]^2$.

Acids, Bases and pH

The pH Scale is a Measure of the Hydrogen Ion Concentration

The **concentration of hydrogen ions** in a solution can vary enormously, so those wise chemists of old decided to express the concentration on a **logarithmic scale**.

$$pH = -\log_{10} [H^+]$$

[H$^+$] is the concentration of hydrogen ions in a solution, measured in moldm^{-3}.

The pH scale normally goes from **0** (very acidic) to **14** (very alkaline). **pH 7** is regarded as being **neutral**.

You Can Calculate pH From Hydrogen Ion Concentration...

If you know the **hydrogen ion concentration** of a solution, you can calculate its **pH** by sticking the numbers into the **formula**.

A solution of hydrochloric acid has a hydrogen ion concentration of 0.01 moldm^{-3}. What is the pH of the solution?

$$pH = -\log_{10} [H^+] = -\log_{10} [0.01] = 2$$

Use the 'log' button on your calculator for this.

...Or Hydrogen Ion Concentration From pH

If you've got the **pH** of a solution, and you want to know its **hydrogen ion concentration**, then you need the **inverse** of the pH formula:

$$[H^+] = 10^{-pH}$$

Now you can use this formula to find [H$^+$].

A solution of sulfuric acid has a pH of 1.52.
What is the hydrogen ion concentration of this solution?

$$[H^+] = 10^{-pH} = 10^{-1.52} = 0.03 \text{ moldm}^{-3} = 3 \times 10^{-2} \text{ moldm}^{-3}$$

For Strong Monoprotic Acids Hydrogen Ion Concentration = Acid Concentration

Hydrochloric acid and nitric acid (HNO$_3$) are **strong acids** so they ionise fully.
They're also **monoprotic**, so each mole of acid produces **one mole of hydrogen ions**.
This means the H$^+$ concentration is the **same** as the acid concentration.

Monoprotic means that each molecule of an acid will release 1 proton when it dissociates.

So for 0.1 moldm^{-3} hydrochloric acid, [H$^+$] is **0.1 moldm^{-3}**. Its pH = $-\log_{10}$ [H$^+$] = $-\log_{10}$ 0.1 = **1.0**.
Here are a couple of examples for you:

1) Calculate the pH of 0.05 moldm^{-3} nitric acid.   [H$^+$] = 0.05 $\Rightarrow$ pH = $-\log_{10}$ 0.05 = **1.30**

2) If a solution of hydrochloric acid has a pH of 2.45, what is the hydrogen ion concentration of the acid?   [H$^+$] = 10$^{-2.45}$ = **3.55 $\times$ 10^{-3} moldm^{-3}**

For Strong Diprotic Acids Hydrogen Ion Concentration = Twice Acid Concentration

Each molecule of a **strong diprotic acid** releases **2 protons** when it dissociates.
Sulfuric acid is an example of a **strong diprotic acid**: H$_2$SO$_{4(l)}$ + water $\rightarrow$ 2H$^+_{(aq)}$ + SO$_4^{2-}{}_{(aq)}$

There's more about diprotic acids on page 102.

So, **diprotic acids** produce **two moles of hydrogen ions** for **each mole of acid**, meaning that the H$^+$ concentration is twice the concentration of the acid.

So for 0.1 moldm^{-3} sulfuric acid, [H$^+$] is **0.2 moldm^{-3}**. Its pH will be: pH = $-\log_{10}$ [H$^+$] = $-\log_{10}$ 0.2 = **0.70.**
Here's an example:

3) Calculate the pH of 0.25 moldm^{-3} sulfuric acid.   [H$^+$] = 2 $\times$ 0.25 = 0.5 $\Rightarrow$ pH = $-\log_{10}$ 0.5 = **0.30**

Acids, Bases and pH

Use K_w to Find the pH of a Base

Sodium hydroxide (NaOH) and potassium hydroxide (KOH) are **strong bases** that **fully ionise** in water — they donate **one mole of OH⁻ ions** per mole of base.

This means that the concentration of OH⁻ ions is the **same** as the **concentration of the base**. So for 0.02 moldm⁻³ sodium hydroxide solution, [OH⁻] is also **0.02 moldm⁻³**.

But to work out the **pH** you need to know **[H⁺]** — luckily this is linked to **[OH⁻]** through the **ionic product of water**, K_w:

$$K_w = [H^+][OH^-]$$

So if you know [OH⁻] for a **strong aqueous base** and K_w at a certain temperature, you can work out **[H⁺]** and then the **pH**.

> The value of K_w at 298 K is 1.0×10^{-14} mol²dm⁻⁶. Find the pH of 0.1 moldm⁻³ NaOH at 298 K.
>
> $$[OH^-] = 0.1 \text{ moldm}^{-3} \Rightarrow [H^+] = \frac{K_w}{[OH^-]} = \frac{1.0 \times 10^{-14}}{0.1} = 1.0 \times 10^{-13} \text{ moldm}^{-3}$$
>
> So pH = $-\log_{10} 1.0 \times 10^{-13}$ = **13.0**

Practice Questions

Q1 Which substance is acting as a Brønsted-Lowry base in the following equation? $HX + H_2O \rightarrow H_3O^+ + X^-$

Q2 Explain what is meant by the term 'strong acid' and give an example of one.

Q3 Define K_w and give its value at 298 K.

Q4 Define pH.

Exam Questions

1 a)   How did Brønsted and Lowry define: (i) an acid,   (ii) a base.   [2 marks]
   b)   Show, by writing appropriate equations, how HSO_4^- can behave as:
        (i) a Brønsted-Lowry acid,   (ii) a Brønsted-Lowry base.   [2 marks]

2 Hydrocyanic acid (HCN) is a weak acid. Define the term 'weak acid' and write
   a balanced equation for the equilibrium that occurs when HCN dissolves in water.   [4 marks]

3 Nitric acid, HNO_3, is a strong monoprotic acid.
   a)   Explain what is meant by a monoprotic acid.   [1 mark]
   b)   Find the concentration of a solution of nitric acid that has a pH value of 0.55.
        Give your answer to 2 decimal places.   [2 marks]

4 At 298 K, the value of K_w is 1.00×10^{-14} mol²dm⁻⁶.
   Find the pH of a 0.125 moldm⁻³ solution of sodium hydroxide at this temperature.
   Give your answer to 1 decimal place.   [3 marks]

5 A flask of water is heated to 30 °C. At 30 °C, the value of K_w is 1.47×10^{-14} mol²dm⁻⁶.
   a)   Calculate the hydrogen ion concentration of the water in the flask.   [2 marks]
   b)   Calculate the pH of the water at this temperature.   [1 mark]
   Give your answers to 2 decimal places.

Acids and bases — the Julie Andrews and Marilyn Manson of the chemistry world...

Don't confuse strong acids with concentrated acids, or weak acids with dilute acids. Strong and weak are to do with how much an acid ionises, whereas concentrated and dilute are to do with the number of moles of acid you've got per dm³. You can have a strong dilute acid, or a weak concentrated acid. And it works just the same way with bases too.

More pH Calculations

Those calculators should be nicely warmed up by now. You'll be needing that log function key again too.

To Find the pH of a Weak Acid you Use K_a (the Acid Dissociation Constant)

Weak acids **don't** ionise fully in solution, so the [H$^+$] **isn't** the same as the acid concentration. This makes it a **bit trickier** to find their pH. You have to use yet another **equilibrium constant**, K_a.

For a weak aqueous acid, HA, you get the following equilibrium: $HA_{(aq)} \rightleftharpoons H^+_{(aq)} + A^-_{(aq)}$

As only a **tiny amount** of HA dissociates, you can assume that $[HA]_{start} = [HA]_{equilibrium}$.

So if you apply the equilibrium law, you get: $K_a = \dfrac{[H^+][A^-]}{[HA]}$

You can also assume that **all** the H$^+$ ions come from the **acid**, so [H$^+$] = [A$^-$].

So $K_a = \dfrac{[H^+]^2}{[HA]}$ ← The units of K_a are moldm^{-3}.

Here's an example of how to use K_a to find the **pH** of a weak acid:

1) Find the pH of a 0.02 moldm^{-3} solution of propanoic acid (CH_3CH_2COOH) at 298K. K_a for propanoic acid at this temperature is 1.30×10^{-5} moldm^{-3}.

First write an expression for K_a for the weak acid, then rearrange it to find [H$^+$]^2:

$K_a = \dfrac{[H^+]^2}{[CH_3CH_2COOH]} \Rightarrow [H^+]^2 = K_a[CH_3CH_2COOH] = (1.30 \times 10^{-5}) \times 0.02 = 2.60 \times 10^{-7}$

Take the square root of this number to find [H$^+$]: $[H^+] = \sqrt{2.60 \times 10^{-7}} = 5.10 \times 10^{-4}$ moldm^{-3}

Now use [H$^+$] to find the pH of the acid: $pH = -\log_{10}[H^+] = -\log_{10} 5.10 \times 10^{-4} = \mathbf{3.29}$

You Might Have to Find the Concentration or K_a of a Weak Acid

You don't need to know anything new for this type of calculation. You usually just have to find **[H$^+$]** from the pH, then fiddle around with the **K_a expression** to find the missing bit of information.

2) The pH of an ethanoic acid (CH_3COOH) solution is 3.02 at 298 K. Calculate the molar concentration of this solution. The K_a of ethanoic acid is 1.75×10^{-5} moldm^{-3} at 298 K.

Use the pH of the acid to find [H$^+$]: $[H^+] = 10^{-pH} = 10^{-3.02} = 9.55 \times 10^{-4}$ moldm^{-3}

Now write an expression for K_a, and rearrange it to equal [CH_3COOH]:

$K_a = \dfrac{[H^+]^2}{[CH_3COOH]} \Rightarrow [CH_3COOH] = \dfrac{[H^+]^2}{K_a} = \dfrac{(9.55 \times 10^{-4})^2}{1.75 \times 10^{-5}} = \mathbf{0.0521}$ **moldm^{-3}**

3) A solution of 0.162 moldm^{-3} HCN has a pH of 5.05 at 298 K. What is the value of K_a for HCN at 298 K?

Use the pH of the acid to find [H$^+$]: $[H^+] = 10^{-pH} = 10^{-5.05} = 8.91 \times 10^{-6}$ moldm^{-3}

Then write an expression for K_a, and substitute in the values for [H$^+$] and [HCN]:

$K_a = \dfrac{[H^+]^2}{[HCN]} = \dfrac{(8.91 \times 10^{-6})^2}{0.162} = \mathbf{4.90 \times 10^{-10}}$ **moldm^{-3}**

More pH Calculations

$pK_a = -log_{10} K_a$ and $K_a = 10^{-pK_a}$

pK_a is calculated from K_a in exactly the same way as pH is calculated from $[H^+]$ — and vice versa.

So if an acid has a K_a value of 1.50×10^{-7}, then $pK_a = -log_{10}(1.50 \times 10^{-7}) = 6.82$.

Notice how pK_a values aren't annoyingly tiny like K_a values.

And if an acid has a pK_a value of 4.32, then $K_a = 10^{-4.32} = 4.79 \times 10^{-5}$.

Just to make things that bit more complicated, there might be a **pK_a** value in a 'find the pH' type of question.
If so, you need to convert it to K_a so that you can use the **K_a expression**.

4) Calculate the pH of 0.050 moldm^{-3} methanoic acid (HCOOH).
Methanoic acid has a pK_a of 3.75 at 298 K.

$K_a = 10^{-pK_a} = 10^{-3.75} = 1.78 \times 10^{-4}$ moldm^{-3} ⟵ First you have to convert the pK_a to K_a.

$K_a = \dfrac{[H^+]^2}{[HCOOH]}$ ⟹ $[H^+]^2 = K_a[HCOOH] = 1.78 \times 10^{-4} \times 0.050 = 8.9 \times 10^{-6}$

⟹ $[H^+] = \sqrt{8.9 \times 10^{-6}} = 2.98 \times 10^{-3}$ moldm^{-3} pH $= -\log 2.98 \times 10^{-3} = $ **2.53**

Sometimes you have to give your answer as a **pK_a** value. In this case, you just work out the K_a value as usual and then convert it to **pK_a** — and Bob's your pet hamster.

Practice Questions

Q1 Explain what is meant by the term 'weak acid' and give an example of one.

Q2 Write an equation for the reversible dissociation of the weak acid HA.

Q3 Explain how to find the pH of a weak acid.

Q4 Define pK_a.

Exam Questions

1 The value of K_a for the weak acid HA, at 298 K, is 5.60×10^{-4} moldm^{-3}.
   a) Write an expression for K_a for the weak acid HA. [1 mark]
   b) Calculate the pH of a 0.280 moldm^{-3} solution of HA at 298 K. [3 marks]

2 The pH of a 0.150 moldm^{-3} solution of a weak monoprotic acid, HX, is 2.65 at 298 K.
   a) Calculate the value of K_a for the acid HX at 298 K. [4 marks]
   b) Calculate pK_a for this acid. [1 mark]

3 Hydrocyanic acid is a weak monoprotic acid with the formula HCN.
   At 298 K it has a K_a of 4.9×10^{-10} moldm^{-3}.
   a) Write an expression for K_a for hydrocyanic acid. [1 mark]
   b) At 298 K, a solution of hydrocyanic acid has a pH of 4.15.
      Find the concentration of the solution in moldm^{-3}. [3 marks]
   c) Find the pK_a of hydrocyanic acid at 298 K. [1 mark]

4 At 298 K, the weak monoprotic acid HF has a pK_a of 3.16.
   Find the pH of a 0.5 moldm^{-3} solution of HF at 298 K. [5 marks]

My mate had a red Ka — but she drove it into a lamppost...

Strong acids have high K_a values and weak acids have low K_a values. For pK_a values, it's the other way round — the stronger the acid, the lower the pK_a. If something's got p in front of it, like pH, pK_w or pK_a, it tends to mean $-\log_{10}$ of whatever. Not all calculators work the same way, so make sure you know how to work logs out on your calculator.

pH Curves, Titrations and Indicators

If you add alkali to an acid, the pH changes in a squiggly sort of way.

Use **Titration** to Find the **Concentration** of an **Acid** or **Alkali**

Titrations allow you to find out **exactly** how much alkali is needed to **neutralise** a quantity of acid.

1) You measure out some **acid** of known concentration using a pipette and put it in a flask, along with some **appropriate indicator**.

2) First do a rough titration — add the **alkali** to the acid using a **burette** fairly quickly to get an approximate idea where the solution changes colour (the **end point**). Give the flask a regular **swirl**.

3) Now do an **accurate** titration. Run the alkali in to within 2 cm³ of the end point, then add it **drop by drop**. If you don't notice exactly when the solution changes colour you've **overshot** and your result won't be accurate.

4) **Record** the amount of alkali needed to **neutralise** the acid. It's best to **repeat** this process a few times, making sure you get very similar answers each time (within about 0.1 cm³ of each other).

You can also find out how much **acid** is needed to neutralise a quantity of **alkali**. It's exactly the same process as above, but you add **acid to alkali** instead.

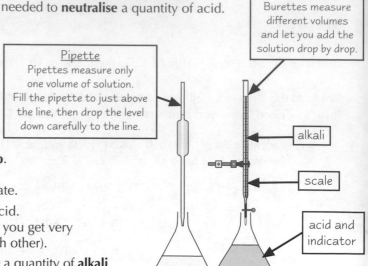

__Pipette__
Pipettes measure only one volume of solution. Fill the pipette to just above the line, then drop the level down carefully to the line.

__Burette__
Burettes measure different volumes and let you add the solution drop by drop.

alkali

scale

acid and indicator

pH Curves Plot pH Against Volume of Acid or Base Added

The graphs below show the pH curves for the **different combinations** of **strong and weak** monoprotic acids and bases.

strong acid/strong base

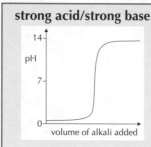

The pH starts around **1**, as there's an excess of **strong acid**.

It finishes up around pH **13**, when you have an excess of **strong base**.

strong acid/weak base

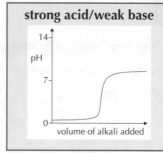

The pH starts around **1**, as there's an excess of **strong acid**.

It finishes up around pH **9**, when you have an excess of **weak base**.

weak acid/strong base

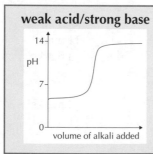

The pH starts around **5**, as there's an excess of **weak acid**.

It finishes up around pH **13**, when you have an excess of **strong base**.

weak acid/weak base

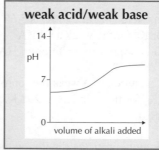

The pH starts around **5**, as there's an excess of **weak acid**.

It finishes up around pH **9**, when you have an excess of **weak base**.

All the graphs apart from the weak acid/weak base graph have a bit that's almost vertical — this is the **equivalence point** or **end point**. At this point, a tiny amount of base causes a sudden, big change in pH — it's here that all the acid is just **neutralised**.

You don't get such a sharp change in a **weak acid/weak base** titration. If you used an indicator for this type of titration, its colour would change very **gradually**, and it would be very tricky to see the exact end point. So you're usually better off using a **pH meter** for this type of titration.

If you titrate a **base** with an **acid** instead, the **shapes** of the curves **stay the same**, but they **flip over**:

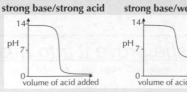
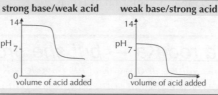

pH Curves, Titrations and Indicators

pH Curves can Help you Decide which Indicator to Use

When you use an **indicator**, you need it to change colour exactly at the **end point** of your titration. So you need to pick one that changes colour over a **narrow pH range** that lies **entirely** on the **vertical part** of the **pH curve**.

So for this titration you'd want an indicator that changed colour somewhere between pH 7 and pH 11.

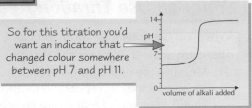

Methyl orange and **phenolphthalein** are **indicators** that are often used for acid-base titrations. They each change colour over a **different pH range**:

Name of indicator	Colour at low pH	Approx. pH of colour change	Colour at high pH
Methyl orange	red	3.1 – 4.4	yellow
Phenolphthalein	colourless	8.3 – 10	pink

For a **strong acid/strong alkali** titration, you can use **either** of these indicators — there's a rapid pH change over the range for **both** indicators.

For a **strong acid/weak alkali** only **methyl orange** will do. The pH changes rapidly across the range for methyl orange, but not for phenolphthalein.

For a **weak acid/strong alkali**, **phenolphthalein** is the stuff to use. The pH changes rapidly over phenolphthalein's range, but not over methyl orange's.

For **weak acid/weak alkali** titrations there's no sharp pH change, so **no** indicator will work.

Practice Questions

Q1 Explain what 'equivalence point' means.

Q2 Why are indicators not used for weak acid/weak alkali titrations?

Q3 What indicator could you use for a weak acid/strong alkali titration?

Q4 Sketch the pH curve for a strong acid/weak alkali titration.

Exam Questions

1   A known volume of acid is titrated with a known volume of alkali.
    The pH of the solution is followed throughout the titration using a pH meter.

    a)   Suggest an acid and an alkali that may have been used in
         this titration.   [2 marks]

    b)   Suggest a suitable indicator for use in this titration.   [1 mark]

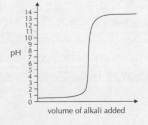

2   a)   Sketch a graph to show the shape of the pH curve for the titration of a weak acid with a weak base.
         Assume that both solutions have a concentration of 0.1 moldm^{-3}.   [2 marks]

    b)   What will the pH of the solution be at the equivalence point?   [1 mark]

3   Look at the three pH curves below labelled X, Y and Z.

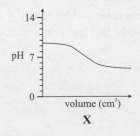
X

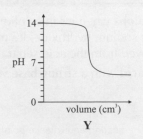
Y

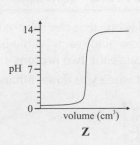
Z

    a)   Which of these curves represents the titration of sodium hydroxide with ethanoic acid?   [1 mark]
    b)   Write an equation for the reaction between sodium hydroxide and ethanoic acid.   [2 marks]
    c)   Suggest an indicator that you could use for this titration, and explain your choice.   [2 marks]

Try learning this stuff drop by drop...

Titrations involve playing with big bits of glassware that you're told not to break as they're really expensive — so you instantly become really clumsy. If you manage not to smash the burette, you'll find it easier to get accurate results if you use a dilute acid or alkali — drops of dilute acid and alkali contain fewer particles so you're less likely to overshoot.

Titration Calculations

Now you know how to do a titration, here's what you can do with the results...

You Can Use Titration Results to Calculate Concentrations

When you've done a titration you can use your results to calculate the **concentration** of your acid or base.

There are a few things you can do to make sure your titration **results** are as **accurate** as possible:

1) Measure the neutralisation volume as precisely as you possibly can. This will usually be to the **nearest 0.05 cm³**.
2) It's a good idea to **repeat** the titration at least three times and take a **mean average** titre value. That'll help you to make sure your answer is **reliable**.
3) Don't use any **anomalous** (unusual) results — as a rough guide, all your results should be within 0.2 cm³ of each other.

If you use a **pH meter** rather than an indicator, you can draw a pH curve of the titration and use it to work out how much acid or base is needed for neutralisation.

You do this by finding the **equivalence point** (the mid-point of the line of rapid pH change) and drawing a **vertical line downwards** until it meets the x-axis. The value at this point on the x-axis is the volume of acid or base needed.

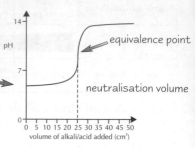

Here's an Example Calculation...

Here's an example of how you can use the neutralisation volume to calculate the concentration of the acid or base:

Example: 25 cm³ of 0.5 moldm⁻³ HCl was needed to neutralise 35 cm³ of NaOH solution. Calculate the concentration of the sodium hydroxide solution.

First write a **balanced equation** and decide **what you know** and what you **need to know**:

$$HCl + NaOH \rightarrow NaCl + H_2O$$
$$25\ cm^3 \quad 35\ cm^3$$
$$0.5\ moldm^{-3} \quad ?$$

Now work out how many **moles of HCl** you have:

$$\text{Number of moles of HCl} = \frac{\text{concentration} \times \text{volume (cm}^3)}{1000} = \frac{0.5 \times 25}{1000} = 0.0125 \text{ moles}$$

You should remember this formula from AS — you divide by 1000 to get the volume from cm³ to dm³.

From the equation, you know 1 mole of HCl neutralises 1 mole of NaOH.

So 0.0125 moles of HCl must neutralise **0.0125** moles of NaOH.

Now it's a doddle to work out the **concentration of NaOH**.

This is just the formula above, rearranged.

$$\text{Concentration of NaOH}_{(aq)} = \frac{\text{moles of NaOH} \times 1000}{\text{volume (cm}^3)} = \frac{0.0125 \times 1000}{35} = \textbf{0.36 moldm}^{-3}$$

A Diprotic Acid Releases Two Protons When it Dissociates

A **diprotic acid** is one that can release **two protons** when it's in solution. **Ethanedioic acid** (HOOC-COOH) is diprotic. When ethanedioic acid reacts with a **base** like sodium hydroxide, it's **neutralised**. But the reaction happens in **two stages**, because the **two protons** are removed from the acid **separately**.

This means that when you titrate **ethanedioic acid** with a **strong base** you get a pH curve with two **equivalence points**:

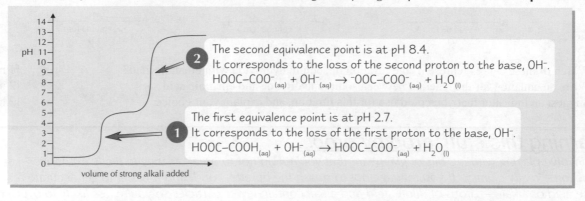

The second equivalence point is at pH 8.4. It corresponds to the loss of the second proton to the base, OH⁻.
$$HOOC–COO^-_{(aq)} + OH^-_{(aq)} \rightarrow {}^-OOC–COO^-_{(aq)} + H_2O_{(l)}$$

The first equivalence point is at pH 2.7. It corresponds to the loss of the first proton to the base, OH⁻.
$$HOOC–COOH_{(aq)} + OH^-_{(aq)} \rightarrow HOOC–COO^-_{(aq)} + H_2O_{(l)}$$

Titration Calculations

You Can Find the **Concentration** of a **Diprotic Acid** From Titration Results Too

You can calculate the concentration of a **diprotic** acid from titration data in the same way as you did for a monoprotic acid.

Example: 25 cm³ of ethanedioic acid, $C_2H_2O_4$, was completely neutralised by 20 cm³ of 0.1 moldm⁻³ NaOH solution. Calculate the concentration of the ethanedioic acid solution.

Write a **balanced equation** and decide what you know and what you need to know:

$$C_2H_2O_4 + 2NaOH \rightarrow Na_2C_2O_4 + 2H_2O$$

| 25 cm³ | 20 cm³ |
| ? | 0.1 moldm⁻³ |

Now work out how many **moles of NaOH** you have:

$$\text{Number of moles of NaOH} = \frac{\text{concentration} \times \text{volume (cm}^3)}{1000} = \frac{0.1 \times 20}{1000} = 0.002 \text{ moles}$$

Because it's a diprotic acid, you need twice as many moles of base as moles of acid.

You know from the equation that you need 2 moles of NaOH to neutralise 1 mole of $C_2H_2O_4$.

So 0.002 moles of NaOH must neutralise (0.002 ÷ 2) = **0.001 moles of $C_2H_2O_4$**.

Now find the **concentration of $C_2H_2O_4$**.

$$\text{Concentration of } C_2H_2O_4 = \frac{\text{moles of } C_2H_2O_4 \times 1000}{\text{volume (cm}^3)} = \frac{0.001 \times 1000}{25} = \textbf{0.04 moldm}^{-3}$$

Practice Questions

Q1 Write a balanced equation for the reaction between the strong monoprotic acid HNO_3 and NaOH.

Q2 How many moles of NaOH would you need to neutralise one mole of a monoprotic acid?

Q3 What is a diprotic acid?

Q4 How many moles of NaOH would you need to neutralise one mole of a diprotic acid?

Exam Questions

1 A student performed a titration with 25 cm³ of hydrochloric acid, adding 0.1 moldm⁻³ sodium hydroxide from a burette. The student's results are shown in the table below.

	Titration 1	Titration 2	Titration 3
Titre Volume (cm³ of NaOH)	25.6	25.65	25.55

   a) Write a balanced equation for the reaction. [1 mark]
   b) (i) Calculate the average titre of sodium hydroxide that was needed to neutralise the hydrochloric acid. [1 mark]
      (ii) Use this to find the number of moles of sodium hydroxide that were needed to neutralise the acid. [2 marks]
   c) Find the concentration of the hydrochloric acid. [2 marks]

2 Sulfuric acid is a diprotic acid. 25 cm³ of this acid is needed to neutralise 35.65 cm³ of 0.1 moldm⁻³ sodium hydroxide.
   a) Write a balanced equation for the reaction. [1 mark]
   b) Calculate:
      (i) The number of moles of sodium hydroxide present in the 35.65 cm³ sample. [2 marks]
      (ii) The number of moles of sulfuric acid needed to neutralise the sodium hydroxide. [1 mark]
      (iii) The concentration of the sulfuric acid used. [2 marks]

3 What volume of 0.1 moldm⁻³ hydrochloric acid would be needed to neutralise 10 cm³ of 0.25 moldm⁻³ sodium hydroxide? [6 marks]

Diprotic acids — twice the titration fun...

And that's a whole lot of calculation-based merriment. Don't forget, if it's a diprotic acid that you're using, you need twice as many moles of NaOH to neutralise it as you would a monoprotic acid. But when it comes down to it, it's the same story as any other chemistry calculation — write out the equation, compare the number of moles, and put the values into the right formula.

Buffer Action

How can a solution resist becoming more acidic if you add acid to it? Why would it want to? Here's where you find out...

Buffers Resist Changes in pH

A **buffer** is a solution that **resists** changes in pH when **small** amounts of acid or alkali are added.

A buffer **doesn't** stop the pH from changing completely — it does make the changes **very slight** though.
Buffers only work for small amounts of acid or alkali — put too much in and they'll go "Waah" and not be able to cope.
You get **acidic buffers** and **basic buffers**.

Acidic Buffers are Made from a Weak Acid and one of its Salts

Acidic buffers have a pH of less than 7 — they're made by mixing a **weak acid** with one of its **salts**.
Ethanoic acid and **sodium ethanoate** ($CH_3COO^-Na^+$) is a good example:

The ethanoic acid is a **weak acid**, so it only **slightly** dissociates: $CH_3COOH_{(aq)} \rightleftharpoons H^+_{(aq)} + CH_3COO^-_{(aq)}$.

But the salt **fully** dissociates into its ions when it dissolves: $CH_3COONa_{(s)} + water \rightarrow CH_3COO^-_{(aq)} + Na^+_{(aq)}$.

So in the solution you've got heaps of **undissociated ethanoic acid molecules**, and heaps of **ethanoate ions** from the salt.

When you alter the **concentration** of H^+ or OH^- **ions** in the buffer solution the **equilibrium position** moves
to **counteract** the change (this is down to **Le Chatelier's principle** — see page 93). Here's how it all works:

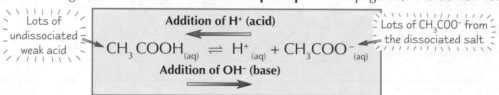

Lots of undissociated weak acid

Addition of H⁺ (acid)

$$CH_3COOH_{(aq)} \rightleftharpoons H^+_{(aq)} + CH_3COO^-_{(aq)}$$

Addition of OH⁻ (base)

Lots of CH_3COO^- from the dissociated salt

If you add a **small** amount of **acid** the **H^+ concentration** increases. Most of the extra H^+ ions
combine with CH_3COO^- ions to form CH_3COOH. This shifts the equilibrium to the **left**,
reducing the H^+ concentration to close to its original value. So the **pH** doesn't change.

The large number of CH_3COO^- ions make sure that the buffer can cope with the addition of acid.

There's no problem doing this as there's loads of spare CH_3COOH molecules.

If a **small** amount of **base** (e.g. NaOH) is added, the **OH^- concentration** increases.
Most of the extra OH^- ions react with H^+ ions to form water — removing H^+ ions from the solution.
This causes more CH_3COOH to **dissociate** to form H^+ ions — shifting the equilibrium to the **right**.
The H^+ concentration increases until it's close to its original value, so the **pH** doesn't change.

Basic Buffers are Made from a Weak Base and one of its Salts

Basic buffers have a pH greater than 7 — and they're made by mixing a **weak base** with one of its **salts**.
A solution of **ammonia** (NH_3, a weak base) and **ammonium chloride** (NH_4Cl, a salt of ammonia) acts as a **basic** buffer.
The **salt** fully dissociates in solution: $NH_4Cl_{(aq)} \rightarrow NH_4^+_{(aq)} + Cl^-_{(aq)}$.

Some of the NH_3 molecules will also react with water molecules: $NH_{3\,(aq)} + H_2O_{(aq)} \rightleftharpoons NH_4^+_{(aq)} + OH^-_{(aq)}$.

So the solution will contain loads of **ammonium ions** (NH_4^+), and lots of **ammonia** molecules too.

The **equilibrium position** of this reaction
can move to **counteract** changes in pH:

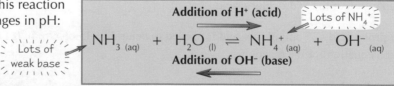

Lots of weak base

Addition of H⁺ (acid)

$$NH_{3\,(aq)} + H_2O_{(l)} \rightleftharpoons NH_4^+_{(aq)} + OH^-_{(aq)}$$

Addition of OH⁻ (base)

Lots of NH_4^+

If a small amount of **base** is added, the OH^- concentration **increases**, making the solution more **alkaline**.
Most of the extra OH^- ions will react with the NH_4^+ ions, to form NH_3 and H_2O. So the equilibrium
will shift to the **left**, removing OH^- ions from the solution, and stopping the pH from changing much.

If a small amount of **acid** is added, the H^+ concentration **increases**, making the solution more **acidic**.
- Some of the H^+ ions react with **OH^-** ions to make **H_2O**. When this happens the equilibrium position
  moves to the right to replace the OH^- ions that have been used up.
- Some of the H^+ ions react with **NH_3** molecules to form NH_4^+: $NH_3 + H^+ \rightleftharpoons NH_4^+$
These reactions will **remove** most of the extra H^+ ions that were added — so the pH **won't** change much.

Buffer Action

Here's How to Calculate the pH of a Buffer Solution

Calculating the **pH** of an acidic buffer isn't too tricky. You just need to know the K_a of the weak acid and the **concentrations** of the weak acid and its salt. Here's how to go about it:

Example: A buffer solution contains 0.40 moldm^{-3} methanoic acid, HCOOH, and 0.6 moldm^{-3} sodium methanoate, HCOO$^-$Na$^+$. For methanoic acid, $K_a = 1.6 \times 10^{-4}$ moldm^{-3}. What is the pH of this buffer?

Firstly, write the expression for K_a of the weak acid:

$$HCOOH_{(aq)} \rightleftharpoons H^+_{(aq)} + HCOO^-_{(aq)} \Rightarrow K_a = \frac{\left[H^+_{(aq)}\right] \times \left[HCOO^-_{(aq)}\right]}{\left[HCOOH_{(aq)}\right]}$$

Remember — these all have to be equilibrium concentrations.

Then rearrange the expression and stick in the data to calculate $[H^+_{(aq)}]$:

$$\left[H^+_{(aq)}\right] = K_a \times \frac{\left[HCOOH_{(aq)}\right]}{\left[HCOO^-_{(aq)}\right]}$$

$$\Rightarrow \left[H^+_{(aq)}\right] = 1.6 \times 10^{-4} \times \frac{0.4}{0.6} = 1.07 \times 10^{-4} \text{ moldm}^{-3}$$

You have to make a **few assumptions** here:
- HCOO$^-$Na$^+$ is fully dissociated, so assume that the equilibrium concentration of HCOO$^-$ is the same as the initial concentration of HCOO$^-$Na$^+$.
- HCOOH is only slightly dissociated, so assume that its equilibrium concentration is the same as its initial concentration.

Finally, convert $[H^+_{(aq)}]$ to pH: $\quad pH = -\log_{10}[H^+_{(aq)}] = -\log_{10}(1.07 \times 10^{-4}) = \mathbf{3.97}$ And that's your answer.

Buffers are Really Handy

Jeremy resolved to use a shampoo with a buffer in future.

Most **shampoos** contain a pH 5.5 buffer — it counteracts the alkaline soap in the shampoo. The soap might get your hair squeaky clean, but alkalis don't make your hair look shiny.

Biological washing powders contain buffers too. They keep the pH at the right level for the enzymes to work best.

There are also lots of **biological buffer** systems in our bodies, making sure all our tissues are kept at the **right pH**. For example, it's vital that **blood** stays at a pH very near to 7.4, so it contains a buffer system.

Practice Questions

Q1 What's a buffer solution?

Q2 How can a mixture of ethanoic acid and sodium ethanoate act as a buffer?

Q3 Describe how to make a basic buffer.

Q4 Give two uses of buffer solutions.

Exam Questions

1   A buffer solution contains 0.40 moldm^{-3} benzoic acid, C$_6$H$_5$COOH, and 0.20 moldm^{-3} sodium benzoate, C$_6$H$_5$COO$^-$Na$^+$. At 25 °C, K_a for benzoic acid is 6.4×10^{-5} moldm^{-3}.

   a)   Calculate the pH of the buffer solution.   [3 marks]

   b)   Explain the effect on the buffer of adding a small quantity of dilute sulfuric acid.   [3 marks]

2   A buffer was prepared by mixing solutions of butanoic acid, CH$_3$(CH$_2$)$_2$COOH, and sodium butanoate, CH$_3$(CH$_2$)$_2$COO$^-$Na$^+$, so that they had the **same** concentration.

   a)   Write a chemical equation to show butanoic acid acting as a weak acid.   [1 mark]

   b)   Given that K_a for butanoic acid is 1.5×10^{-5} moldm^{-3}, calculate the pH of the buffer solution.   [3 marks]

Old buffers are often resistant to change...

So that's how buffers work — add acid, and the equilibrium moves one way, add alkali and it moves the other. That way you don't end up changing the concentration of hydrogen ions in the solution, so the pH doesn't change either. There's a pleasing simplicity about it that I find rather elegant. Like a fine wine with a nose of berry and undertones of... OK, I'll shut up now.

Naming Organic Compounds

There are zillions of organic compounds. And loads of them are pretty similar — just with slight, but crucial differences. There's no way you could memorise all their names, so chemists have devised a clever way of naming each of them.

Nomenclature *is a Fancy Word for* Naming *Organic Compounds*

You can name **any** organic compound using **rules** of nomenclature.
Here's how to use the rules to name a **branched alcohol**.

1) Count the carbon atoms in the **longest continuous chain** — this gives you the stem:

> *Don't forget — the longest carbon chain may be bent.*

Number of carbons	1	2	3	4	5	6	7	8	9	10
Stem	meth-	eth-	prop-	but-	pent-	hex-	hept-	oct-	non-	dec-

2) The **main functional group** of the molecule usually gives you the end of the name (the **suffix**).

Homologous series	Prefix or Suffix	Example
alkanes	-ane	Propane $CH_3CH_2CH_3$
branched alkanes	-yl	methylpropane $CH_3CH(CH_3)CH_3$
cycloalkanes	cyclo-	cyclohexane C_6H_{12}
alkenes	-ene	propene $CH_3CH=CH_2$
haloalkanes	chloro- bromo- iodo-	chloroethane CH_3CH_2Cl
alcohols	-ol	ethanol CH_3CH_2OH

The longest chain is 5 carbons, so the stem is **pent-**

The main functional group is **-OH**, so the compound's name is going to be based on "**pentanol**".

3) Number the carbons in the **longest** carbon chain so that the carbon with the main functional group attached has the lowest possible number. If there's more than one longest chain, pick the one with the **most side-chains**.

4) Write the carbon number that the functional group is on **before the suffix**.

5) Any side-chains or less important functional groups are added as prefixes at the start of the name. Put them in **alphabetical** order, with the **number** of the carbon atom each is attached to.

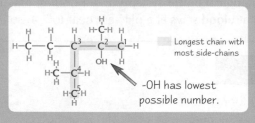

Longest chain with most side-chains

-OH has lowest possible number.

–OH is on carbon-2, so it's some sort of "**pentan-2-ol**".

6) If there's more than one **identical** side-chain or functional group, use **di-** (2), **tri-** (3) or **tetra-** (4) before that part of the name — but ignore this when working out the alphabetical order.

> There's an ethyl group on carbon-3, and methyl groups on carbon-2 and carbon-4, so it's **3-ethyl-2,4-dimethylpentan-2-ol**

You Can Name Haloalkanes *Using These* Rules

Haloalkanes are just alkanes where one or more hydrogens have been swapped for a **halogen**. You can name them using the rules above.

dichloromethane   trichloromethane   2-iodopropane   2-bromo-2-chloro-1, 1, 1-trifluoroethane

Naming Organic Compounds

Naming **Alkenes** — Look at the Position of the **Double Bond**

Alkenes have at least one **double bond** in their carbon chain.
For alkenes with more than three carbons, you need to say which carbon the double bond starts from.

Example 1) The longest chain is **5** carbons, so the stem of the name is **pent-**.

2) The functional group is **C=C**, so it's **pentene**.

3) Number the carbons from right to left (so the double bond starts on the lowest possible number). The first carbon in the double bond is **carbon 2**. So this is **pent-2-ene**.

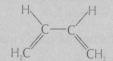

Here are some more examples:

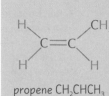

propene CH$_2$CHCH$_3$

Pent-2-ene CH$_3$CHCHCH$_2$CH$_3$

buta-1,3-diene CH$_2$CHCHCH$_2$

If the alkene has two double bonds, the suffix becomes diene. When there's more than one double bond, the stem of the name usually gets an extra 'a' too (e.g. but<u>a</u>- or pent<u>a</u>-not but- or pent-). And you might see the numbers written first, e.g. 1,3-butadiene.

Alcohols can be **Primary, Secondary** or **Tertiary**

1) The alcohol homologous series has the **general formula C$_n$H$_{2n+1}$OH**.

2) An alcohol is **primary**, **secondary** or **tertiary**, depending on which carbon atom the hydroxyl group **–OH** is bonded to...

Primary (1°) E.g.	Secondary (2°) E.g.	Tertiary (3°) E.g.

Propan-1-ol

R–alkyl group

Propan-2-ol

2-methylpropan-2-ol

Practice Questions

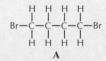

Q1 Name the haloalkane shown in the diagram on the right.

Q2 How many carbon atoms are there in 2-methylpropane? Draw its structure.

Q3 Write out the general formula for an alcohol, and draw an example of a tertiary alcohol.

Exam Questions

1 A diene is reacted with hydrogen bromide, HBr, to give a mixture of haloalkanes, including haloalkanes A and B:

    A   **B**

   a) Name haloalkane A. [1 mark]

   b) Name haloalkane B. [1 mark]

   c) Draw and name the diene molecule that both haloalkanes were formed from. [2 marks]

2 A secondary alcohol has the molecular formula C$_4$H$_{10}$O.
   It can be dehydrated to form two different alkenes, both of which have the molecular formula C$_4$H$_8$.

   a) Draw the structure of the secondary alcohol, and name it. [2 marks]

   b) Give the names of the two different alkenes that can be formed by dehydrating the alcohol. [2 marks]

It's as easy as 1,2,3-trichloropent-2-ene...

The best thing to do now is find some random alkanes, alkenes, haloalkanes and alcohols, and work out their names using the rules. Then have a bash at it the other way around — read the name and draw the compound. It might seem a wee bit tedious now, but come the exam, you'll be thanking me. Doing the exam questions will give you some good practice too.

Formulas and Isomers

If you thought the last page was a bit wordy, this one's all about how a molecule's atoms are actually arranged, and how you can represent its shape on paper.

There are **Loads of Ways** of **Representing** Organic Compounds

TYPE OF FORMULA	WHAT IT SHOWS YOU	FORMULA FOR BUTAN-1-OL
General formula	An algebraic formula that can describe **any member** of a family of compounds.	$C_nH_{2n+1}OH$ (for all alcohols)
Empirical formula	The **simplest ratio** of atoms of each element in a compound (cancel the numbers down if possible). (So ethane, C_2H_6, has the empirical formula CH_3.)	$C_4H_{10}O$
Molecular formula	The **actual** number of atoms of each element in a molecule.	$C_4H_{10}O$
Structural formula	Shows the atoms **carbon by carbon**, with the attached hydrogens and functional groups.	$CH_3CH_2CH_2CH_2OH$
Displayed formula	Shows how all the atoms are **arranged**, and all the bonds between them.	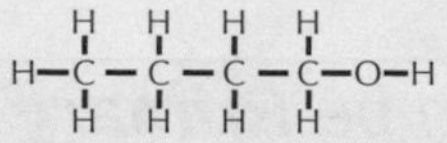
Skeletal formula	Shows the **bonds** of the carbon skeleton **only**, with any functional groups. The hydrogen and carbon atoms aren't shown. This is handy for drawing large complicated structures, like cyclic hydrocarbons.	

A functional group is a reactive part of a molecule — it gives it most of its chemical properties.

A **homologous series** is a bunch of organic compounds which have the **same general formula**.
Alkanes are a homologous series. Each member of a homologous series differs by $-CH_2-$.

Structural Isomers have different **Structural Arrangements** of Atoms

In structural isomers, the atoms are **connected** in different ways. But they still have the **same molecular formula**. There are **three types** of structural isomers:

CHAIN ISOMERS

Chain isomers have different arrangements of the **carbon skeleton**. Some are **straight chains** and others **branched** in different ways.

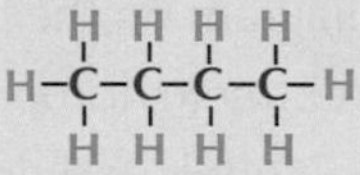

butane 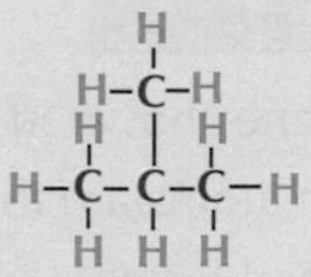   **methylpropane**

POSITIONAL ISOMERS

Positional isomers have the **same skeleton** and the **same atoms or groups of atoms** attached. The difference is that the **functional group** is attached to a **different carbon atom**.

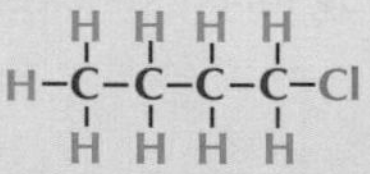

1-chlorobutane 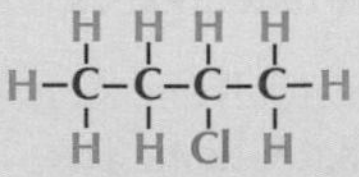   **2-chlorobutane**

FUNCTIONAL GROUP ISOMERS

Functional group isomers have the same atoms arranged into **different functional groups**.

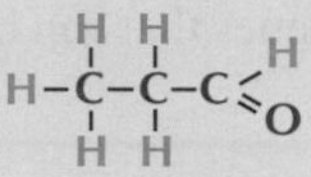

propanal 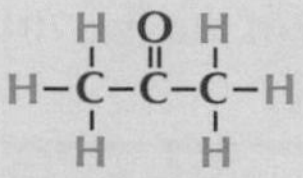   **propanone** 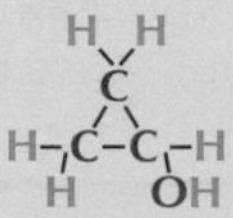   **cyclopropanol**

These are all functional group isomers of C_3H_6O.

Formulas and Isomers

E/Z isomerism is a Type of Stereoisomerism

The astronauts liked to arrange things differently in space.

1) **Stereoisomers** have the same structural formula but a **different arrangement** in space.
   (Just bear with me for a moment... that will become clearer, I promise.)

2) You can **twist** and **rotate** a molecule any way you like around a **single bond**.
   But a **double bond** has a **fixed position** — you **can't** rotate the rest of the molecule around it.

3) Because of the **lack of rotation** around the double bond, some **alkenes** have stereoisomers.

4) Stereoisomers happen when each double-bonded carbon atom has two **different atoms** or **groups** attached to it. Then you get an 'E-isomer' and a 'Z-isomer'.

   For example, the double-bonded carbon atoms in but-2-ene each have an **H** and a **CH_3** group attached.

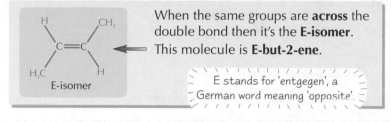

When the same groups are **across** the double bond then it's the **E-isomer**. This molecule is **E-but-2-ene**.

E stands for 'entgegen', a German word meaning 'opposite'.

When the same groups are **both above** or **both below** the double bond then it's the **Z-isomer**. This molecule is **Z-but-2-ene**.

Z stands for 'zusammen', the German for 'together'.

Practice Questions

Q1 Explain what the term 'general formula' means.

Q2 Give both the empirical formula and the molecular formula of methylpropane.

Q3 Explain why 1,1-dichloropropane and 1,2-dichloropropane are positional isomers.

Q4 What are stereoisomers?

Exam Questions

1   A chemist has samples of three chemicals, labelled A, B and C.
   a) Chemical A is a straight chain alcohol with the molecular formula $C_5H_{11}OH$.
      Give the names of two positional isomers that fit this description. [2 marks]
   b) Chemical B is an alkene with the molecular formula C_5H_{10}.
      Give the names of a pair of stereoisomers with this molecular formula. [2 marks]
   c) Chemical C is a hydrocarbon with the molecular formula C_3H_6.
      Give the names of two functional group isomers with this molecular formula. [2 marks]

2   The ethers are a homologous series of hydrocarbon molecules.
   The structural formulas of three members of the series are shown below.

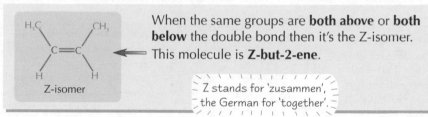

   CH_3OCH_3 $CH_3CH_2OCH_3$ $CH_3CH_2OCH_2CH_3$
   Methoxymethane   **Ethoxymethane**   **Ethoxyethane**

   a) Give the empirical formula of ethoxyethane [1 mark]
   b) Write the general formula of the ether family. [1 mark]
   c) Use your general formula to predict the empirical formula of an ether molecule with ten carbon atoms. [1 mark]
   d) Draw the displayed formula of a functional group isomer of methoxymethane. [1 mark]

Human structural isomers...

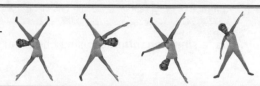

Optical Isomerism

As if E/Z isomerism wasn't exciting enough, take a deep breath and prepare for stereoisomerism part two — optical isomerism...

Optical Isomers are Mirror Images of Each Other

Optical isomerism is another type of stereoisomerism. Stereoisomers have the **same structural formula**, but have their atoms arranged differently in **space**.

A **chiral** (or **asymmetric**) carbon atom is one that has **four different groups** attached to it. It's possible to arrange the groups in two different ways around the carbon atom so that two different molecules are made — these molecules are called **enantiomers** or **optical isomers**.

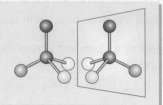

The enantiomers are **mirror images** and no matter which way you turn them, they can't be **superimposed**.

> If the molecules can be superimposed, they're <u>achiral</u> — and there's no optical isomerism.

You have to be able to draw optical isomers. But first you have to identify the chiral centre...

Example

<u>Locating the chiral centre:</u>
Look for the carbon atom with four different groups attached. Here it's the carbon with the four groups H, OH, COOH and CH₃ attached.

chiral centre

2-hydroxypropanoic acid

<u>Drawing isomers:</u>
Once you know the chiral carbon, draw one enantiomer in a tetrahedral shape. Don't try to draw the full structure of each group — it gets confusing. Then draw a mirror image beside it.

enantiomers of 2-hydroxypropanoic acid

Optical Isomers Rotate Plane-Polarised Light

Optical isomers are **optically active** — they **rotate plane-polarised light**. One enantiomer rotates it in a **clockwise** direction, and the other rotates it in an **anticlockwise** direction.

> Normal light vibrates in all directions, but plane-polarised light only vibrates in one direction.

Most Optical Isomers are Produced as Racemates in the Laboratory

A **racemate** (or **racemic mixture**) contains **equal quantities** of each enantiomer of an optically active compound.

Racemates **don't** show any optical activity — the two enantiomers **cancel** each other's light-rotating effect.
Chemists often react two **achiral** things together and get a **racemic** mixture of a **chiral** product.
This is because when two molecules react there's an **equal chance** of forming each of the enantiomers.

Look at the reaction between butane and chlorine:

Butane + Cl₂ → HCl + Enantiomer 1 or Enantiomer 2 → Enantiomer 1 | Enantiomer 2

A **chlorine atom** replaces one of the H groups, to give **2-chlorobutane**.
Either of the H groups can be replaced, so the reaction produces a **mixture** of the **two possible enantiomers**.
Each hydrogen has a **fifty-fifty chance** of being replaced, so the two optical isomers are formed in **equal amounts**.

You can modify a reaction to produce a **single enantiomer** using chemical methods, but it's **difficult** and **expensive**.

Optical Isomerism

Drugs work by changing **chemical reactions** that are taking place in the body.
Most drugs do this by binding to an **active site** — usually on an **enzyme** or a specific **receptor** molecule.

A drug must be **exactly** the right **shape** to fit into the correct active site — only one enantiomer will do.
The other enantiomer could fit into a different enzyme, and might cause **harmful side-effects** or have **no effect** at all.
So if a drug happens to be chiral, it must be made so that it only contains one enantiomer.

Thalidomide is an example of a drug whose optical isomers have different effects.

1) Thalidomide was developed in the 1950s as a **sedative** drug. **Early tests** seemed to show that it had **few side effects**, and so it was launched as a prescription drug.

2) Thalidomide also proved to be an effective **anti-sickness** drug. It was widely prescribed to **pregnant women** in the late 1950s and early 1960s to combat **morning sickness**.

3) At the same time babies began to be born with **malformed limbs** — thalidomide was shown to be causing these symptoms, and its use was **banned**.

4) The drug had **not been tested thoroughly** enough, particularly in relation to its possible **effects on the developing foetus**.

5) Further studies demonstrated that the two **optical isomers** have **different effects** in the body. One is an effective **anti-sickness** drug. The other can cause **birth defects**. In this case even if a **pure dose** of the safe isomer is given, the body will **convert** some of it to the harmful isomer.

6) More recent research has shown that thalidomide can be useful for treating **other diseases**, such as some types of cancer. Patients can now be treated with thalidomide, but **very strict controls** are in place to ensure that it is **never** taken by anyone who is pregnant.

Practice Questions

Q1 What's a chiral molecule?

Q2 What's a racemic mixture?

Q3 The displayed formula of 2-methylbutan-1-al is shown on the right. Explain why the carbon atom marked with a * is a chiral centre.

Q4 Explain why most optically active pharmaceutical drugs only contain one isomer.

Exam Questions

1 There are sixteen possible structural isomers of the compound $C_3H_6O_2$, four of which show stereoisomerism.

a) Explain the meaning of the term *stereoisomerism*. [2 marks]

b) Draw a pair of stereoisomers of $C_3H_6O_2$, with hydroxyl groups. Label them *E* and *Z*. [3 marks]

c) (i) There are two chiral isomers of $C_3H_6O_2$. Draw the enantiomers of one of the chiral isomers. [2 marks]
(ii) What structural feature in the molecule gives rise to optical isomerism? [1 mark]
(iii) State how you could distinguish between the enantiomers. [2 marks]

2 Parkinson's disease involves a deficiency of dopamine. It is treated by giving patients a single pure enantiomer of DOPA (dihydroxyphenylalanine), a naturally occurring amino acid, which is converted to dopamine in the brain.

a) DOPA is a chiral molecule. Its structure is shown on the right. Mark the structure's chiral centre. [1 mark]

b) A DOPA racemate was synthesised in 1911, but today a natural form of the pure DOPA enantiomer is isolated from fava beans for use as a pharmaceutical.
(i) Explain the meaning of the term *racemate*. [1 mark]
(ii) Suggest two reasons why the pure DOPA enantiomer is used in preference to the racemate. [2 marks]

Time for some quiet reflection...

This isomer stuff's not all bad — you get to draw little pretty pictures of molecules. If you're having difficulty picturing them as 3D shapes, you could always make some models with matchsticks and plasticine®. It's easier to see the mirror image structure with a solid version in front of you. And if you become a famous artist, you can sell them for millions...

Aldehydes and Ketones

Aldehydes and ketones are closely related compounds, with one wee, but crucial, difference...

Aldehydes and Ketones contain a Carbonyl Group

Aldehydes and ketones are both **carbonyl compounds** so they both contain the **carbonyl** functional group, **C=O**. The difference is, they've got their carbonyl groups in **different positions**.

'R' represents a carbon chain of any length.

Aldehydes have their carbonyl group at the **end** of the carbon chain. Their names end in **–al**.

methanal   propanal

Ketones have their carbonyl group in the middle of the carbon chain. Their names end in **–one**, and often have a number to show which **carbon** the carbonyl group is on.

propanone   pentan-2-one

There are a Few Tests to Distinguish Between Aldehydes and Ketones

They all work on the idea that an **aldehyde** can be **easily oxidised** to a carboxylic acid, but a ketone can't.

As an aldehyde is oxidised, another compound is **reduced** — so a reagent is used that **changes colour** as it's reduced.

TOLLENS' REAGENT

Tollens' reagent is a **colourless** solution of **silver nitrate** dissolved in **aqueous ammonia**.

If it's heated in a test tube with an aldehyde, a **silver mirror** forms after a few minutes.

$$\underset{\text{colourless}}{Ag(NH_3)_2^+}{}_{(aq)} + e^- \longrightarrow \underset{\text{silver}}{Ag_{(s)}} + 2NH_{3(aq)}$$

The aldehyde/ketone will be flammable, so heat the mixture in a water bath rather than over a flame.

FEHLING'S SOLUTION OR BENEDICT'S SOLUTION

Fehling's solution is a **blue** solution of complexed **copper(II) ions** dissolved in **sodium hydroxide**.

If it's heated with an aldehyde the copper(II) ions are reduced to a **brick-red precipitate** of **copper(I) oxide**.

$$\underset{\text{blue}}{Cu^{2+}}{}_{(aq)} + e^- \longrightarrow \underset{\text{brick-red}}{Cu^+}{}_{(s)}$$

Benedict's solution is exactly the same as Fehling's solution except the copper(II) ions are dissolved in **sodium carbonate** instead. You still get a **brick-red precipitate** of copper(I) oxide though.

You can Reduce Aldehydes and Ketones Back to Alcohols

In AS Chemistry you saw how **primary alcohols** can be **oxidised** to produce **aldehydes** and **carboxylic acids**, and how **secondary alcohols** can be **oxidised** to make **ketones**. Using a **reducing agent** you can **reverse** these reactions.

NaBH$_4$ (sodium tetrahydridoborate(III) or sodium borohydride) dissolved in **water with methanol** is usually the reducing agent used. But in equations, **[H]** is often used to indicate a hydrogen from a reducing agent.

1) Reducing an **aldehyde** to a **primary alcohol**.

Here's the reaction mechanism:

The H⁻ ions (:Ḧ) come from the **reducing agent**.

The H⁺ ions usually come from water. Sometimes **a weak acid** is added as a source of H⁺.

2) Reducing a **ketone** to a **secondary alcohol**.

The reaction mechanism for ketones is the same as for aldehydes.

These are **nucleophilic addition** reactions. The H⁻ ion acts as a **nucleophile** and **adds** on to the δ⁺ carbon atom.

Aldehydes and Ketones

Hydrogen Cyanide will React with Carbonyls by Nucleophilic Addition

Hydrogen cyanide reacts with carbonyl compounds to produce **hydroxynitriles** (molecules with a CN and an OH group).
It's a **nucleophilic addition reaction** — a **nucleophile** attacks the molecule, causing an extra group to be **added**.

Hydrogen cyanide's a **weak acid** — it partially dissociates in water to form **H⁺** ions and **CN⁻** ions. $HCN \rightleftharpoons H^+ + CN^-$

1) The CN⁻ group **attacks** the partially positive carbon atom and **donates** a pair of electrons. Both electrons from the double bond transfer to the oxygen.

2) H⁺ (from either hydrogen cyanide or water) bonds to the oxygen to form the **hydroxyl group** (OH).

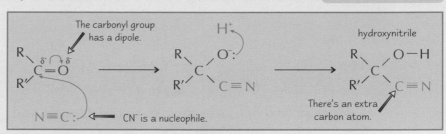

Do a Risk Assessment Before Combining Any Chemicals

A risk assessment involves **reviewing the hazards** of the reacting chemicals, the products and any conditions needed, such as heat. You don't have to wrap yourself in cotton wool, but you do have to take **all reasonable precautions** to reduce the risk of an accident.

Here's a bit of a risk assessment for reacting hydrogen cyanide with a carbonyl compound (above):
Hydrogen cyanide is a highly toxic gas. So to reduce the risk, a solution of acidified potassium cyanide is used instead. The cyanide ions needed for the reaction are formed in the solution. Even so, the reaction should be done in a fume cupboard. Also, the organic compounds are flammable, so if you need to heat them use a water bath or electric mantle.

Practice Questions

Q1 What's the difference between the structures of aldehydes and ketones?

Q2 Describe a chemical test that lets you distinguish between aldehydes and ketones.

Q3 Explain why the reduction of an aldehyde or a ketone is a nucleophilic addition reaction.

Q4 Why is acidified KCN preferred to HCN as a reagent in the reaction: $CH_3CHO + HCN \rightarrow CH_3CH(OH)CN$?

Exam Question

1 The compound C_3H_6O can exist as an aldehyde and a ketone.

   a) Draw and name the carbonyl isomers of C_3H_6O.   [4 marks]

   b) Hydrogen cyanide, HCN, will react with C_3H_6O carbonyl compounds to form a compound with the molecular formula C_4H_6ON.

     (i)   Name the type of mechanism that this reaction proceeds by.   [1 mark]

     (ii)   Draw the different products formed when the carbonyl compound is an aldehyde and a ketone. Indicate which product will be produced as a racemic mixture and explain why.   [4 marks]

   c) The aldehyde C_3H_6O can be reduced to an alcohol, C_3H_7OH.

     (i)   Write an equation for the reaction. Show the structures of the aldehyde and alcohol clearly.   [1 mark]

     (ii)   Suggest suitable reagent(s) and conditions for the reaction.   [2 marks]

2 There are two straight-chain carbonyl compounds with the molecular formula C_4H_8O.

   a) Name the two compounds.   [2 marks]

   b) Name a reagent that could be used to distinguish between the isomers and give the expected result for each.   [3 marks]

   c) During the above test one of the isomers is converted to a carboxylic acid. Name the organic product of the reaction.   [1 mark]

There ain't nobody nowhere gonna oxidise me...

You've got to be a dab hand at recognising functional groups from a mile off. The carbonyl group's just the first of many. Make sure you know how aldehydes differ from ketones and what you get when you oxidise or reduce them both. The mechanisms are a pain to learn, but you've just got to do it. Keep trying to write them out from memory till you can.

Carboxylic Acids and Esters

Carboxylic acids are much more interesting than cardboard boxes — as you're about to discover...

Carboxylic Acids contain –COOH

A <u>carboxyl</u> group contains a <u>carbonyl</u> group and a <u>hydroxyl</u> group.

Carboxylic acids contain the **carboxyl** functional group **–COOH**.
To name them, you find and name the longest alkane chain, take off the 'e' and add **'–oic acid'**.

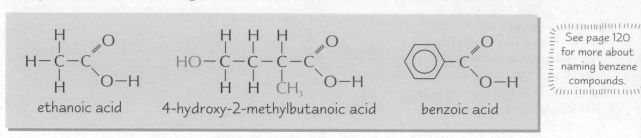

ethanoic acid   4-hydroxy-2-methylbutanoic acid   benzoic acid

See page 120 for more about naming benzene compounds.

The carboxyl group is always at the **end** of the molecule and when naming it's more important than other functional groups — so all the other functional groups in the molecule are numbered starting from this carbon.

Carboxylic Acids are *Weak Acids*

Carboxylic acids are **weak acids** — in water they partially dissociate into a **carboxylate ion** and an **H⁺ ion**.

The equilibrium lies to the left because most of the molecules don't dissociate.

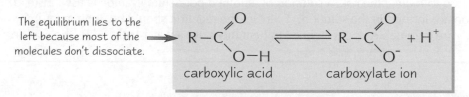

carboxylic acid   carboxylate ion

Pete was hoping for an A⁻ in his carboxylic acid practical, but he ended up with an H⁺.

Carboxylic Acids React with *Carbonates* to Form *Carbon Dioxide*

Carboxylic acids react with **carbonates**, CO_3^{2-} or **hydrogencarbonates**, HCO_3^- to form a **salt**, **carbon dioxide** and **water**.

ethanoic acid   sodium ethanoate
$$2CH_3COOH_{(aq)} + Na_2CO_{3(s)} \rightarrow 2CH_3COONa_{(aq)} + H_2O_{(l)} + CO_{2(g)}$$
$$CH_3COOH_{(aq)} + NaHCO_{3(s)} \rightarrow CH_3COONa_{(aq)} + H_2O_{(l)} + CO_{2(g)}$$

In these reactions, carbon dioxide fizzes out of the solution.

Carboxylic Acids React with *Alcohols* to form *Esters*

There's more on esters at the top of the next page.

If you heat a **carboxylic acid** with an **alcohol** in the presence of a **strong acid catalyst**, you get an ester. It's called an **esterification** reaction. Concentrated sulfuric acid is usually used as the acid catalyst.

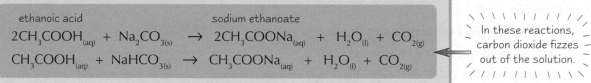

carboxylic acid   alcohol   ester   water

This oxygen comes from the alcohol.

It's also a condensation reaction as it releases water.

Here's how ethanoic acid reacts with ethanol to make the ester, ethyl ethanoate:

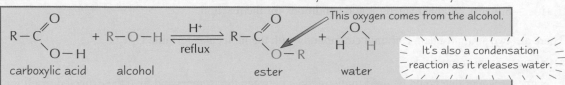

ethanoic acid   ethanol   ethyl ethanoate   water

Carboxylic Acids and Esters

Esters have the Functional Group –COO–

You've just seen that an ester is formed by reacting an alcohol with a carboxylic acid. Well, the **name** of an **ester** is made up of **two parts** — the **first** bit comes from the **alcohol**, and the **second** bit from the carboxylic acid.

1) Look at the **alkyl group** that came from the **alcohol**. This is the first bit of the ester's name.

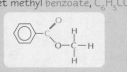

This is an **ethyl** group.

2) Now look at the part that came from the carboxylic acid. Swap its '-oic acid' ending for 'oate' to get the second bit of the name.

This came from ethanoic acid, so it is an ethanoate.

3) Put the two parts together.

It's **ethyl** ethanoate
$CH_3COOCH_2CH_3$

The name's written the opposite way round from the formula.

This goes for molecules with benzene rings too — react methanol with benzoic acid, and you get methyl benzoate, $C_6H_5COOCH_3$.

If either of the carbon chains is **branched** you need to name the attached groups too.

For an ester, number the carbons starting from the C atoms in the C–O–C bond.

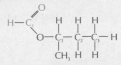

1-methylpropyl methanoate
$HCOOCH(CH_3)CH_2CH_3$

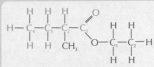

ethyl 2-methylbutanoate
$CH_3CH_2CH(CH_3)COOCH_2CH_3$

Esters are Used as *Food Flavourings, Perfumes, Solvents* and *Plasticisers*

1) Esters have a **sweet smell** — it varies from gluey sweet for smaller esters to a fruity 'pear drop' smell for the larger ones. This makes them useful in perfumes. The food industry uses esters to **flavour** things like drinks and sweets too.

The nice fragrances and flavours of lots of flowers and fruits come from esters.

2) Esters are **polar** liquids so lots of **polar organic compounds** will dissolve in them. They've also got quite **low boiling points**, so they **evaporate easily** from mixtures. This makes them good solvents in **glues** and **printing inks**.

3) Esters are used as **plasticisers** — they're added to plastics during polymerisation to make the plastic more **flexible**. Over time, the plasticiser molecules escape though, and the plastic becomes brittle and stiff.

Practice Questions

Q1 Why is this compound called 3,3,2-trimethylbutanoic acid and not 2,2,3-trimethylbutanoic acid?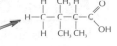

Q2 Name the catalyst used in an esterification reaction between an alcohol and a carboxylic acid.

Q3 What properties of esters make them good solvents?

Q4 Name the alcohol and carboxylic acid that can be used to make propyl butanoate.

Exam Questions

1 The structures of substances X and Y are shown on the right:
a) Write a balanced equation for the reaction between substance X and sodium carbonate, Na_2CO_3. [3 marks]
b) Substance Y can be synthesised from substance X in a single step process. Give the name of the other reagent necessary for this synthesis and name the type of reaction. [2 marks]

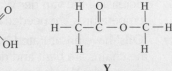

X   Y

2 Ethyl butanoate is an ester that is made by reacting ethanol and butanoic acid.
a) Draw the structure of ethyl butanoate. [1 mark]
b) Ethyl butanoate is used as a solvent in some adhesives. Name another possible use for ethyl butanoate. [1 mark]

3 3-methylbutyl ethanoate is the ester responsible for the odour of pear essence.
a) Write an equation for the formation of this ester from an alcohol and a carboxylic acid. [1 mark]
b) Name the carboxylic acid used. [1 mark]
c) What conditions would be needed for this reaction? [2 marks]

Ahh... the sweet smell of success...

I bet your apple-flavoured sweets have never been near a nice rosy apple — it's all the work of esters. And for that matter, I reckon prawn cocktail crisps have never met a prawn, or a cocktail either. None of it's real. And as for potatoes... But maybe nothing is real. Perhaps you're just a mind in a jar, being conned by an evil genius into thinking things are real.

More on Esters

OK, brace yourself. There's some pure unadulterated lard coming up.

Esters are **Hydrolysed** to Form **Alcohols**

Hydrolysis is when a substance is split up by water — but using just water is often really slow, so an **acid** or an **alkali** is often used to speed it up. There are two types of hydrolysis of esters — **acid hydrolysis** and **base hydrolysis**. With both types you get an **alcohol**, but the second product in each case is different.

ACID HYDROLYSIS

Acid hydrolysis splits the ester into an **acid** and an **alcohol** — it's the reverse of the reaction on page 114.

You have to **reflux** the ester with a **dilute acid**, such as hydrochloric or sulfuric. For example:

As it's a reversible reaction, you need to use lots of water to push the equilibrium over to the right.

ethyl ethanoate + H_2O ⇌ (reflux, H^+) ethanoic acid + ethanol

BASE HYDROLYSIS

This time you have to **reflux** the ester with a **dilute alkali**, such as sodium hydroxide. You get a **carboxylate ion** and an **alcohol**. For example:

ethyl ethanoate + OH^- ⇌ (reflux) ethanoate + ethanol

Fats and Oils are Esters of Glycerol and Fatty Acids

Fatty acids are long chain **carboxylic acids**. They combine with glycerol (propane-1,2,3-triol) to make fats and oils. The fatty acids can be **saturated** (no double bonds) or **unsaturated** (with C=C double bonds).

Most of a fat or oil is made from fatty acid chains — so it's these that give them many of their properties.

- **'Fats'** have mainly **saturated** hydrocarbon chains — they fit neatly together, increasing the van der Waals forces between them. This means higher temperatures are needed to melt them and they're **solid** at room temperature.

- **'Oils'** have unsaturated hydrocarbon chains — the double bonds mean the chains are bent and don't pack together well, decreasing the effect of the van der Waals forces. So they're easier to melt and are **liquids** at room temperature.

Oils and Fats can be **Hydrolysed** to Make **Glycerol**, **Soap** and **Fatty Acids**

Like any ester, you can **hydrolyse** oils and fats by heating them with **sodium hydroxide**.

And you'll never guess what the sodium salt produced is — **a soap**.

fat + $3NaOH$ → glycerol + $3\ CH_3(CH_2)_{16}COO^-Na^+$ (sodium salt (soap))

Next, you can convert the sodium salt to a **fatty acid** by adding an acid such as HCl.

$$CH_3(CH_2)_{16}COO^-Na^+ + H^+ \rightarrow CH_3(CH_2)_{16}COOH + Na^+$$
sodium salt (soap)   fatty acid

More on Esters

Biodiesel is a Mixture of Methyl Esters of Fatty Acids

1) Vegetable oils, e.g. rapeseed oil, make good vehicle fuels, but you can't burn them directly in engines.
2) The oils must be converted into **biodiesel** first. This involves reacting them with **methanol**, using **potassium hydroxide** as a **catalyst**.
3) You get a mixture of **methyl esters** of fatty acids — this is biodiesel.

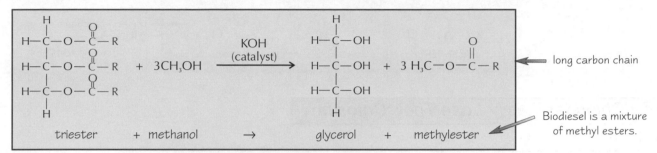

long carbon chain

Biodiesel is a mixture of methyl esters.

triester + methanol → glycerol + methylester

Biodiesel isn't Usually 100% Carbon Neutral

1) Biodiesel can be thought of as **carbon neutral**, because when crops grow they absorb the same amount of CO_2 as they produce when they're burned.
2) But it's not quite as simple as that — energy is **used** to make the fertilizer to grow the crops, and it's used in planting, harvesting and converting the oil. If this energy comes from fossil fuels, then the process **won't** be carbon neutral overall.

Practice Questions

Q1 Describe two different ways of hydrolysing an ester to make an alcohol.

Q2 Draw the structure of a fat.

Q3 What two products do you get when you hydrolyse an oil or fat with sodium hydroxide?

Q4 What does biodiesel consist of?

Exam Questions

1 Compound C, shown on the right, is found in raspberries.

a) Name compound C. [1 mark]

b) Draw and name the structures of the products formed when compound C is refluxed with dilute sulfuric acid. What kind of reaction is this? [5 marks]

c) If compound C is refluxed with excess sodium hydroxide, a similar reaction occurs. What is the difference between the products formed in this reaction and the products of the reaction described in b)? [1 mark]

2 When a vegetable oil was refluxed with concentrated aqueous sodium hydroxide, the products were propane-1,2,3-triol and a salt.

a) Draw the structure of propane-1,2,3-triol. [1 mark]

b) The salt was treated with an excess of hydrochloric acid, and oleic acid, $CH_3(CH_2)_7CH=CH(CH_2)_7COOH$, was produced. Write an equation for the formation of this acid from its salt. [1 mark]

c) Describe a simple chemical test that you could use to distinguish a solution of oleic acid from a solution of $CH_3(CH_2)_{16}COOH$. [2 marks]

Sodium salts — it's all good, clean fun...

I bet you never knew that you could get from something you fry your chips in to something you wash your hands with in one small leap. There are lots of yukkily-complicated structures to learn on these pages — you might find it easier if you think through where the ester breaks in each case and where the atoms of the other reactant add on. Keep at it...

Acyl Chlorides

Acyl chlorides are easy to make and are good starting points for making other types of molecule.

Acyl Chlorides have the Functional Group –COCl

Acyl (or acid) chlorides have the functional group **COCl** — their general formula is $C_nH_{2n-1}OCl$. All their names end in **–oyl chloride**.

ethanoyl chloride   4-hydroxy-2,3-dimethylpentanoyl chloride

The carbon atoms are numbered from the end with the acyl functional group. (This is the same as with carboxylic acids.)

Acyl Chlorides Easily Lose Their Chlorine

Acyl chlorides react with...

...WATER
A **vigorous** reaction with cold water, producing a **carboxylic acid**.

ethanoyl chloride   ethanoic acid

...ALCOHOLS
A **vigorous** reaction at room temperature, producing an **ester**.

ethanoyl chloride   methyl ethanoate

This irreversible reaction is a much easier, faster way to produce an ester than esterification (page 114).

...AMMONIA
A **violent** reaction at room temperature, producing an **amide**.

ethanoyl chloride   ethanamide

See pages 123 to 125 for amines and amides.

...AMINES
A **violent** reaction at room temperature, producing an **N-substituted amide**.

ethanoyl chloride   N-methylethanamide

Each time, **Cl** is **substituted** by an oxygen or nitrogen group and misty fumes of **hydrogen chloride** are given off.

Acyl Chlorides and Acid Anhydrides React in the Same Way

An **acid anhydride** is made from two identical carboxylic acid molecules. If you know the name of the carboxylic acid, they're easy to name — just take away '**acid**' and add '**anhydride**'.

ethanoic acid   ethanoic anhydride

You need to know the reactions of **water**, **alcohol**, **ammonia** and **amines** with acid anhydrides. Luckily, they're almost the same as those of acyl chlorides — the reactions are just **less vigorous** and you get a **carboxylic acid** formed instead of HCl.

e.g. $(CH_3CO)_2O_{(l)} + H_2O_{(l)} \rightarrow 2CH_3COOH_{(aq)}$
ethanoic anhydride + water → ethanoic acid

Acyl Chlorides

Acyl Chloride Reactions are **Nucleophilic Addition-Elimination**

In acyl chlorides, both the chlorine and the oxygen atoms draw electrons **towards** themselves, so the carbon has a slight **positive** charge — meaning it's easily attacked by **nucleophiles**.

Here's the mechanism for a **nucleophilic addition-elimination** reaction between ethanoyl chloride and methanol:

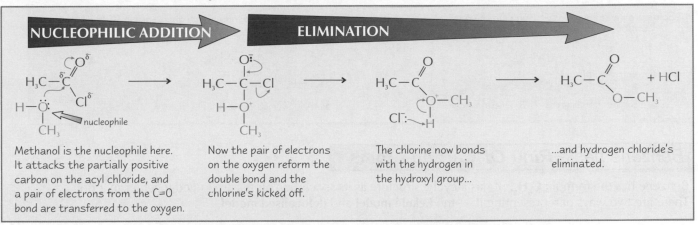

NUCLEOPHILIC ADDITION   **ELIMINATION**

Methanol is the nucleophile here. It attacks the partially positive carbon on the acyl chloride, and a pair of electrons from the C=O bond are transferred to the oxygen.

Now the pair of electrons on the oxygen reform the double bond and the chlorine's kicked off.

The chlorine now bonds with the hydrogen in the hydroxyl group...

...and hydrogen chloride's eliminated.

The other reactions of acyl chlorides that you need to know all work in exactly the same way.

You just need to change the nucleophile to water (H_2O:), ammonia ($\ddot{N}H_3$) or an amine (e.g. $CH_3\ddot{N}H_2$).

Ethanoic Anhydride is Used for the **Manufacture** of **Aspirin**

Aspirin is an **ester** — it's made by reacting **salicylic acid** (which has an alcohol group) with **ethanoic anhydride** or **ethanoyl chloride**.

Ethanoic anhydride is used in industry because:

- it's **cheaper** than ethanoyl chloride.
- it's **safer** to use than ethanoyl chloride as it's **less corrosive**, reacts **more slowly** with water, and **doesn't** produce dangerous **hydrogen chloride** fumes.

salicylic acid   ethanoic anhydride   aspirin   ethanoic acid

Practice Questions

Q1 Write the equation for the reaction between ethanoyl chloride and ammonia.

Q2 Draw the structure of ethanoic anhydride.

Q3 What part of an acyl chloride is attacked by nucleophiles?

Q4 Give TWO reasons why ethanoic anhydride is preferred to ethanoyl chloride when producing aspirin.

Exam Questions

1 Ethanoyl chloride and ethanoic anhydride both react with methanol.
  a) Write equations for both reactions and name the organic product that is formed in both reactions. [3 marks]
  b) Give an observation that could be made for the reaction with ethanoyl chloride that would not occur with ethanoic anhydride. [1 mark]
  c) Ethanoic acid can also be used with methanol to prepare the organic product named in (a). Give one advantage of using ethanoyl chloride. [1 mark]

2 Ethanoyl chloride and ethylamine react together at room temperature.
  a) Write an equation for this reaction and name the organic product. [2 marks]
  b) Draw a mechanism for this reaction. [4 marks]

Anhydrides — aren't they some sort of alien robot...

I could easily lose my mind doing this stuff, let alone a little chlorine particle, and what's worse is I think all those hydrogen chloride fumes are getting to me... I feel kind of dizzy... my head hurts... and I want to lie down and sleep... can't.. keep.. eyes... open... zzzzzzzzzzzzzzzzzzzzzzzzzzzzzzzzzzzzzz...

Aromatic Compounds

Yep, it's another section of organic chemistry, I'm afraid — and it kicks off with the beautiful benzene ring.

Aromatic Compounds are Derived from Benzene

Aliphatic hydrocarbons have no rings of carbon atoms.

Arenes or **aromatic compounds** contain a **benzene ring**.
They're named in two ways. There isn't an easy rule — you just have to learn these examples:

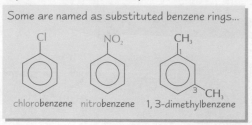

Some are named as substituted benzene rings...

chlorobenzene   nitrobenzene   1, 3-dimethylbenzene

...while others are named as compounds with a phenyl group (C_6H_5) attached.

phenol   2-methylphenol   phenylamine

Benzene has a Ring Of Carbon Atoms

Benzene has the formula C_6H_6. It has a cyclic structure as its six carbon atoms are joined together in a ring. There are two ways of representing it — the **Kekulé model** and **delocalised model**.

The Kekulé Model Came First

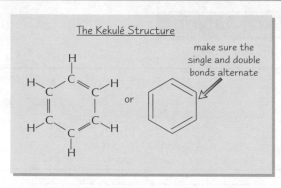

The Kekulé Structure

make sure the single and double bonds alternate

or

1) This was proposed by German chemist Friedrich August Kekulé in 1865. He came up with the idea of a **ring** of C atoms with **alternating single** and **double** bonds between them.

2) He later adapted the model to say that the benzene molecule was constantly **flipping** between two forms (**isomers**) by switching over the double and single bonds.

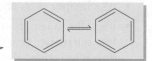

3) If the Kekulé model was correct, you'd expect there to always be three bonds with the length of a **C–C bond** (147 pm) and three bonds with the length of a **C=C bond** (135 pm).

4) However **X-ray diffraction studies** have shown that all the carbon-carbon bonds in benzene have the **same length** — (140 pm) so they're **between** the length of a single bond and a double bond.

5) So the Kekulé structure **can't** be quite right, but it's still used today as it's useful for drawing reaction mechanisms.

The Delocalised Model Replaced Kekulé's Model

The bond-length observations are explained by the delocalised model.

1) In this model, each carbon donates an electron from its **p-orbital**. These electrons combine to form a ring of **delocalised electrons**.

2) All the bonds in the ring are the same, so they are the **same length**.

3) The electrons in the rings are said to be **delocalised** because they don't belong to a specific carbon atom. They are represented as a circle in the ring of carbons rather than as double or single bonds

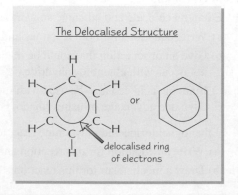

The Delocalised Structure

or

delocalised ring of electrons

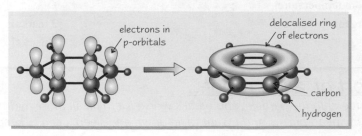

electrons in p-orbitals

delocalised ring of electrons

carbon

hydrogen

Benzene is a planar (flat) molecule — it's got a ring of carbon atoms with their hydrogens sticking out all on a flat plane.

Aromatic Compounds

Enthalpy Changes *Give More Evidence for* Delocalisation

1) Cyclohexene has **one** double bond. When it's hydrogenated, the enthalpy change is **–120 kJmol⁻¹**. If benzene had three double bonds (as in the Kekulé structure), you'd expect it to have an enthalpy of hydrogenation of –360 kJmol⁻¹.

2) But the **experimental** enthalpy of hydrogenation of benzene is **–208 kJmol⁻¹** — far **less exothermic** than expected.

3) Energy is put in to break bonds and released when bonds are made. So **more energy** must have been put in to break the bonds in benzene than would be needed to break the bonds in the Kekulé structure.

$$\text{cyclohexene} + H_2 \rightarrow \quad \Delta H^{\ominus}_{hydrogenation} = -120 \text{ kJmol}^{-1}$$

$$\text{Kekulé structure of benzene} + 3H_2 \rightarrow \quad \text{predicted } \Delta H^{\ominus}_{hydrogenation} = -360 \text{ kJmol}^{-1}$$

$$\text{actual } \Delta H^{\ominus}_{hydrogenation} = -208 \text{ kJmol}^{-1}$$

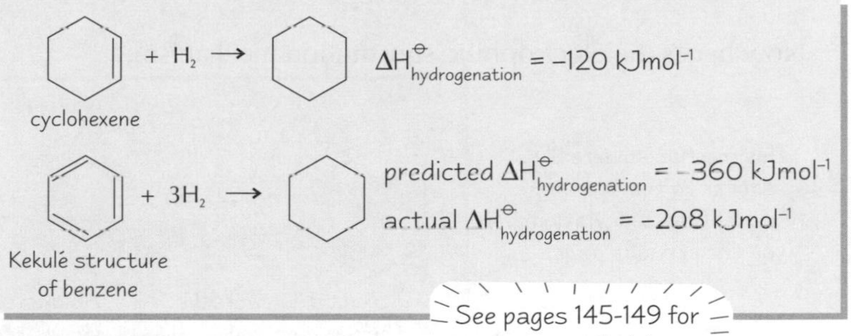

See pages 145-149 for more on enthalpies.

4) This difference indicates that benzene is **more stable** than the Kekulé structure would be. This is thought to be due to the **delocalised ring of electrons**.

Arenes Undergo Electrophilic Substitution *Reactions*

The benzene ring is a region of **high electron density**, so it attracts **electrophiles**. As the benzene ring's so stable, it tends to undergo **electrophilic substitution** reactions, which preserve the delocalised ring.

You need to know two **electrophilic substitution mechanisms** for benzene — **Friedel-Crafts acylation** (shown below) and the **nitration reaction** on the next page.

Friedel-Crafts Acylation *Reactions Produce* Phenylketones

Many **useful** chemicals such as **dyes** and **pharmaceuticals** contain benzene rings, but because benzene is so **stable**, it's fairly **unreactive**. **Friedel-Crafts acylation** reactions are used to add an **acyl group** (**–C(=O)–R**) to the benzene ring. Once an acyl group has been added, the side chains can be modified by **further reactions** to produce **useful products**.

An electrophile has to have a pretty strong **positive charge** to be able to attack the stable benzene ring — most just **aren't polarised enough**. But some can be made into **stronger electrophiles** using a catalyst called a **halogen carrier**. Friedel-Crafts acylation uses an **acyl chloride** (see page 118) as an electrophile and a **halogen carrier** such as AlCl₃.

Here's how the AlCl₃ makes the acyl chloride electrophile stronger:

AlCl₃ accepts a **lone pair of electrons** from the acyl chloride. As the lone pair of electrons is pulled away, the **polarisation** in the acyl chloride **increases** and it forms a **carbocation**. This makes it a much, much **stronger electrophile**, and gives it a strong enough charge to **react** with the **benzene ring**.

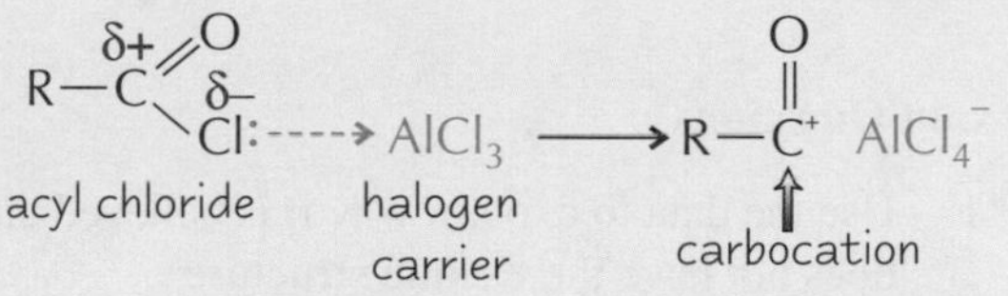

And here's how the electrophile is substituted into the benzene ring. The mechanism has **two steps**:

1) **Electrons** in the benzene ring are **attracted** to the positively charged **carbocation**. Two electrons from the benzene **bond** with the carbocation. This **partially breaks** the **delocalised ring** and gives it a **positive charge**.

2) The **negatively charged AlCl₄⁻** ion is attracted to the **positively charged ring**. One **chloride ion** breaks away from the aluminium chloride ion and **bonds** with the **hydrogen ion**. This **removes the hydrogen** from the ring forming **HCl**. It also allows the catalyst to reform.

The reactants need to be **heated under reflux** in a **non-aqueous** environment for the reaction to occur.

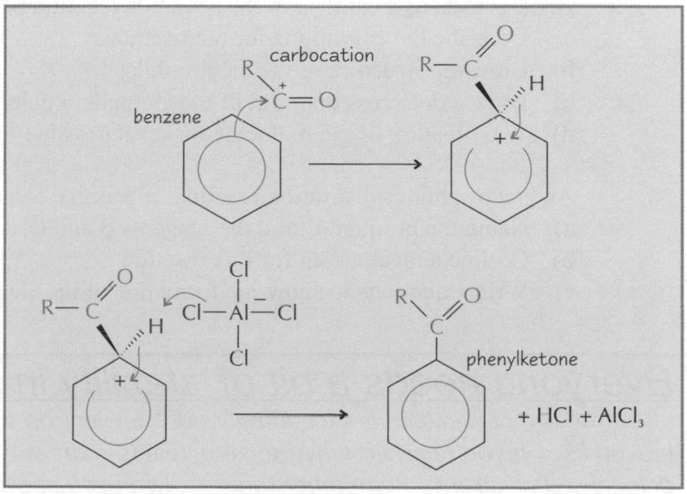

Aromatic Compounds

Nitration is Used in the Manufacture of Explosives and Dyes

When you warm **benzene** with **concentrated nitric** and **sulfuric acids**, you get **nitrobenzene**.
Sulfuric acid acts as a **catalyst** — it helps to make the nitronium ion, NO_2^+, which is the electrophile.

$$HNO_3 + H_2SO_4 \rightarrow H_2NO_3^+ + HSO_4^- \implies H_2NO_3^+ \rightarrow NO_2^+ + H_2O$$

Now here's the electrophilic substitution mechanism:

This mechanism's really similar to the one for Friedel-Crafts acylation on the previous page.

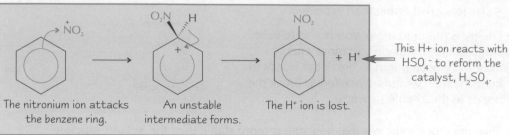

The nitronium ion attacks the benzene ring.

An unstable intermediate forms.

The H^+ ion is lost.

This H+ ion reacts with HSO_4^- to reform the catalyst, H_2SO_4.

If you only want one NO_2 group added (**mononitration**), you need to keep the temperature **below 55 °C**.
Above this temperature you'll get lots of substitutions.

Nitration reactions are really useful

1) Nitro compounds can be **reduced** to form **aromatic amines** (see page 123). These are used to manufacture **dyes** and **pharmaceuticals**.

2) Nitro compounds **decompose violently** when heated, so they are used as **explosives** — such as 2,4,6-trinitromethylbenzene (**trinitrotoluene — TNT**).

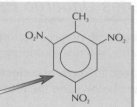

Practice Questions

Q1 Explain why electrophiles are attracted to benzene.

Q2 Which halogen carrier is used in the Friedel-Crafts acylation reaction?

Q3 What is used as a catalyst in the nitration of benzene?

Q4 What type of compound is this?

Exam Questions

1   Use the data to explain why it is believed that benzene does not have the Kekulé structure.

$$\bigcirc + H_2 \rightarrow \bigcirc \quad \Delta H = -120 \text{ kJ mol}^{-1}$$

$$\bigcirc + 3H_2 \rightarrow \bigcirc \quad \Delta H = -208 \text{ kJ mol}^{-1} \quad \text{[2 marks]}$$

2   $AlCl_3$ is used as a catalyst in the acylation reaction between benzene and ethanoyl chloride.
    a) Describe the conditions for the reaction.   [2 marks]
    b) Draw the structure of the electrophile.   [1 mark]
    c) Draw a dot-cross diagram of the aluminium chloride molecule, $AlCl_3$.   [1 mark]
    d) What feature of the molecule makes it possible for $AlCl_3$ to accept Cl^-?   [1 mark]

3   An electrophilic substitution reaction of benzene is summarised in the diagram:
    a) Name the product A, and the reagents B and C, and give the conditions D.   [4 marks]
    b) Outline a mechanism for this reaction.   [3 marks]
    c) Write equations to show the formation of the electrophile.   [2 marks]

$$\bigcirc \xrightarrow[D]{B + C} \bigcirc$$

benzene   A

Everyone needs a bit of stability in their life...

The structure of benzene is exceedingly odd — even top scientists struggled to find out what its molecular structure looked like. If you're asked why benzene reacts in the way it does, it's bound to be something to do with the ring of delocalised electrons. Remember there's a hydrogen at every point on the benzene ring — it's easy to forget they're there.

Amines Mainly

This is a-mean, fishy-smelling topic...

Amines are Organic Derivatives of **Ammonia**

If one or more of the **hydrogens** in **ammonia** (NH_3) is replaced with an organic group, you get an **amine**.

If **one** hydrogen is **replaced** with an organic group, you get a **primary amine** — if **two** are replaced, it's a **secondary amine**, **three** means it's a **tertiary amine** and if all **four** are replaced it's called a **quaternary ammonium ion**.

methylamine (primary amine)   dimethylamine (secondary amine)   trimethylamine (tertiary amine)   tetramethylamine ion (quaternary ammonium ion)   phenylamine (primary amine)

aliphatic amines   aromatic amine

Because quaternary ammonium ions are **positively charged**, they will hang around with any negative ions that are near. The complexes formed are called **quaternary ammonium salts** — like **tetramethylammonium chloride** $((CH_3)_4N^+Cl^-)$. **Small amines** smell similar to **ammonia**, with a **slightly 'fishy'** twist. **Larger amines** smell very **'fishy'**. (Nice.)

Quaternary Ammonium Salts are Used as **Cationic Surfactants**

Cationic surfactants are used in things like **fabric conditioners** and **hair products**. They are **quaternary ammonium salts** with at least one long hydrocarbon chain.

The **positively charged** part (ammonium ion) will bind to negatively charged surfaces such as hair and fibre. This gets rid of **static**.

An Amine Has a **Lone Pair of Electrons** That Can Form **Dative Covalent Bonds**

1) Amines act as **bases** because they **accept protons**. There's a **lone pair of electrons** on the **nitrogen** atom that forms a **dative covalent (coordinate) bond** with an H^+ ion.

2) The **strength** of the **base** depends on how **available** the nitrogen's lone pair of electrons is. The more **available** the **lone pair** is, the more likely the amine is to **accept a proton**, and the **stronger** a base it will be. A **lone pair** of electrons will be **more available** if its **electron density** is **higher**.

3) **Primary aliphatic amines** are **stronger** bases than **ammonia**, which is a **stronger** base than **aromatic amines**. Here's why:

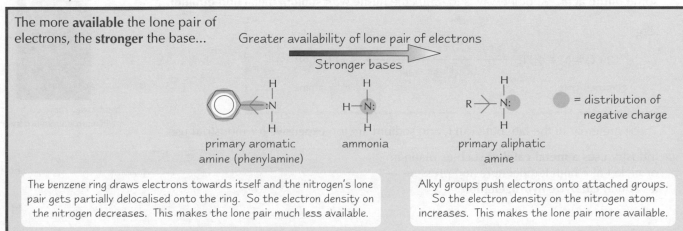

The more **available** the lone pair of electrons, the **stronger** the base...

Greater availability of lone pair of electrons

Stronger bases

primary aromatic amine (phenylamine)   ammonia   primary aliphatic amine

= distribution of negative charge

The benzene ring draws electrons towards itself and the nitrogen's lone pair gets partially delocalised onto the ring. So the electron density on the nitrogen decreases. This makes the lone pair much less available.

Alkyl groups push electrons onto attached groups. So the electron density on the nitrogen atom increases. This makes the lone pair more available.

Amines Mainly

You know what amines are, you know they act as bases and that they turn up in conditioner — but how are they made?

Aliphatic Amines are made from Haloalkanes or Nitriles

There are **two** ways to produce aliphatic amines — either from **haloalkanes** or by **reducing nitriles**. The method for producing **aromatic amines** is different again — as you'll see on the next page.

YOU CAN HEAT A HALOALKANE WITH AMMONIA

Amines can be made by heating a **haloalkane** with **excess ammonia**.

E.g.

$$2\,NH_3 + CH_3CH_2Br \longrightarrow CH_3CH_2NH_2 + NH_4Br$$

ammonia   ethylamine

You'll get a **mixture** of primary, secondary and tertiary amines, and quaternary ammonium salts, as more than one hydrogen is likely to be substituted.

You can separate the products using **fractional distillation**.

You need to know the mechanism for this reaction:

Ammonia attacks the carbon in the haloalkane. The halogen is released.

ammonia + haloalkane → alkylammonium salt ...then... alkylammonium salt ⇌ primary amine + ammonium salt

A second ammonia molecule donates its lone pair of electrons to a hydrogen, which breaks off from the alkylammonium salt.

As long as there's some more of the haloalkane, **further substitutions** can take place. They keep happening until you get a **quaternary ammonium salt**, which can't react any further as it has no lone pair of electrons.

Amines undergo the same mechanism as ammonia (see above).

primary amine → (haloalkane) → secondary amine → (haloalkane) → tertiary amine → (haloalkane) → quaternary ammonium ion

OR YOU CAN REDUCE A NITRILE

You can reduce a nitrile to an amine using **LiAlH$_4$** (a strong reducing agent), followed by some **dilute acid**. Another way is to reflux the nitrile with **sodium** metal and **ethanol**.

E.g.

$$CH_3C\equiv N + 4[H] \xrightarrow[\text{(2) dilute acid}]{\text{(1) LiAlH}_4} CH_3CH_2NH_2$$

ethanenitrile   ethylamine

I can't afford LiAlH$_4$...

Becky was reduced to tears by lithium aluminium hydride.

These are great in the lab, but LiAlH$_4$ and sodium are too **expensive** for industrial use.

Industry uses a **metal catalyst** such as platinum or nickel at a high temperature and pressure — it's called **catalytic hydrogenation**.

$$CH_3C\equiv N + 2H_2 \xrightarrow[\text{high temp. \& pressure}]{Ni} CH_3CH_2NH_2$$

ethanenitrile   ethylamine

Amines Mainly

Aromatic Amines are made by Reducing a Nitro Compound

Aromatic amines are produced by **reducing** a nitro compound, such as **nitrobenzene**.
There are **two steps** to the method:

1) First you need to heat a mixture of a **nitro compound**, **tin metal** and **concentrated hydrochloric acid** under **reflux** — this makes a salt. For example, if you use nitrobenzene, the salt formed is $C_6H_5NH_3^+Cl^-$.

2) Then to turn the salt into an **aromatic amine**, you need to add an alkali, such as **sodium hydroxide** solution.

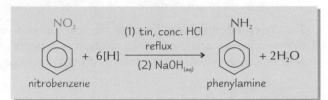

And now for a weeny bit about amides...

Amides are Carboxylic Acid Derivatives

Amides contain the functional group **–CONH$_2$**.
The **carbonyl group** pulls electrons away from the NH$_2$ group, so amides behave differently from amines.

one of the hydrogens is replaced with an alkyl group

amide   N-substituted amide

Practice Questions

Q1 Explain how amines and ammonia can form dative (co-ordinate) covalent bonds.

Q2 Explain why amines and ammonia can act as nucleophiles.

Q3 What conditions are needed to reduce nitrobenzene to phenylamine?

Q4 Haloalkanes react with amines to produce a mixture of primary, secondary and tertiary amines.
How can they be separated?

Exam Questions

1 a) Explain how methylamine, CH_3NH_2, can act as a base. [1 mark]
   b) Methylamine is a stronger base than ammonia, NH_3. However, phenylamine, $C_6H_5NH_2$, is a weaker base than ammonia. Explain these differences in base strength. [4 marks]

2 The reaction between an amine and a haloalkane produces a mixture of products.
   a) Write an equation for the formation of a quaternary ammonium salt from ethylamine and bromomethane. [2 marks]
   b) What condition would help to ensure that the quaternary ammonium salt would be a major product? [1 mark]
   c) Name a use for quaternary ammonium salts. [1 mark]

3 a) Propylamine can be synthesised from bromopropane.
      Suggest a disadvantage of this synthesis route. [1 mark]
   b) Propylamine can also be synthesised from propanenitrile.
      (i) Suggest suitable reagents for its preparation in a laboratory. [2 marks]
      (ii) Why is this method not suitable for industrial use? [1 mark]
      (iii) What reagents and conditions are used in industry? [2 marks]

4 Ethylamine can react with bromoethane to form a compound of molecular formula $C_4H_{11}N$.
   a) Write an equation for the reaction and name the organic product. [2 marks]
   b) Name and outline a mechanism for the reaction. [4 marks]

You've got to learn it — amine it might come up in your exam...

Rotting fish smells so bad because the flesh releases diamines as it decomposes. Is it fish that smells of amines or amines that smell of fish — it's one of those chicken or egg things that no one can answer. Well, enough philosophical pondering — we all know the answer to the meaning of life. It's 42. Now make sure you know the answers to the questions above.

Amino Acids and Proteins

Wouldn't it be nice if you could go to sleep with this book under your pillow and when you woke up you'd know it all.

Amino Acids have an **Amino Group** and a **Carboxyl** Group

An amino acid has a **basic amino group** (NH_2) and an **acidic carboxyl group** (COOH). This makes them **amphoteric** — they've got both acidic and basic properties.

They're **chiral molecules** because the carbon has **four** different groups attached. So a solution of a single amino acid enantiomer will **rotate polarised light** — see page 110.

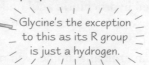
Glycine's the exception to this as its R group is just a hydrogen.

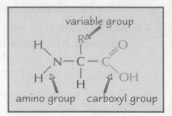
variable group
amino group   carboxyl group

Amino Acids Have **Common** and **Systematic Names**

Amino acids often have a **common** and a **systematic name**. You just have to **learn** the common names, but you can **work out** the systematic names.

Systematic naming

1) Find the **longest carbon chain** that includes the carboxylic acid group and write down its name.

2) **Number the carbons** in the chain starting with the carbon in the carboxylic acid group as number 1.

3) Write down the **positions of any NH_2 groups** and show that they are NH_2 groups with the word '**amino**'.

$$NH_2$$
amino group

$$H_3C \underset{3}{} — \underset{2}{CH} — \underset{1}{COOH}$$

propanoic acid

2-aminopropanoic acid
(alanine is its common name)

Amino Acids Can Exist As **Zwitterions**

A zwitterion is a **dipolar ion** — it has both a **positive** and a **negative charge** in different parts of the molecule. Zwitterions only exist near an amino acid's **isoelectric point**. This is the **pH** where the **average overall charge** on the amino acid is zero. It's different for different amino acids — it depends on their R-group.

In conditions more **acidic** than the isoelectric point, the $-NH_2$ group is likely to be **protonated**.	At the isoelectric point, both the carboxyl group and the amino group are likely to be ionised — forming an ion called a **zwitterion**.	In conditions more **basic** than the isoelectric point, the $-COOH$ group is likely to **lose** its proton.
low pH	zwitterion	high pH

Paper Chromatography is used to **Separate Out** a Mixture of Amino Acids

If you have a solution containing a **mixture** of amino acids, you can use **paper chromatography** to **separate** them out and **identify** the different amino acids that make up your solution.

1) Draw a **pencil line** near the bottom of a piece of chromatography paper and put a **concentrated spot** of the mixture of amino acids on it.

2) Dip the bottom of the paper (not the spot) into a solvent.

3) As the solvent spreads up the paper, the different amino acids move with it, but at **different rates**, so they separate out.

4) You can **identify** each amino acid by comparing how far it moves to how far the solvent moves.

5) Amino acids aren't coloured — so you have to spray **ninhydrin solution** on the paper to turn them purple.

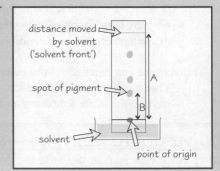
distance moved by solvent ('solvent front')
spot of pigment
solvent
point of origin
A
B

Amino Acids and Proteins

Proteins are Condensation Polymers of Amino Acids

Proteins are made up of **lots** of amino acids joined together by **peptide links**. The chain is put together by **condensation** reactions and broken apart by **hydrolysis** reactions.

Here's how two amino acids join together to make a **dipeptide**:

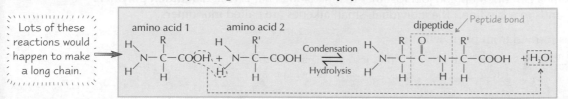

Lots of these reactions would happen to make a long chain.

Proteins are really polyamides — the monomers are joined by amide groups. In proteins these are called peptide links.

To break up the protein (**hydrolyse** it) you need to use pretty harsh conditions. **Hot aqueous 6 M hydrochloric acid** is added, and the mixture is heated under reflux for 24 hours. The final mixture is then neutralised.

Hydrogen Bonds Help Keep Proteins in Shape

1) Proteins are **not** just straight chains of amino acids. Instead, the chains **fold** or **twist** due to **intermolecular forces**.

2) **Hydrogen bonding** is one type of force that holds proteins in shape. Hydrogen bonds exist between polar groups — e.g. –OH and –NH$_2$.

3) Their 3-dimensional shape is **vital** to how proteins **function**. For example, changing the shape of a protein that acts as an **enyzme** can stop it working.

4) Factors such as **heat** and **pH** can affect hydrogen bonding and so can **change the shape** of proteins.

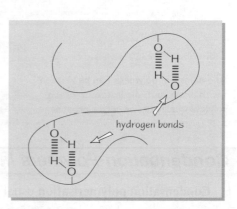

hydrogen bonds

Practice Questions

Q1 Draw the structure of a typical amino acid using R to represent the variable group.

Q2 Draw the zwitterion that would be formed from your amino acid.

Q3 What type of bond links the two amino acids in a dipeptide?

Q4 Outline the role of hydrogen bonds in proteins.

Exam Questions

1 Aspartic acid is an amino acid of molecular formula C$_4$H$_7$O$_4$N. It has two carboxylic acid groups and one amine group. The carbon chain is unbranched.
   a) Draw the structure of aspartic acid and label any chiral carbons with an asterisk *. [2 marks]
   b) Give the systematic name for aspartic acid. [1 mark]

2 The diagram on the right shows the structure of an amino acid. 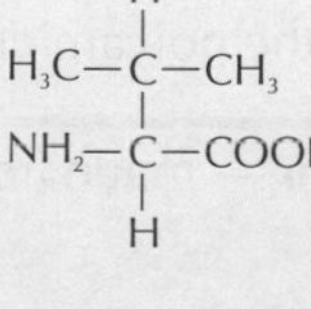
   a) Draw the structure of the zwitterion of this amino acid. [1 mark]
   b) Write an equation showing what happens when an excess of aqueous acid is added to the zwitterion. Draw the structure of the product. [1 mark]

3 The amino acid serine is otherwise known as 2-amino-3-hydroxypropanoic acid.
   a) Draw the structure of serine. [1 mark]
   b) When two amino acids react together, a dipeptide is formed.
      (i) Explain the meaning of the term dipeptide. [2 marks]
      (ii) Draw the structures of the two dipeptides that are formed when serine and glycine react together. The structure of glycine is shown on the right. [2 marks] 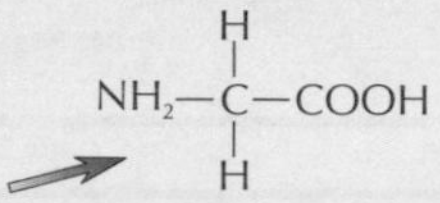
   c) The dipeptides formed can be hydrolysed to give the original amino acids again. Give the reagent(s) and conditions for this reaction. [2 marks]

My top three tides of all time — high tide, low tide and peptide...

The word zwitterion is such a lovely word — it flutters off your tongue like a butterfly. This page isn't too painful — another organic structure, a nice experiment and some stuff on proteins. There's even a two-for-the-price-of-one equation — forwards it's condensation; backwards it's hydrolysis. Remember to learn the conditions for the hydrolysis though.

Polymers

Polymers are long molecules made by joining lots of little molecules together. They're made using addition or condensation polymerisation, as you're about to see.

Addition Polymers are Formed from Alkenes

Addition polymerisation is a free radical addition reaction.

The double bonds in alkenes can open up and join together to make long chains called **addition polymers**. It's like they're holding hands in a big line. The individual, small alkenes are called **monomers**.

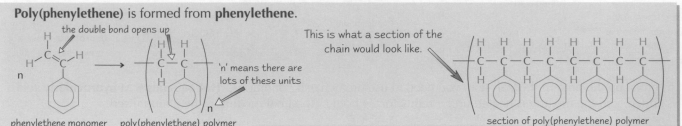

Poly(phenylethene) is formed from **phenylethene**.

the double bond opens up

'n' means there are lots of these units

This is what a section of the chain would look like.

phenylethene monomer   poly(phenylethene) polymer

section of poly(phenylethene) polymer

Poly(chloroethene) is formed from **chloroethene**.

The names of polymers can be written with or without the brackets — e.g. poly(chloroethene) or polychloroethene.

chloroethene monomer   poly(chloroethene) polymer

Polychloroethene's also known as PVC — Poly(vinyl chloride) or Pretty Vulgar Clothing.

Condensation Polymers Include Polyamides, Polyesters and Polypeptides

1) **Condensation polymerisation** usually involves two different types of monomer.

2) Each monomer has at least **two functional groups**. Each functional group reacts with a group on another monomer to form a link, creating polymer chains.

3) Each time a link is formed, a small molecule is lost (water) — that's why it's called **condensation** polymerisation.

4) Examples of condensation polymers include **polyamides**, **polyesters** and **polypeptides** (or proteins, see page 127).

Reactions Between Dicarboxylic Acids and Diamines Make Polyamides

The **carboxyl** groups of **dicarboxylic acids** react with the **amino** groups of **diamines** to form **amide links**.

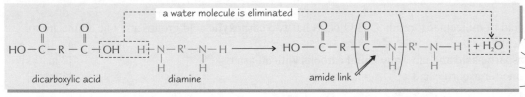

a water molecule is eliminated

dicarboxylic acid   diamine   amide link $+ H_2O$

Dicarboxylic acids and diamines have functional groups at each end of the molecule, so long chains can form.

You need to know how the polyamides **nylon 6,6** and **KEVLAR®** link together.

Example   **Nylon 6,6** — made from **1,6-diaminohexane** and **hexanedioic acid**.

1,6-diaminohexane   hexanedioic acid   nylon 6,6 $+ 2nH_2O$

Example   **KEVLAR®** — made from **benzene-1,4-diamine** and **benzene-1,4-dicarboxylic acid**.

benzene-1,4-diamine   benzene-1,4-dicarboxylic acid   Kevlar $+ 2nH_2O$

Polymers

Reactions Between *Dicarboxylic Acids* and *Diols* Make *Polyesters*

The **carboxyl** groups of dicarboxylic acids can also react with the **hydroxyl** groups of **diols** to form **ester links**.

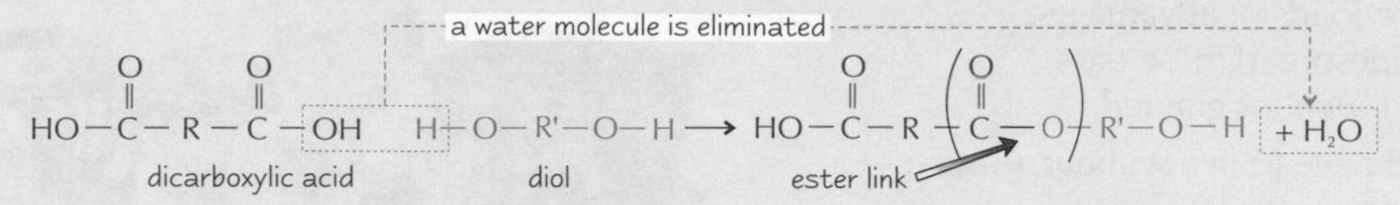

Polymers joined by **ester links** are called **polyesters** — an example is **Terylene™**.

Example | **Terylene (PET)** — formed from **benzene-1,4-dicarboxylic acid** and **ethane-1,2-diol**.

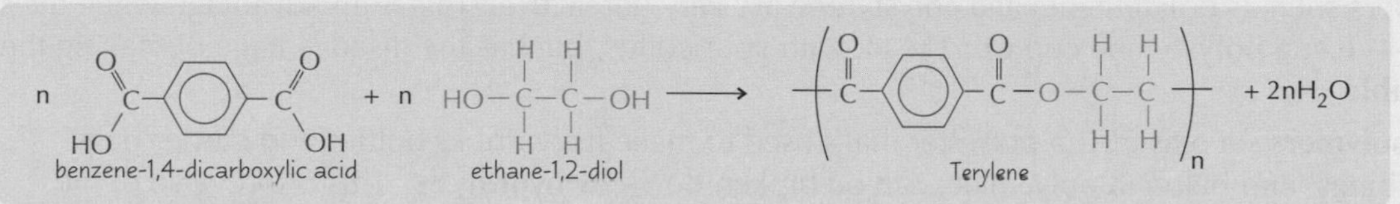

Practice Questions

Q1 What type of polymer is: a) polyethene, b) polyester?

Q2 What molecule is eliminated when a polyester is made?

Q3 Which type of condensation polymer is made when a dicarboxylic acid reacts with a diamine?

Q4 Name an example of a polyester.

Exam Questions

1 The diagram shows three repeating units of a polymer.

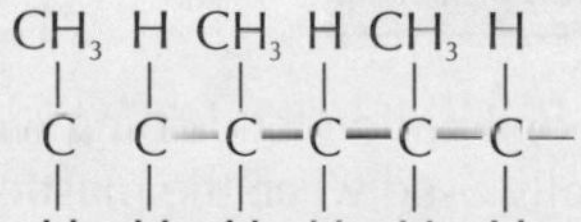

a) Name the polymer.   [1 mark]
b) Name the **type** of polymerisation reaction by which this polymer would be formed.   [1 mark]
c) Write an equation for the formation of this polymer from its monomers.   [1 mark]

2 a) Nylon 6,6 is the most commonly produced nylon. A section of the polymer is shown below:

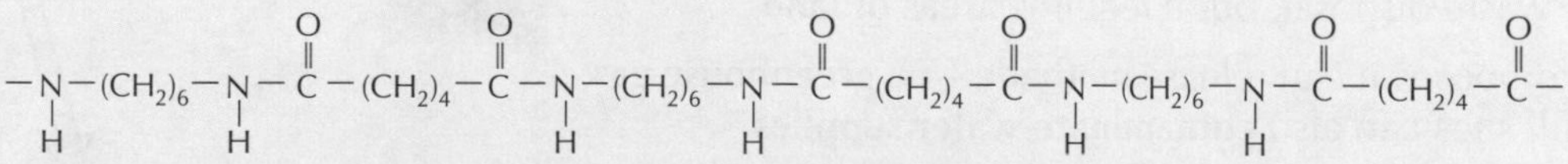

   (i) Draw the structural formulae of the monomers from which nylon 6,6 is formed.
        It is not necessary to draw the carbon chains out in full.   [2 marks]

   (ii) Suggest why this nylon polymer is called nylon 6,6.   [1 mark]

   (iii) Give a name for the linkage between the monomers in this polymer.   [1 mark]

b) A polyester is formed by the reaction between the monomers hexanedioic acid and 1,6-hexanediol.

   (i) Draw the repeating unit for the polyester.   [1 mark]

   (ii) Explain why this is an example of condensation polymerisation.   [1 mark]

Never miss your friends again — form a polymer...

These polymers are really useful — KEVLAR® is used for bulletproof vests and nylon's used for parachutes. And someone had to invent them. Just think, you could be the next mad inventor, working for the biggest secret agency in the world. And you'd have a really fast car, which obviously would turn into a yacht with the press of a button...

More About Polymers

Polymers are amazingly useful. But they have one big drawback...

Polymers — *Useful* but Difficult to *Get Rid Of*

1) Synthetic polymers have loads of **advantages**, so they're incredibly widespread these days — we take them pretty much for granted.

   Just imagine what you'd have to live without ⟶ if there were no polymers...

   (Okay... I could live without the polystyrene head, but the rest of this stuff is pretty useful.)

2) **Addition polymers** such as poly(ethene) and polystyrene are very unreactive. This is an advantage when they are being used — e.g. a polystyrene cup won't react with your coffee, but has the disadvantage of making them **non-biodegradable**.

3) **Condensation polymers** such as PET (a polyester that's used to make fizzy drinks bottles and carrier bags amongst other things) and nylon (a polyamide) can be broken down by **hydrolysis**. This is a reverse of the reaction by which they were made. These polymers are **biodegradable**, although the process is **very slow**.

Waste Plastics Have to be Disposed Of

It's estimated that in the UK we throw away over **3 million tons** of plastic (i.e. synthetic polymers) every year. Because plastics either take a **very long time** to biodegrade or are **non-biodegradable**, the question of what to do with all those plastic objects when we've finished using them is an important one.

The options are **burying**, **burning** or sorting for **reusing** or **recycling**. None of these methods is an ideal solution — they all have **advantages** and **disadvantages** associated with them.

Waste Plastics can be Buried

1) **Landfill** is one option for dealing with waste plastics. It is generally used when the plastic is:
   - difficult to separate from other waste,
   - not in sufficient quantities to make separation financially worthwhile,
   - too difficult technically to recycle.

2) Landfill is a relatively **cheap** and **easy** method of waste disposal, but it requires **areas of land**.

3) As the waste decomposes it can release **methane** — a **greenhouse gas**. **Leaks** from landfill sites can also **contaminate water supplies**.

4) The **amount of waste** we generate is becoming more and more of a problem, so there's a need to **reduce** landfill as much as possible.

Waste Plastics can be Burned

1) Waste plastics can be **burned** and the heat used to generate **electricity**.

2) This process needs to be carefully **controlled** to reduce **toxic** gases. For example, polymers that contain **chlorine** (such as **PVC**) produce **HCl** when they're burned — this has to be removed.

3) So, waste gases from the combustion are passed through **scrubbers** which can **neutralise** gases such as HCl by allowing them to react with a **base**.

4) But, the waste gases, e.g. carbon dioxide, will still contribute to the **greenhouse effect**.

Rex and Dirk enjoy some waist plastic.

More About Polymers

Waste Plastics can be **Recycled**

Because many plastics are made from non-renewable **oil-fractions**, it makes sense to recycle plastics as much as possible. There's more than one way to recycle plastics. After **sorting** into different types:

- some plastics (poly(propene), for example) can be **melted** and **remoulded**,
- some plastics can be **cracked** into **monomers**, and these can be used to make more plastics or other chemicals.

Plastic products are usually marked to make sorting easier. The different numbers show different polymers

e.g.  = PVC, and = poly(propene)

Environmental Pressure May **Change** Our Use of Plastics

Plastics are so **useful** that it's very unlikely we'll stop using them anytime soon, despite the disposal problems. However, **greater awareness** of the problems may change how we use plastics.

For example, supermarkets have been criticised for the amount of plastic waste they create. They've responded by:

1) Trying to **reduce** the amount of **plastic packaging** on food items.

2) Employing scientists to **develop** plastics that **biodegrade quickly**, e.g. food packaging that biodegrades in just one year.

3) Encouraging us to **reuse plastics** by selling 'bags for life' that can be used over and over again or by charging for carrier bags.

Practice Questions

Q1 Why isn't poly(ethene) biodegradable?

Q2 Give one advantage and one disadvantage of burning as a method of disposing of waste plastics.

Q3 Name a polymer that can be melted and remoulded.

Q4 Explain why burning PVC produces HCl gas, but burning poly(ethene) doesn't.

Exam Questions

1 One of the products of burning waste PVC is HCl gas.
The gas can be removed by passing it through scrubbers containing other substances.

a) Choose a substance from the following list that could be used to neutralise the HCl produced and write an equation for the neutralisation reaction.

$CaCO_3$ $\qquad$ NH_4Cl $\qquad$ CH_3COOH. [3 marks]

b) Name an environmental problem that a product of this reaction may contribute to. [1 mark]

2 The diagram on the right shows sections of two polymers.

a) Name:
   i) each type of polymer. [2 marks]
   ii) the monomers that each of these polymers is formed from. [2 marks]

b) Which of these polymers is biodegradable? [1 mark]

c) Give one advantage and one disadvantage of landfill as a disposal method for waste plastic. [2 marks]

A $\left[\begin{array}{c} O \\ \| \\ C-N-(CH_2)_5 \\ | \\ H \end{array} \begin{array}{c} O \\ \| \\ C-N-(CH_2)_5 \\ | \\ H \end{array} \right]$ B $\left(\begin{array}{cccc} CH_3 & H & CH_3 & H \\ | & | & | & | \\ C-C-C-C \\ | & | & | & | \\ H & H & H & H \end{array} \right)$

Phil's my recycled plastic plane — but I don't know where to land Phil...

You might have noticed that all this recycling business is a hot topic these days. And not just in the usual places, such as Chemistry books. No, no, no... recycling even makes it regularly onto the news as well. This suits examiners just fine — they like you to know how useful and important chemistry is. So learn this stuff, pass your exam, and do some recycling.

Organic Synthesis and Analysis

In your exam you may be asked to suggest a pathway for the synthesis of a particular molecule. These pages contain a summary of some of the reactions you should know.

Chemists use Synthesis Routes to Get from One Compound to Another

Chemists have got to be able to make one compound from another. It's vital for things like **designing medicines**. It's also good for making imitations of **useful natural substances** when the real things are hard to extract.

These reactions are covered elsewhere in the book, so check back for extra details.

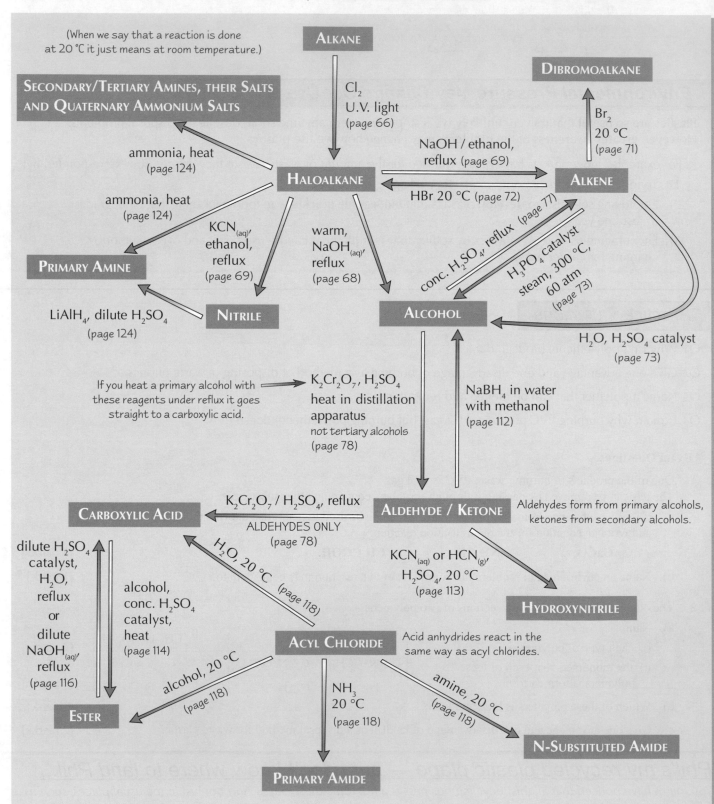

(When we say that a reaction is done at 20 °C it just means at room temperature.)

ALKANE

Cl₂ U.V. light (page 66)

DIBROMOALKANE

Br₂ 20 °C (page 71)

SECONDARY/TERTIARY AMINES, THEIR SALTS AND QUATERNARY AMMONIUM SALTS

ammonia, heat (page 124)

HALOALKANE

NaOH / ethanol, reflux (page 69)

HBr 20 °C (page 72)

ALKENE

ammonia, heat (page 124)

KCN₍aq₎, ethanol, reflux (page 69)

warm, NaOH₍aq₎, reflux (page 68)

conc. H₂SO₄, reflux (page 77)

H₃PO₄ catalyst, steam, 300 °C, 60 atm (page 73)

PRIMARY AMINE

LiAlH₄, dilute H₂SO₄ (page 124)

NITRILE

ALCOHOL

H₂O, H₂SO₄ catalyst (page 73)

If you heat a primary alcohol with these reagents under reflux it goes straight to a carboxylic acid.

K₂Cr₂O₇, H₂SO₄ heat in distillation apparatus not tertiary alcohols (page 78)

NaBH₄ in water with methanol (page 112)

CARBOXYLIC ACID

K₂Cr₂O₇ / H₂SO₄, reflux ALDEHYDES ONLY (page 78)

ALDEHYDE / KETONE

Aldehydes form from primary alcohols, ketones from secondary alcohols.

H₂O, 20 °C (page 118)

KCN₍aq₎ or HCN₍g₎, H₂SO₄, 20 °C (page 113)

HYDROXYNITRILE

dilute H₂SO₄ catalyst, H₂O, reflux or dilute NaOH₍aq₎, reflux (page 116)

alcohol, conc. H₂SO₄ catalyst, heat (page 114)

ACYL CHLORIDE

Acid anhydrides react in the same way as acyl chlorides.

alcohol, 20 °C (page 118)

NH₃ 20 °C (page 118)

amine, 20 °C (page 118)

ESTER

PRIMARY AMIDE

N-SUBSTITUTED AMIDE

Organic Synthesis and Analysis

Synthesis Route for Making **Aromatic Compounds**

There are not so many of these reactions to learn — so make sure you know all the itty-bitty details.
If you can't remember any of the reactions, look back to the relevant pages and take a quick peek over them.

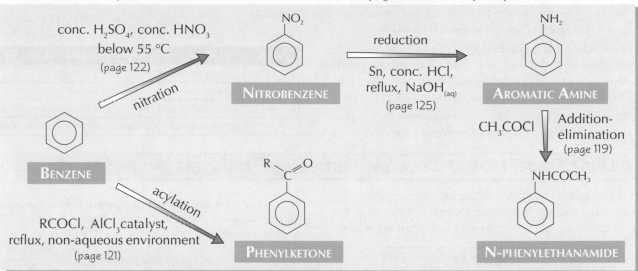

Make Sure You Give **All** The Information

If you're asked how to make one compound from another in the exam, make sure you include:

1) any **special procedures**, such as refluxing.
2) the **conditions** needed, e.g. high temperature or pressure, or the presence of a catalyst.
3) any **safety** precautions, e.g. do it in a fume cupboard.

If there are things like hydrogen chloride or hydrogen cyanide around, you really don't want to go breathing them in.
Stuff like bromine and strong acids and alkalis are corrosive, so you don't want to splash them on your bare skin.

Here are a few other bits and bobs you need to know...

You Can **Test** Whether You've Got a **Primary**, **Secondary** or **Tertiary** Alcohol

To find out which sort of alcohol you've got, you warm it with the **oxidising** agent **acidified potassium dichromate(VI)**.
Then you watch for a colour change...

PRIMARY – the orange dichromate slowly turns green as an **aldehyde** forms (then eventually a carboxylic acid).

SECONDARY – the orange dichromate slowly turns green as a **ketone** forms.

TERTIARY – nothing happens — boring, but easy to remember.

> The colour change is the orange dichromate(VI) ion, $Cr_2O_7^{2-}$, being reduced to the green chromium(III) ion, Cr^{3+}.

The test shows the **same result** for **primary** and **secondary** alcohols — you have to perform **another test** to find out if you have produced an aldehyde or a ketone.

> You can distinguish between an **aldehyde** and a **ketone** using **Tollens' reagent** (an aldehyde forms a **silver mirror**, **no reaction** with a **ketone**), or with either **Fehling's solution** or **Benedict's solution** (an aldehyde gives a **brick red precipitate**, **no reaction** with a **ketone**).
> There's more about these tests on page 112.

Use **Bromine** to Test for **Unsaturation**

To see if a compound is unsaturated (contains any double bonds), all you need to do is shake it with **orange bromine solution**.

The solution will quickly **decolourise** if the substance is unsaturated.

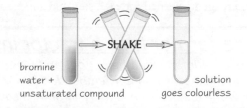

bromine water + unsaturated compound → SHAKE → solution goes colourless

Organic Synthesis and Analysis

Use NaOH and Silver Nitrate Solution to Find Out What Haloalkane You've Got

At AS you learned that you could identify **halide ions** using **silver nitrate solution**.
Well, you can use this test to distinguish between different haloalkanes.

1) You first need to get the halogen off the haloalkane and into solution as a halide ion — so add **sodium hydroxide**.

X is a halogen

$$CH_3CH_2X + OH^- \rightarrow CH_3CH_2OH + X^-$$

2) Now add **nitric acid** (to get rid of any ions that may interfere with the test), and then **silver nitrate solution**. And — ta da — you get a precipitate of a silver halide. The **colour** of the precipitate tells you what halide it is.

3) If you're not sure which colour your precipitate is, you can test its solubility in **ammonia solution**.

Halide	With silver nitrate	Solubility of precipitate in ammonia solution
F⁻	no precipitate	—
Cl⁻	white precipitate	dissolves in dilute ammonia solution
Br⁻	cream precipitate	dissolves in concentrated ammonia solution
I⁻	yellow precipitate	insoluble in concentrated ammonia solution

And Don't Forget How to Calculate Percentage Yield and Atom Economy

Examiners love to squidge calculations into organic synthesis questions. Make sure that you've not forgotten how to calculate **percentage yield** and **atom economy**:

You can use **masses in grams** or **relative molecular masses**. In your calculation, use the M_r of each product/reactant, multiplied by the number of moles of it given in the balanced equation.

$$\% \text{ yield} = \frac{\text{actual yield}}{\text{theoretical yield}} \times 100$$

The theoretical yield is how much product you expect to get from the reactants (you can work this out from the balanced equation).

$$\% \text{ atom economy} = \frac{\text{mass of desired product}}{\text{total mass of reactants}} \times 100$$

The atom economy is the proportion of reactant atoms that become part of the desired product. A low atom economy means a reaction is very wasteful — and wasteful reactions aren't sustainable.

Practice Questions

Q1 Give two reagents that when combined with an alcohol will give an ester.
Q2 What type of organic product would be formed from the reaction between a haloalkane and ammonia?
Q3 What reagents are needed to change nitrobenzene into phenylamine?
Q4 Which type of alcohol cannot be oxidised by acidified potassium dichromate?

Exam Questions

1 Describe a chemical test that would distinguish between each of the following pairs of compounds.

a) A (pentagon) B $\underset{H}{\overset{H}{>}}C=C\underset{H}{\overset{C_3H_7}{<}}$ [3 marks]

b) C $\underset{H}{\overset{H}{>}}C=C\underset{H}{\overset{CH_2CH_2COCH_3}{<}}$ D $\underset{H}{\overset{H}{>}}C=C\underset{H}{\overset{CH_2CH_2CHO}{<}}$ [3 marks]

2 The diagram on the right shows a possible reaction pathway for the two-step synthesis of a ketone from a haloalkane.

Haloalkane P $\xrightarrow[\text{NaOH}]{\text{Step 1}}$ Alcohol Q $\xrightarrow{\text{Step 2}}$ Ketone R

a) Give the name of alcohol Q. [1 mark]

Here is the balanced equation for Step 1 of the synthesis: $CH_3CHClCH_3 + NaOH \rightarrow CH_3CH(OH)CH_3 + NaCl$

b) Calculate the % atom economy for the production of Alcohol Q in Step 1. Give your answer to 1 d.p. [3 marks]
c) 24.9 g of Alcohol Q was produced from 39.25 g of Haloalkane P. Calculate the percentage yield of Step 1. [4 marks]
d) Give the other reagents and the conditions needed to carry out Step 2. [2 marks]

3 Ethyl methanoate is one of the compounds responsible for the smell of raspberries. Outline, with reaction conditions, how it could be synthesised in the laboratory from methanol. [7 marks]

4 How would you synthesise propanol starting with propane? State the reaction conditions and reagents needed for each step and any particular safety considerations. [8 marks]

I saw a farmer turn a tractor into a field once — now that's impressive...

There's loads of information here. Tons and tons of it. But you've covered pretty much all of it before, so it shouldn't be too hard to make sure it's firmly embedded in your head. If it's not, you know what to do — go back over it again. Then cover the diagrams up, and try to draw them out from memory. And then make a table with the tests in.

Mass Spectrometry

A mass spectrometer ionises organic molecules and then breaks lots of the ions up into fragments. Not good news for the molecule, but very handy for scientists. A mass spectrum can tell you the molecular mass, molecular formula, empirical formula, structural formula and your horoscope for the next fortnight.

Mass Spectrometry Can Tell You the M_r of a Compound

1) **Mass spectrometry** can be used to find the **relative molecular mass**, M_r, of a compound.

2) Electrons in the spectrometer **bombard** the sample molecules and **break electrons off**, forming ions.
A **mass spectrum** is produced, showing the **relative amounts** of ions with different mass-to-charge ratios.

3) To find the M_r of a compound you look at the **molecular ion peak** (the **M peak**).

4) This peak is due to the molecular ion, which is formed by the loss of **one electron**.
E.g. for pentane: $C_5H_{10} \rightarrow [C_5H_{10}]^+ + e^-$

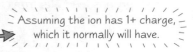
Assuming the ion has 1+ charge, which it normally will have.

5) The **mass/charge value** of the molecular ion peak is the molecular mass:

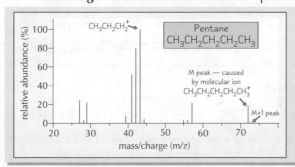

Here's the mass spectrum of pentane. Its M peak is at 72 — so the compound's M_r is 72.

For most <u>organic compounds</u> the M peak is the one with the second highest mass/charge ratio. The smaller peak to the right of the M peak is called the M+1 peak — it's caused by the presence of the carbon isotope ^{13}C.

The Molecular Ion can be Broken into Smaller Fragments

The bombarding electrons make some of the molecular ions break up into **fragments**.
The fragments that are ions show up on the mass spectrum, making a **fragmentation pattern**.
Fragmentation patterns can be used to identify **molecules** and even their **structures**.

Fragment (consider revising)

For propane, the molecular ion is $CH_3CH_2CH_3^+$, and the fragments it breaks into include CH_3^+ ($M_r = 15$) and $CH_3CH_2^+$ ($M_r = 29$).

Only the **ions** show up on the mass spectrum — the **free radicals** are 'lost'.

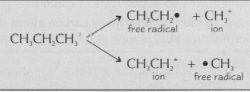

$$CH_3CH_2CH_3 \nearrow CH_3CH_2^{\bullet} + CH_3^+$$
free radical / ion

$$\searrow CH_3CH_2^+ + {\bullet}CH_3$$
ion / free radical

Mr Clippy's grammar advice also applies to chemistry.

To work out the structural formula, you've got to work out what **ion** could have made each peak from its **m/z value**.
(You assume that the m/z value of a peak matches the **mass** of the ion that made it.)

Example

Use this mass spectrum to work out the structure of the molecule:

It's only the m/z values you're interested in — ignore the heights of the bars for now.

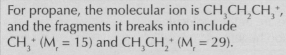

Some common groups to look for:

Fragment	Molecular Mass
CH_3	15
C_2H_5	29
C_3H_7	43
OH	17

1. Identify the fragments

This molecule's got a peak at 15 m/z, so it's likely to have a **CH₃ group**.

It's also got a peak at 17 m/z, so it's likely to have an **OH group**.

Other ions are matched to the peaks here:

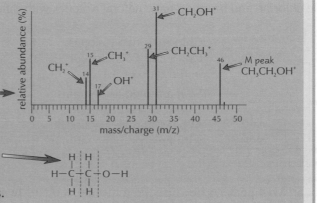

2. Piece them together to form a molecule with the correct M_r

Ethanol has all the fragments on this spectrum.

Ethanol's **molecular mass** is 46. This should be the same as the m/z value of the M peak — it is.

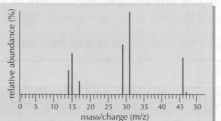

Mass Spectrometry

More *Stable* Fragments Give *Higher Peaks*

1) Some fragment ions are **more stable** than others, so they're more likely to form. That means they'll be more abundant, and will have **higher peaks** on the mass spectrum.

2) The mass spectrum of ethanol on the previous page shows that the **most abundant peak** (the one with the highest bar) has a mass of **31** — this represents the fragment ion **CH₂OH⁺**.

3) Two very stable fragment ions are **carbocations** and **acylium ions** — you'd expect high peaks for them.

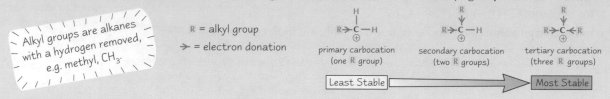

A carbocation is an ion with a **positively charged carbon**. Carbocations are relatively stable ions because alkyl groups feed **electrons** towards the positive charge. Carbocations with more alkyl groups are more stable.

Alkyl groups are alkanes with a hydrogen removed, e.g. methyl, CH₃.

R = alkyl group
→ = electron donation

primary carbocation (one R group)   secondary carbocation (two R groups)   tertiary carbocation (three R groups)

Least Stable ⟶ Most Stable

The **acylium ion (RCO⁺)** is often formed from aliphatic ketones. It has the resonance forms:

$$R - C \equiv \overset{\oplus}{\underset{..}{O}} \longleftrightarrow R - \overset{\oplus}{C} = \overset{..}{\underset{..}{O}}$$

If a molecule has resonance forms, it changes between them very quickly.

Resonating between two or more different structures helps **stabilise** what would otherwise be an unstable ion. So there's a good chance that a tall peak at m/z = **43** will be CH₃CO⁺, and one at m/z = **57** will be CH₃CH₂CO⁺. *As long as you could get such a fragment from the molecule — a peak at 43 may also be C₃H₇⁺.*

Use the *M+1 Peak* to Work Out the *Number* of *Carbon Atoms* in a Molecule

The peak **1 unit** to the right of the M peak is called the **M+1 peak**. It's mostly due to the **carbon-13 isotope** which exists naturally and makes up about 1.1% of carbon.

You can use the M+1 peak to find out **how many carbon atoms** there are in the molecule. There's even a handy formula for this (as long as the molecule's not too huge):

$$\frac{\text{height of M+1 peak}}{\text{height of M peak}} \times 100 = \text{number of carbon atoms in compound}$$

EXAMPLE

If a molecule has a molecular peak with a relative abundance of 44.13%, and an M+1 peak with a relative abundance of 1.41%, how many carbon atoms does it contain?

(1.41 ÷ 44.13) x 100 = 3.195   So the molecule contains 3 carbon atoms.

An *M+2 Peak* Tells Us the Compound Contains *Chlorine* or *Bromine*

If a molecule's got either **chlorine** or **bromine** in it you'll get an **M+2 peak** as well as an M peak and an M+1 peak. It's because chlorine and bromine have natural isotopes with different masses and they all show up on the spectrum.

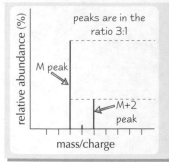

Chlorine's two isotopes, **Cl-35** and **Cl-37**, occur in the ratio of **3:1**.

So if a molecule contains chlorine, it will give an **M peak** and an M+2 peak with heights in the ratio **3:1**.

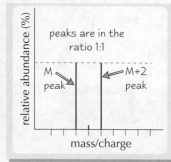

Bromine's got two isotopes, **Br-79** and **Br-81**, that occur in **equal amounts**.

So if a molecule contains **bromine**, the M peak and M+2 peak will both have the **same height**.

If you spot an **M+4 peak** it's because the molecule contains **two** atoms of the **halogen**. (This changes the ratios of the M and M+2 peaks though.)

Mass Spectrometry

And Here's One More Example...

So, a mass spectrum can give you loads of information. Have a look at this one.
It's of an **oxygen-containing organic compound**:

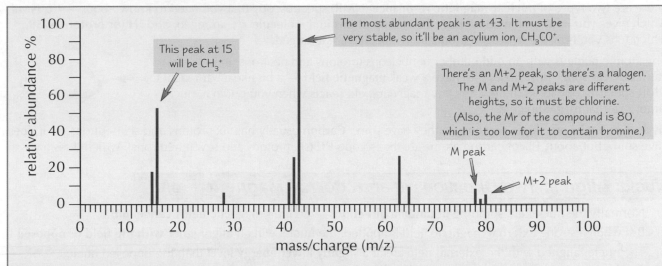

This peak at 15 will be CH_3^+

The most abundant peak is at 43. It must be very stable, so it'll be an acylium ion, CH_3CO^+.

There's an M+2 peak, so there's a halogen. The M and M+2 peaks are different heights, so it must be chlorine.
(Also, the Mr of the compound is 80, which is too low for it to contain bromine.)

M peak

M+2 peak

Adding ^{35}Cl and ^{37}Cl in turn to the fragment of mass 43 (CH_3CO^+) gives the correct M_r values of **78** and **80**.
So the formula of the molecule is **CH_3COCl (ethanoyl chloride)**.

The masses of the fragments at 63 and 65 also differ by **2 units**. So they're likely to contain
the ^{35}Cl and ^{37}Cl isotopes. They're due to the fragments $CO^{35}Cl^+$ and $CO^{37}Cl^+$.

Practice Questions

Q1 What is meant by the term 'molecular ion'?

Q2 Why do compounds containing chlorine or bromine have two molecular ion peaks?

Q3 Explain why the M+1 peak is formed and why it is always small.

Q4 What would be the mass of the fragment $CH_3CH_2CH_2^+$?

Exam Question

1   On the right is the mass spectrum of a haloalkane.
    Use the spectrum to answer the questions below.

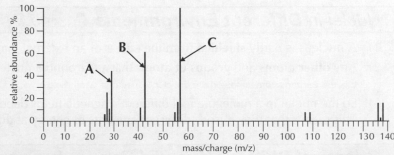

a) Does the molecule contain chlorine or bromine? Explain how you know.   [2 marks]

b) What are the two main molecular masses of this haloalkane?   [1 mark]

c) Give the formulae of the fragments labelled A, B and C.   [3 marks]

d) Suggest a molecular formula for this molecule and draw its structure. Explain your suggestion.   [4 marks]

Unstable fragments — that's my brain on a Monday morning...

So mass spectrometry's a bit like weighing yourself, then taking bits off your body, weighing them separately, then trying to
work out how they all fit together. Luckily you won't get anything as complicated as a body, and you won't need to cut
yourself up either. Good news all round. Learn this page and watch out for the M+1 and M+2 peaks in the exam.

NMR Spectroscopy

NMR isn't the easiest of things, so ingest this information one piece at a time — a bit like eating a bar of chocolate (but this isn't so yummy).

NMR *Gives You Information About a Molecule's* Structure

There are two types of **nuclear magnetic resonance** (**NMR**) spectroscopy that you need to know about — **^{13}C NMR**, which gives you information about how the **carbon atoms** in a molecule are arranged, and **^1H** (or **proton**) **NMR**, which tells you how the **hydrogen atoms** in a molecule are arranged.

Any atomic nucleus with an **odd** number of nucleons (protons and neutrons) in its nucleus has a **nuclear spin**. This causes it to have a weak **magnetic field** — a bit like a bar magnet. NMR spectroscopy looks at how this tiny magnetic field reacts when you put in a much larger external magnetic field.

Hydrogen nuclei are **single protons**, so they have spin. **Carbon** usually has six protons and six neutrons, so it **doesn't** have spin. But about 1% of carbon atoms are the isotope **^{13}C** (six protons and seven neutrons), which does have spin.

Nuclei *Align in Two Directions in an* External Magnetic Field

1) Normally the nuclei are spinning in **random directions** — so their magnetic fields **cancel out**.

2) But when a strong **external** magnetic field is applied, the nuclei will all align either **with the field** or **opposed to it**.

3) The nuclei aligned with the external field are at a **slightly lower energy level** than the opposed nuclei.

4) **Radio waves** of the right frequency can give the nuclei that are aligned with the external magnetic field enough energy to flip up to the higher energy level. The nuclei opposed to the external field, can **emit** radio waves and flip down to the lower energy level.

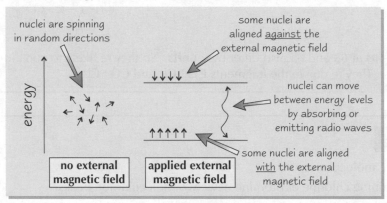

5) To start with, there are more nuclei **aligned** with the external field, so there will be an **overall absorption** of energy. NMR spectroscopy **measures** this **absorption**.

Nuclei in *Different Environments Absorb* Different Amounts of Energy

1) A nucleus is partly **shielded** from the effects of an external magnetic field by its **surrounding electrons**.

2) Any **other atoms** and **groups of atoms** that are around a nucleus will also affect the amount of electron shielding.
    E.g. If a carbon atom bonds to a more electronegative atom (like oxygen) the amount of electron shielding around its nucleus will decrease.

3) So the nuclei in a molecule feel different magnetic fields depending on their **environments**. This means that they absorb **different amounts** of energy at **different frequencies.**

Chemical Shift *is Measured Relative to* Tetramethylsilane

1) Nuclei in different environments absorb energy of **different frequencies**. NMR spectroscopy measures these differences relative to a **standard substance** — the difference is called the **chemical shift (δ)**.

2) The standard substance is **tetramethylsilane (TMS)**, $Si(CH_3)_4$. This molecule has 12 hydrogen atoms all in **identical environments**, so it produces a **single** absorption peak, well away from most other absorption peaks.

> Tetramethylsilane is also inert (so it doesn't react with the sample), non-toxic, and volatile (so it's easy to remove from the sample).

3) Chemical shift is measured in **parts per million** (or **ppm**) relative to TMS. So the single peak produced by TMS is given a **chemical shift value of 0**.

4) You'll often see a peak at $\delta = 0$ on spectra because TMS is added to the test compound for calibration purposes.

NMR Spectroscopy

¹³C NMR Tells You How Many Different Carbon Environments a Molecule Has

If you have a sample of a chemical that contains **carbon** atoms, you can use a ¹³C NMR spectrum of the molecule to help work out what it is. The spectrum will have **one peak** on it for each **carbon environment** in the molecule.

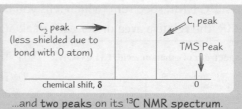

There are **two carbon atoms** in a molecule of **ethanol**:

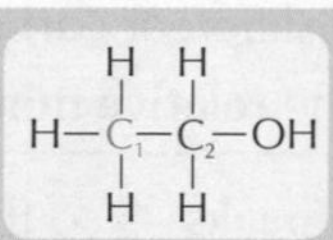

Because they are bonded to **different** atoms, each has a **different** amount of **electron shielding** — so there are **two carbon environments** in the ethanol molecule...

C_2 peak (less shielded due to bond with O atom)

C_1 peak

TMS Peak

chemical shift, δ

...and **two peaks** on its ¹³C NMR spectrum.

¹³C NMR spectra are usually much **simpler** than ¹H NMR spectra — they have fewer, sharper peaks. Here are some more examples:

ETHANE

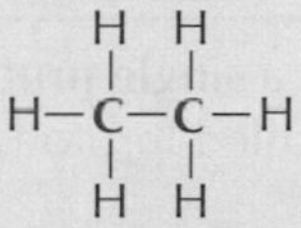

Both carbon nuclei in ethane have the **same environment**. Each C has 3 Hs and a CH₃ group attached. So ethane has one peak on its ¹³C NMR spectrum.

¹³C NMR spectrum of ethane

chemical shift, δ

PROPANONE

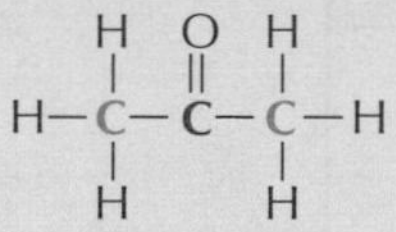

In propanone there are **two different carbon environments**. The **end carbons** both have the same environment — they each have 3 Hs and a COCH₃ attached. The **centre carbon** has 2 CH₃ groups and an O attached by a double bond. So propanone's ¹³C NMR spectrum has **2 peaks**.

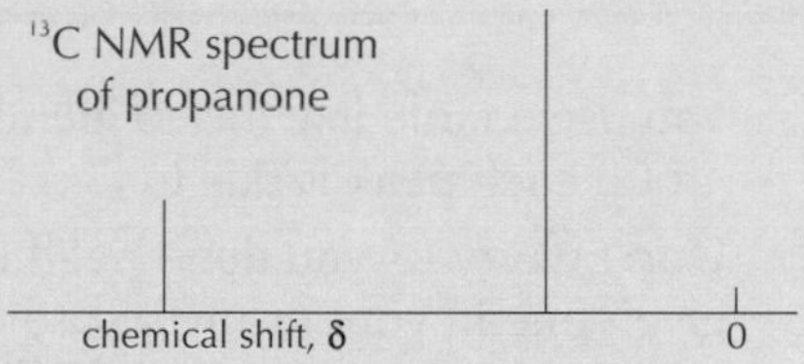

¹³C NMR spectrum of propanone

chemical shift, δ

CYCLOHEXANE-1,3-DIOL

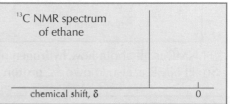

In cyclohexane-1,3-diol there are **four different carbon environments**. If you think about the symmetry of the molecule you can see why this is. So cyclohexane-1,3-diol's ¹³C NMR spectrum will have **4 peaks**.

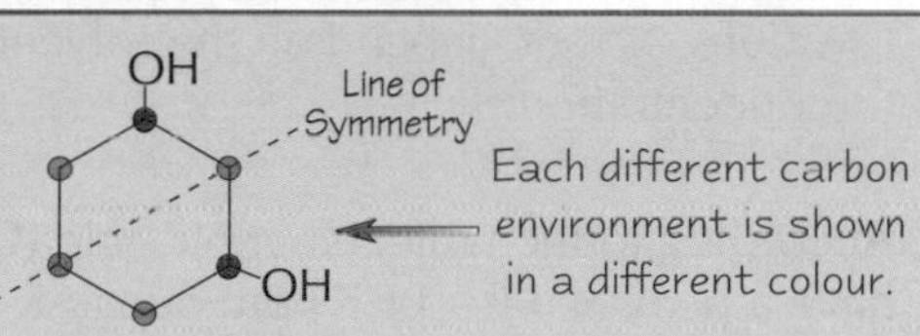

Each different carbon environment is shown in a different colour.

Practice Questions

Q1 Which part of the electromagnetic spectrum is absorbed in NMR spectroscopy?

Q2 What is a chemical shift?

Q3 Explain what the number of peaks in a ¹³C NMR shows.

Exam Questions

1 Draw the molecular structures of each of the following compounds and predict the number of peaks in the ¹³C NMR spectrum of each compound.
   a) ethyl ethanoate   [2 marks]
   b) 1-chloro-2-methylpropane   [2 marks]
   c) 1,3,5-trichlorocyclohexane   [2 marks]

2 The molecule shown on the right is methoxyethane.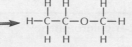
   It has the molecular formula C₃H₈O.
   a) Draw the displayed formulae of the two other possible isomers of C₃H₈O.   [2 marks]
   b) The ¹³C NMR spectrum of one of the three isomers is on the right.
      Deduce which of the three isomers it represents. Explain your answer.   [2 marks]
   c) What is responsible for the peak at 0?   [1 mark]

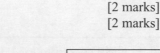

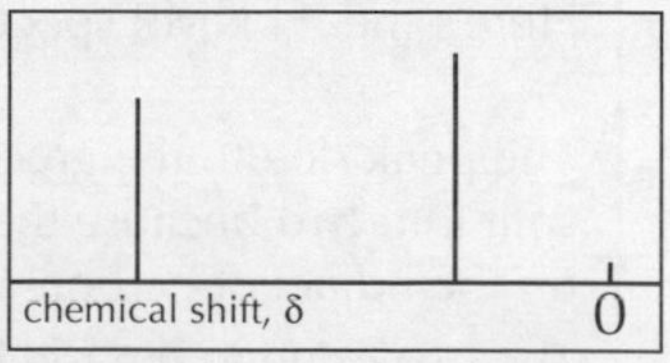

chemical shift, δ

NMR, TMS — IMO this page goes a bit OTT on the TLAs...

The ideas behind NMR are difficult, so don't worry if you have to read these pages quite a few times before they make sense. You've got to make sure you really understand the stuff on these two pages, cos there's more on the next two — and they aren't any easier. Keep bashing away at it though — you'll eventually go "aaah... I get it".

More About NMR

And now that you know the basics, here's the really crunchy bit for you to get your teeth stuck in.

¹H NMR — How Many **Environments** and How Many **Hydrogens** Are In Each

Each peak on a ¹H NMR spectrum is due to one or more hydrogen nuclei (protons) in a **particular environment** — this is similar to a ¹³C NMR spectrum (which tells you the number of different carbon environments).

But, with ¹H NMR, the **relative area** under each peak also tells you the **relative number** of H atoms in each environment.

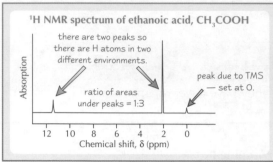

¹H NMR spectrum of ethanoic acid, CH₃COOH

there are two peaks so there are H atoms in two different environments.

peak due to TMS — set at 0.

ratio of areas under peaks = 1:3

Absorption

Chemical shift, δ (ppm)

1) There are **two peaks** — so there are **two environments**.

2) The area ratio is **1:3** — so there's 1 H atom in the environment at δ = 11.5 ppm to every 3 H atoms in the other environment.

3) If you look at the structure of ethanoic acid, this makes sense:

3 H atoms attached to CH₂COOH.

1 H atom attached to COOCH₃.

¹H NMR is all about how **hydrogen nuclei** react to a magnetic field. The nucleus of a hydrogen atom is a **single proton**. So ¹H NMR is also known as '**proton NMR**' — and you might see the hydrogen atoms involved being called 'protons'.

Use a **Table** to Identify the **Hydrogen Atom** Causing the **Chemical Shift**

You use a table like this to **identify** which functional group each peak is due to.

Don't worry — **you don't need to learn it**. You'll be given one in your exam, so use it. The copy you get in your exam may look a little different, and have different values — they depend on the solvent, temperature and concentration.

According to the table, ethanoic acid (CH₃COOH) should have a peak at 11 – 11.7 ppm due to R-CO**O**H, and a peak at 2.0 – 2.9 ppm due to R-CO**C**H₃.

You can see these peaks on ethanoic acid's spectrum above.

Chemical shift, δ (ppm)	Type of H atom
0.8 – 1.3	R – CH₃
1.2 – 1.4	R – CH₂ – R
2.0 – 2.9	R – COCH₃
3.2 – 3.7	halogen – CH₃
3.6 – 3.8	R – CH₂ – Cl
3.3 – 3.9	R – OCH₂ – R
3.3 – 4.0	R – CH₂OH
3.5 – 5.5	R – OH
7.3	⬡ – OH
9.5 – 10	R – CHO
11.0 – 11.7	R – COOH

The hydrogen atoms that cause the shift are highlighted in red. R stands for any alkyl group.

Splitting Patterns Provide More Detail About Structure

The peaks may be **split** into smaller peaks. Peaks always split into the number of hydrogens on the neighbouring carbon, **plus one**. It's called the **n+1 rule**.

Type of Peak	Number of Hydrogens on Adjacent Carbon
Singlet (not split)	0
Doublet (split into two)	1
Triplet (split into three)	2
Quartet (split into four)	3

Here's the ¹H NMR spectrum for **1,1,2-trichloroethane**:

The peak due to the green hydrogens is split into **two** because there's **one hydrogen** on the adjacent carbon atom.

The peak due to the red hydrogen is split into **three** because there are **two hydrogens** on the adjacent carbon atom.

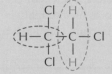

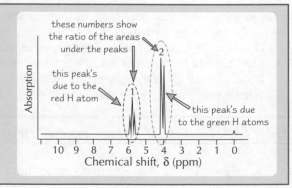

these numbers show the ratio of the areas under the peaks

this peak's due to the red H atom

this peak's due to the green H atoms

Absorption

Chemical shift, δ (ppm)

More About NMR

Put All the **Information Together** to **Predict the Structure**

EXAMPLE Using the spectrum below, and the table of chemical shift data on page 140, predict the structure of the compound.

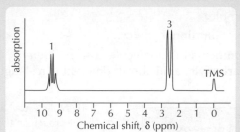

1) The peak at δ = 2.5 ppm is likely to be due to an **R–COCH₃** group, and the peak at δ = 9.5 ppm is likely to be due to an **R–CHO** group.

2) From the areas, there's one proton in the peak at δ = 9.5 ppm, for every three in the peak at δ = 2.5 ppm. This fits nicely with the first bit — so far so good.

3) The quartet's got **three** neighbouring hydrogens, and the doublet's got **one** — so it's likely these two groups are next to each other.

Now you know the molecule has to contain ⟍C– and –C–C–H , all you have to do is fit them together:

Integration Traces Show Areas More Clearly

When the peaks are split, it's not as easy to see the ratio of the **areas**, so an **integration trace** is often shown. The height increases are proportional to the areas.

So the **integration ratio** for this spectrum is **1:2** — this means that there's 1 H atom in the first environment, and 2 H atoms in the second environment.

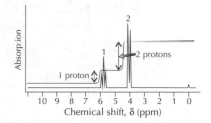

Samples are Dissolved in **Hydrogen-Free Solvents**

1) If a sample has to be dissolved, then a solvent is needed that doesn't contain any ¹H atoms — because these would show up on the spectrum and confuse things.

2) **Deuterated solvents** are often used — their hydrogen atoms have been replaced by **deuterium** (D or ²H). Deuterium's an isotope of hydrogen that's got two nucleons (a proton and a neutron).

3) Because deuterium has an **even number** of nucleons, it doesn't have a spin (so it doesn't create a magnetic field).

4) CCl₄ can also be used — it doesn't contain any ¹H atoms either.

Practice Questions

Q1 What causes peaks to split in proton NMR?
Q2 What causes a triplet of peaks?

Exam Questions

1 The ¹H NMR spectrum on the right is that of an haloalkane. Use the table of chemical shifts on page 140 to do the following:

   a) Predict the environment of the two H atoms with a shift of 3.6 ppm. [1 mark]
   b) Predict the environment of the three H atoms with a shift of 1.3 ppm. [1 mark]
   c) The molecular mass of the molecule is 64.5. Suggest a possible structure and explain your suggestion. [2 marks]
   d) Explain the shapes of the two peaks. [4 marks]

2 For the ¹H NMR spectrum of ethyl ethanoate, CH₃COOC₂H₅, give:
   a) the number of different hydrogen environments. [1 mark]
   b) the integration ratio (from left to right along the molecule, as given above). [1 mark]
   c) the types of peaks you will see (in the same order as the integration ratio). [1 mark]

Never mind splitting peaks — this stuff's likely to cause splitting headaches...

Is your head spinning yet? I know mine is. Round and round like a merry-go-round. It's a hard life when you're tied to a desk trying to get NMR spectroscopy firmly fixed in your head. You must be looking quite peaky by now... so go on, learn this stuff, take the dog around the block, then come back and see if you can still remember it all.

IR Spectroscopy and Chromatography

Chromatography's all about separating mixtures of chemicals — like the old 'felt tip on filter paper' experiment. You need to know about a few different types of chromatography at A2. There's a bit of revision on infrared spectroscopy to do first too...

Infrared Spectroscopy Helps You Identify Organic Molecules

1) In infrared (IR) spectroscopy, a beam of **IR radiation** goes through the sample.

2) The IR energy is absorbed by the **bonds** in the molecules, increasing their **vibrational** energy.

3) **Different bonds** absorb **different wavelengths**. Bonds in different **environments** in a molecule absorb different wavelengths too — so the O–H group in an **alcohol** and the O–H in a **carboxylic acid** absorb different wavelengths.

4) This table shows what **frequencies** different groups absorb —

Functional group	Where it's found	Frequency/ Wavenumber (cm^{-1})	Type of absorption
C–H	most organic molecules	2800 – 3100	strong, sharp
O–H	alcohols	3200 – 3750	strong, broad
O–H	carboxylic acids	2500 – 3300	medium, very broad
C–O	alcohols, carboxylic acids and esters	1100 – 1310	strong, sharp
C=O	aldehydes, ketones, carboxylic acids and esters	1680 – 1800	strong, sharp
C=O	amides	1630 – 1700	medium
C=C	alkenes	1620 – 1680	medium, sharp
N–H	primary amines	3300 – 3500	medium to strong
N–H	amides	about 3500	medium

This tells you what the trough on the graph will look like.

O–H groups tend to have broad absorptions — it's because they take part in hydrogen bonding.

Clark began to regret having an infrared mechanism installed in his glasses.

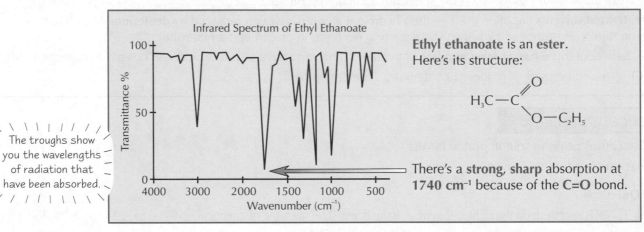

Infrared Spectrum of Ethyl Ethanoate

The troughs show you the wavelengths of radiation that have been absorbed.

Ethyl ethanoate is an **ester**. Here's its structure:

$$H_3C - C \overset{\displaystyle O}{\underset{\displaystyle O - C_2H_5}{}}$$

There's a **strong, sharp** absorption at **1740 cm^{-1}** because of the **C=O** bond.

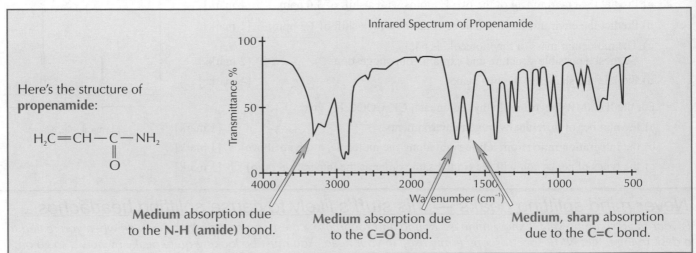

Infrared Spectrum of Propenamide

Here's the structure of **propenamide**:

$$H_2C = CH - \overset{\displaystyle}{\underset{\displaystyle \overset{\|}{O}}{C}} - NH_2$$

Medium absorption due to the **N-H (amide)** bond.

Medium absorption due to the **C=O** bond.

Medium, sharp absorption due to the **C=C** bond.

IR Spectroscopy and Chromatography

*Chromatography is Good for **Separating** and **Identifying** Things*

Chromatography is used to **separate** stuff in a mixture — once it's separated out, you can often **identify** the components.

There are quite a few different types of chromatography — but the ones you need to know about are **column chromatography** and **gas-liquid chromatography** (GLC). All types of chromatography have two phases:

> 1) A **mobile phase** — where the molecules can move. This is always a liquid or a gas.
> 2) A **stationary phase** — where the molecules can't move. This must be a solid, or a liquid held in a solid.

The components in the mixture **separate out** as the mobile phase moves through the stationary phase.

*Column Chromatography is Used To Separate Out **Solutions***

Column chromatography is mostly used for **purifying an organic product.** ← *This needs to be done to separate it from unreacted chemicals or side products.*

1) It involves packing a glass column with a slurry of an absorbent material such as aluminium oxide, coated with water. This is the **stationary phase**.

2) The mixture to be separated is added to the top of the column and allowed to drain down into the slurry. A **solvent** is then run slowly and continually through the column. This solvent is the **mobile phase**.

3) As the mixture is washed through the column, its components **separate out** according to **how soluble** they are in the mobile phase and **how strongly they are adsorbed** onto the stationary phase (**retention).**

4) Each different component will spend some time adsorbed onto the stationary phase and some time dissolved in the mobile phase. The **more** soluble each component is in the mobile phase, the **quicker** it'll pass through the column.

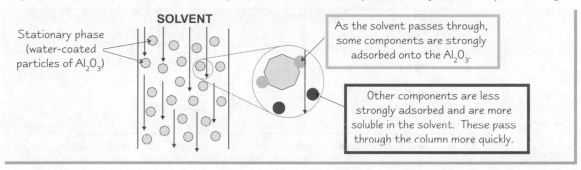

Stationary phase (water-coated particles of Al₂O₃)

SOLVENT

As the solvent passes through, some components are strongly adsorbed onto the Al₂O₃.

Other components are less strongly adsorbed and are more soluble in the solvent. These pass through the column more quickly.

*Gas-Liquid Chromatography is Used To Separate **Mixtures of Volatile Liquids***

1) If you've got a mixture of **volatile liquids** (ones that turn into gases easily), then **gas-liquid chromatography** (GLC) is the way to separate them out so that you can **identify** them.

2) The stationary phase is a **viscous liquid**, such as an oil, which coats the inside of a long tube, which is coiled to save space, and built into an oven. The mobile phase is an **unreactive carrier gas** such as nitrogen.

3) Each component takes a different amount of time from being **injected** into the tube to being **recorded** at the other end. This is the **retention time**.

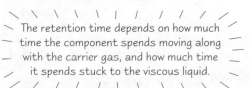
The retention time depends on how much time the component spends moving along with the carrier gas, and how much time it spends stuck to the viscous liquid.

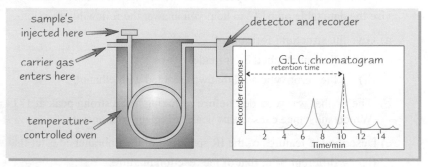
sample's injected here
detector and recorder
carrier gas enters here
temperature-controlled oven
G.L.C. chromatogram
retention time
Recorder response
Time/min

4) Each separate substance will have a unique retention time — so you can use the retention time to **identify** the components of the mixture. (You have to run a known sample under the same conditions for comparison).
For example, if you wanted to know if a mixture contained **octane**, you could run a sample of the **mixture** through the system, then run a sample of **pure octane** through, and see if there's a peak at the **same retention time** on both spectra.

5) The **area** under each peak tells you the relative **amount** of each component that's present in the mixture.

6) GLC can be used to find the **level of alcohol** in **blood** or **urine** — the results are **accurate** enough to be used as evidence in court. It's also used to find the **proportions** of various **esters in oils** used in **paints** — this lets picture restorers know exactly what paint was originally used.

IR Spectroscopy and Chromatography

And now you get this whole page of questions on chromatography and IR Spectra. Keep at it until they're all a doddle.

Practice Questions

Q1 Alcohols and carboxylic acids both contain an OH functional group. Suggest how you could tell the difference between the infrared spectrum of an alcohol and the infrared spectrum of a carboxylic acid.

Q2 What are the two phases in chromatography?

Q3 In column chromatography, explain why the components pass through the column at different rates.

Q4 Give an example of a gas that is used as a carrier gas in gas-liquid chromatography.

Q5 GLC can identify which substances are present and their relative amounts. How is the relative amount determined?

Exam Questions

1   GLC can be used to detect the presence and quantity of alcohol in the blood or urine samples of suspected drink-drivers.

   a) What do the letters GLC stand for?   [1 mark]

   b) Explain how 'retention time' is used to identify ethanol in a sample of blood or urine.   [2 marks]

   c) Why is nitrogen used as the carrier gas?   [1 mark]

2   A molecule with a molecular mass of 74 gives the following IR spectrum.

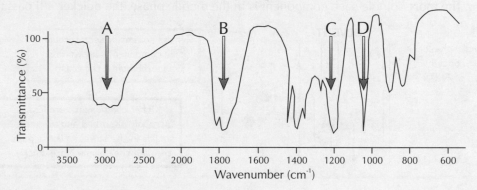

   Use the data table on p142 to help you answer the following questions.

   a) Which functional groups produce the troughs labelled A – D?   [4 marks]

   b) Suggest a molecular formula and name for this molecule. Explain your suggestion.   [3 marks]

3   A student makes an unbranched ester with the molecular formula $C_4H_8O_2$ by refluxing a carboxylic acid with ethanol. The student records an IR spectrum of the reaction mixture before and after refluxing.

   Use the data table on p142 to help you answer the following questions.

   a) Give the structural formula of

   (i)   the unbranched ester made.   [1 mark]

   (ii)   the carboxylic acid that was reacted with ethanol.   [1 mark]

   b) The IR that was recorded **before** refluxing had a strong peak at 1719 cm^{-1}.
       Which substance caused this peak? Explain your answer.   [3 marks]

   c) Describe the features of the IR spectrum of the unbranched ester that would be

   (i)   different from that of the reaction mixture.   [3 marks]

   (ii)   similar to that of the reaction mixture.   [4 marks]

        Explain what causes each of the features that you refer to.

Ooooh — I'm picking up some good vibrations...

Now, I've warned you — infrared glasses are not for fun. They're highly advanced pieces of technology which if placed in the wrong hands could cause havoc and destruction across the Universe. Luckily, you'll be given a data table for IR spectra in the exam, so you don't have to bother learning it. But DO learn how the different types of chromatography work.

Lattice Enthalpies

I'm sure you all remember enthalpies from AS... but here's a quick reminder just in case...

First — A Few Definitions You Should Remember

ΔH is the symbol for **enthalpy change**. Enthalpy change is the **heat** energy transferred in a reaction at **constant pressure**.

$\Delta H^{\ominus}$ means that the enthalpy change was measured under **standard conditions** (**100 kPa and 298 K**).
Exothermic reactions have a **negative ΔH** value, because heat energy is given out.
Endothermic reactions have a **positive ΔH** value, because heat energy is absorbed.

Lattice Enthalpy is a Measure of Ionic Bond Strength

You can calculate the lattice enthalpy of formation using the method on page 149.

Lattice enthalpy can be defined in two ways:

① **Lattice formation enthalpy**: the enthalpy change when **1 mole** of a **solid ionic compound** is **formed** from its **gaseous ions** under standard conditions.

Example:

$Na^+_{(g)} + Cl^-_{(g)} \rightarrow NaCl_{(s)}$ $\Delta H^{\ominus} = -787$ kJ mol^{-1} (exothermic)

$Mg^{2+}_{(g)} + 2Cl^-_{(g)} \rightarrow MgCl_{2(s)}$ $\Delta H^{\ominus} = -2526$ kJ mol^{-1} (exothermic)

② **Lattice dissociation enthalpy**: the enthalpy change when **1 mole** of a **solid ionic compound** is completely **dissociated** into its **gaseous ions** under standard conditions.

Example:

$NaCl_{(s)} \rightarrow Na^+_{(g)} + Cl^-_{(g)}$ $\Delta H^{\ominus} = +787$ kJ mol^{-1} (endothermic)

$MgCl_{2(s)} \rightarrow Mg^{2+}_{(g)} + 2Cl^-_{(g)}$ $\Delta H^{\ominus} = +2526$ kJ mol^{-1} (endothermic)

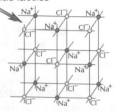

Part of the sodium chloride lattice

Notice that lattice formation enthalpy and lattice dissociation enthalpy are exact opposites.

You Can't Measure Lattice Enthalpy Directly

Lattice enthalpy **can't** be measured directly, but it **can** be calculated with a **Born-Haber cycle** — all will be revealed on the next page. But first you need to know these **definitions**:

Enthalpy change of formation, ΔH_f, is the enthalpy change when **1 mole** of a **compound** is formed from its **elements** in their standard states under standard conditions, e.g. $2C_{(s)} + 3H_{2(g)} + \frac{1}{2}O_{2(g)} \rightarrow C_2H_5OH_{(l)}$

The bond dissociation enthalpy, ΔH_{diss}, is the enthalpy change when all the **bonds of the same type** in **1 mole** of **gaseous molecules** are broken, e.g. $Cl_{2(g)} \rightarrow 2Cl_{(g)}$

Enthalpy change of atomisation of an element, ΔH_{at}, is the enthalpy change when **1 mole** of **gaseous atoms** is formed from an element in its **standard state**, e.g. $\frac{1}{2}Cl_{2(g)} \rightarrow Cl_{(g)}$

Enthalpy change of atomisation of a compound, ΔH_{at}, is the enthalpy change when **1 mole** of a **compound** in its **standard state** is converted to **gaseous atoms**, e.g. $NaCl_{(s)} \rightarrow Na_{(g)} + Cl_{(g)}$

The first ionisation enthalpy, ΔH_{ie1}, is the enthalpy change when **1 mole** of **gaseous 1+ ions** is formed from **1 mole** of **gaseous atoms**, e.g. $Mg_{(g)} \rightarrow Mg^+_{(g)} + e^-$

The second ionisation enthalpy, ΔH_{ie2}, is the enthalpy change when **1 mole** of **gaseous 2+ ions** is formed from **1 mole** of **gaseous 1+ ions**, e.g. $Mg^+_{(g)} \rightarrow Mg^{2+}_{(g)} + e^-$

First electron affinity, ΔH_{ea1}, is the enthalpy change when **1 mole** of gaseous 1– ions is made from **1 mole** of gaseous **atoms**, e.g. $O_{(g)} + e^- \rightarrow O^-_{(g)}$

Second electron affinity, ΔH_{ea2}, is the enthalpy change when **1 mole** of **gaseous 2– ions** is made from **1 mole** of gaseous 1– ions, e.g. $O^-_{(g)} + e^- \rightarrow O^{2-}_{(g)}$

The enthalpy change of hydration, ΔH_{hyd}, is the enthalpy change when **1 mole** of **aqueous ions** is formed from **gaseous ions**, e.g. $Na^+_{(g)} \rightarrow Na^+_{(aq)}$

The enthalpy change of solution, $\Delta H_{solution}$, is the enthalpy change when **1 mole** of **solute** is dissolved in **sufficient solvent** that no further enthalpy change occurs on further dilution, e.g. $NaCl_{(s)} \rightarrow NaCl_{(aq)}$

Lattice Enthalpies

Now you know all your enthalpy change definitions, here's how to use them... Enjoy.

Born-Haber Cycles can be Used to Calculate Lattice Enthalpies

You can't calculate a lattice enthalpy **directly**, so you have to use a **Born-Haber cycle** to figure out what the enthalpy change would be if you took **another, less direct, route**. **Hess's law** says that the **total enthalpy change** of a reaction is always the **same**, no matter which route is taken.

Here's how to draw a Born-Haber cycle for calculating the lattice enthalpy of **NaCl**:

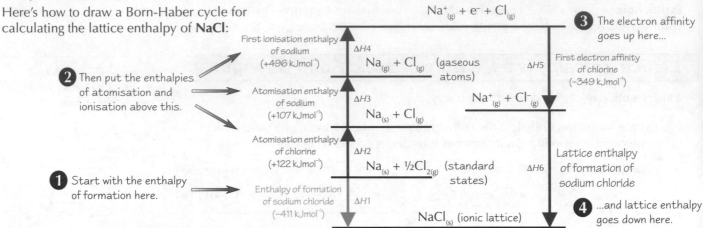

1 Start with the enthalpy of formation here.

2 Then put the enthalpies of atomisation and ionisation above this.

3 The electron affinity goes up here...

4 ...and lattice enthalpy goes down here.

There are **two routes** you can follow to get from the elements in their **standard states** to the **ionic lattice**. The green arrow shows the **direct route** and the purple arrows show the **indirect route**. The enthalpy change for each is the **same**.

From Hess's law: $\Delta H6 = -\Delta H5 - \Delta H4 - \Delta H3 - \Delta H2 + \Delta H1$
$$= -(-349) - (+496) - (+107) - (+122) + (-411) = \textbf{–787 kJmol}^{-1}$$

So the **lattice enthalpy of formation** of sodium chloride is **–787 kJmol^{-1}**.

You need a minus sign if you go the wrong way along an arrow.

Calculations Involving Group 2 Elements are a Bit Different

Born-Haber cycles for compounds containing **Group 2 elements** have a few **changes** from the one above. Make sure you understand what's going on so you can handle whatever compound they throw at you.

Here's the Born-Haber cycle for calculating the lattice enthalpy of **magnesium chloride** ($MgCl_2$).

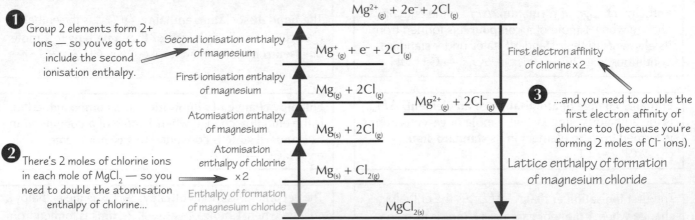

1 Group 2 elements form 2+ ions — so you've got to include the second ionisation enthalpy.

2 There's 2 moles of chlorine ions in each mole of $MgCl_2$ — so you need to double the atomisation enthalpy of chlorine...

3 ...and you need to double the first electron affinity of chlorine too (because you're forming 2 moles of Cl$^-$ ions).

Theoretical Lattice Enthalpies are Often Different from Experimental Values

You can work out a **theoretical lattice enthalpy** by doing some calculations based on the **purely ionic model** of a lattice. The purely ionic model of a lattice assumes that all the ions are **spherical**, and have their charge **evenly distributed** around them.

But the **experimental lattice enthalpy** from the Born-Haber cycle is usually different. This is **evidence** that ionic compounds usually have some **covalent character**.

The positive and negative ions in a lattice **aren't** usually exactly spherical Positive ions **polarise** neighbouring negative ions to different extents, and the **more polarisation** there is, the **more covalent** the bonding will be.

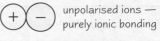
unpolarised ions — purely ionic bonding

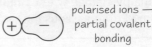
polarised ions — partial covalent bonding

Lattice Enthalpies

Comparing Lattice Enthalpies Can Tell You 'How Ionic' an Ionic Lattice Is

If the experimental and theoretical lattice enthalpies for a compound are **very different**, it shows that the compound has a lot of covalent character.

For example, here are both lattice enthalpy values for some **magnesium halides**.

Compound	Lattice Enthalpy of Formation (kJmol⁻¹)	
	From experimental values in Born-Haber cycle	From theory
Magnesium chloride	−2526	−2326
Magnesium bromide	−2440	−2097
Magnesium iodide	−2327	−1944

1) The **experimental** lattice energies are **bigger** than the theoretical values by a fair bit.

2) This tells you that the **bonding** is **stronger** than the calculations from the ionic model predict.

3) The difference shows that the ionic bonds in the magnesium halides are quite strongly **polarised** and so they have **quite a lot of covalent character**.

Here are some more lattice energies, for **sodium halides** this time:

1) The experimental and theoretical values are a **pretty close match** — so you can say that these compounds fit the 'purely ionic' model very well.

2) This indicates that the structure of the lattice for these compounds is quite close to being **purely ionic**. There's almost no polarisation so they don't have much covalent character.

Compound	Lattice Enthalpy of Formation (kJmol⁻¹)	
	From experimental values in Born-Haber cycle	From theory
Sodium chloride	−787	−766
Sodium bromide	−742	−731
Sodium iodide	−698	−686

Practice Questions

Q1 Define *first ionisation enthalpy* and *first electron affinity*.

Q2 Sketch out Born-Haber cycles for: a) LiF, b) CaCl$_2$.

Q3 Explain why theoretical lattice enthalpies are often different from experimentally determined lattice enthalpies.

Exam Questions

1 Using the data below:
   a) Construct a Born-Haber cycle for potassium bromide (KBr). [4 marks]
   b) Use your Born-Haber cycle to calculate the lattice enthalpy of formation of potassium bromide. [3 marks]

$\Delta H_f^{\ominus}$ [potassium bromide] = −394 kJmol⁻¹ $\Delta H_{at}^{\ominus}$ [bromine] = +112 kJmol⁻¹ $\Delta H_{at}^{\ominus}$ [potassium] = +89 kJmol⁻¹

$\Delta H_{ie1}^{\ominus}$ [potassium] = +419 kJmol⁻¹ $\Delta H_{ea1}^{\ominus}$ [bromine] = −325 kJmol⁻¹

2 The diagram shows part of a Born-Haber cycle for copper(II) chloride.
   a) Identify the enthalpy changes X, Y and Z. [3 marks]
   b) What other enthalpy changes need to be added in order to complete the cycle? [3 marks]

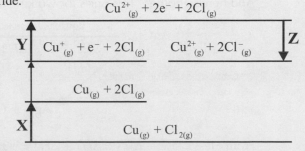

Using Born-Haber cycles — it's just like riding a bike...

All this energy going in and out can get a bit confusing. Remember these simple rules: 1) It takes energy to break bonds, but energy is given out when bonds are made. 2) A negative ΔH means energy is given out (it's exothermic). 3) A positive ΔH means energy is taken in (it's endothermic). 4) Never return to a firework once lit.

Enthalpies of Solution

Once you know what's happening when you stir sugar into your tea, your cuppa'll be twice as enjoyable.

Dissolving Involves Enthalpy Changes

When a solid **ionic lattice** dissolves in water these **two** things happen:

1) The bonds between the ions **break** — this is **endothermic**.
   This enthalpy change is the **lattice enthalpy of dissociation**.

2) Bonds between the ions and the water are **made** — this is **exothermic**.
   The enthalpy change here is called the **enthalpy change of hydration**.

Ionic lattice.

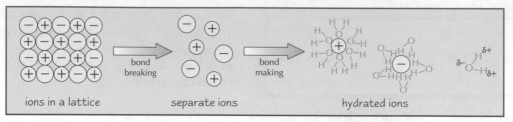

ions in a lattice   separate ions   hydrated ions

Oxygen is more electronegative than hydrogen, so it draws the bonding electrons toward itself, creating a dipole.

3) The **enthalpy change of solution** is the overall effect on the enthalpy of these two things.

Enthalpy Change of Solution can be Calculated

You can work out the enthalpy change of solution using an **enthalpy cycle**.
You just need to know the **lattice dissociation enthalpy** of the compound and the enthalpies of **hydration of the ions**.

Here's how to draw the enthalpy cycle for working out the **enthalpy change of solution** for **sodium chloride**.

1 Put the ionic lattice and the dissolved ions on the top — connect them by the enthalpy change of solution. This is the direct route.

2 Connect the ionic lattice to the gaseous ions by the lattice enthalpy of dissociation. This will be a positive number.

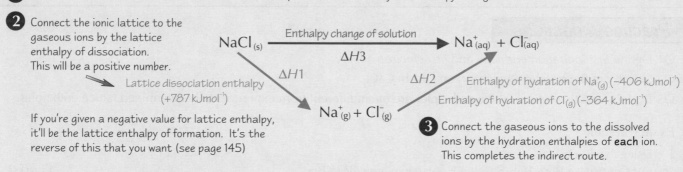

Lattice dissociation enthalpy (+787 kJmol⁻¹)

If you're given a negative value for lattice enthalpy, it'll be the lattice enthalpy of formation. It's the reverse of this that you want (see page 145)

Enthalpy change of solution ΔH3

NaCl(s) → Na⁺(aq) + Cl⁻(aq)

ΔH1   ΔH2   Enthalpy of hydration of Na⁺(g) (−406 kJmol⁻¹)
                     Enthalpy of hydration of Cl⁻(g) (−364 kJmol⁻¹)

Na⁺(g) + Cl⁻(g)

3 Connect the gaseous ions to the dissolved ions by the hydration enthalpies of **each** ion. This completes the indirect route.

From Hess's law: ΔH3 = ΔH1 + ΔH2 = +787 + (−406 + −364) = **+17 kJmol⁻¹**

The enthalpy change of solution is **slightly endothermic**, but this is compensated for by a small increase in **entropy** (see page 150), so sodium chloride still dissolves in water.

And here's another. This one's for working out the **enthalpy change of solution** for **silver chloride**.

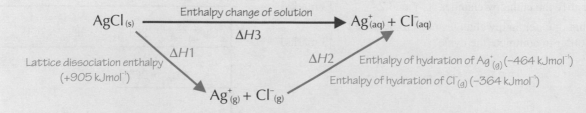

Enthalpy change of solution ΔH3

AgCl(s) → Ag⁺(aq) + Cl⁻(aq)

Lattice dissociation enthalpy (+905 kJmol⁻¹)

ΔH1   ΔH2   Enthalpy of hydration of Ag⁺(g) (−464 kJmol⁻¹)
                     Enthalpy of hydration of Cl⁻(g) (−364 kJmol⁻¹)

Ag⁺(g) + Cl⁻(g)

From Hess's law: ΔH3 = ΔH1 + ΔH2 = +905 + (−464 + −364) = **+77 kJmol⁻¹**

This is much **more endothermic** than the enthalpy change of solution for sodium chloride. There is an **increase in entropy** again, but it's pretty small and not enough to make a difference — so silver chloride is **insoluble** in water.

Enthalpies of Solution

Use *Mean Bond Enthalpies* to Calculate Enthalpy Changes

Energy is **taken in** when bonds are **broken** and **released** when bonds are **formed**, so —

> enthalpy change for a reaction = sum of enthalpies of bonds broken – sum of enthalpies of bonds formed

If you need **more** energy to **break** bonds than is released when bonds are made, it's an **endothermic reaction** and the enthalpy change, ΔH, is **positive**. If **more** energy is **released** than is taken in, it's **exothermic** and ΔH is **negative**.

Example: Calculate the enthalpy change for the following reaction: $N_{2(g)} + 3H_{2(g)} \rightarrow 2NH_{3(g)}$

Bond	Mean bond enthalpy (kJmol⁻¹)
N≡N	945
H—H	436
N—H	391

It's easier to see what's going on if you **sketch out** the molecules involved:

$$N{\equiv}N \ + \ 3\,H{-}H \ \longrightarrow \ 2\,H{-}\underset{\underset{H}{|}}{N}{-}H$$

Now add up the mean bond enthalpies for the reactant **bonds broken**...
$(1 \times N{\equiv}N) + (3 \times H{-}H) = (1 \times 945) + (3 \times 436) = 2253$ kJmol⁻¹

...and for the **new bonds formed** in the products
$(6 \times N{-}H) = (6 \times 391) = 2346$ kJmol⁻¹

So the **enthalpy change for the reaction** = 2253 – 2346 = **–93 kJmol⁻¹**.

You can use this method to calculate the lattice enthalpy of formation — see page 145.

It's negative, so it's exothermic.

Mean Bond Enthalpy Calculations are Only Approximations

The bond enthalpy given above for N–H bonds is **not** exactly right for **every** N–H bond. A given type of bond will **vary in strength** from compound to compound and can even vary **within** a compound. Mean bond enthalpies are the **averages** of these bond enthalpies. Only the bond enthalpies of **diatomic molecules**, such as H_2 and HCl, will always be the same.

So calculations done using **mean bond enthalpies** will never be perfectly accurate. You get much **more exact** results from **experimental data** obtained from the **specific compounds**.

Practice Questions

Q1 Describe the two things that happen when something dissolves.

Q2 Without peeking, draw the enthalpy cycle that you'd use to work out the enthalpy change of solution of NaCl.

Q3 How can mean bond enthalpies be used to calculate enthalpy changes for covalent compounds?

Exam Questions

1   a)   Draw an enthalpy cycle for the enthalpy change of solution of $SrF_{2(s)}$. Label each enthalpy change.   [5 marks]

     b)   Calculate the enthalpy change of solution for SrF_2 from the following data: [2 marks]

       $\Delta H_{latt}^{\ominus} [SrF_{2(s)}] = -2492$ kJmol⁻¹, $\Delta H_{hyd}^{\ominus} [Sr^{2+}_{(g)}] = -1480$ kJmol⁻¹, $\Delta H_{hyd}^{\ominus} [F^{-}_{(g)}] = -506$ kJmol⁻¹

2   Show that the enthalpy change of solution for $MgCl_{2(s)}$ is –122 kJmol⁻¹, given that:

     $\Delta H_{latt}^{\ominus} [MgCl_{2(s)}] = -2526$ kJmol⁻¹, $\Delta H_{hyd}^{\ominus} [Mg^{2+}_{(g)}] = -1920$ kJmol⁻¹, $\Delta H_{hyd}^{\ominus} [Cl^{-}_{(g)}] = -364$ kJmol⁻¹.   [3 marks]

3   a)   Calculate the enthalpy change for $CH_{4(g)} + 2O_{2(g)} \rightarrow CO_{2(g)} + 2H_2O_{(g)}$ using the bond enthalpies in the table below.

Bond	C–H	O=O	C=O	O–H
Mean bond enthalpy (kJmol⁻¹)	412	496	743	463

       [3 marks]

     b)   Give a reason why the enthalpy change calculated by this method is not likely to be as accurate as one calculated using an enthalpy cycle.   [2 marks]

Enthalpy change of solution of the Wicked Witch of the West = 8745 kJmol⁻¹ ...

Compared to the ones on page 146, these enthalpy cycles are an absolute breeze. You've got to make sure the definitions are firmly fixed in your mind though. And don't forget there are two opposite definitions of lattice enthalpy.

Entropy Change

Entropy sounds a bit like enthalpy (to start with at least), but they're not the same thing at all. Read on...

Entropy Tells You How Much Disorder There Is

Entropy is a measure of the **number of ways** that **particles** can be **arranged** and the **number of ways** that the **energy** can be shared out between the particles.

Substances really **like** disorder, so the particles move to try to **increase the entropy**.

There are a few things that affect entropy:

Entropy is represented by the symbol S

PHYSICAL STATE affects Entropy

You have to go back to the good old **solid-liquid-gas** particle explanation thingy to understand this.

Solid particles just wobble about a fixed point — there's **hardly any** randomness, so they have the **lowest entropy**.

Gas particles whizz around wherever they like. They've got the most **random arrangements** of particles, so they have the **highest entropy**.

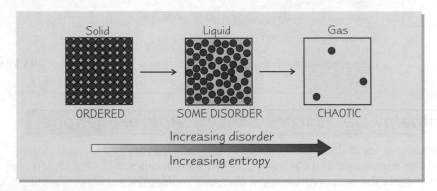

Solid — ORDERED   Liquid — SOME DISORDER   Gas — CHAOTIC

Increasing disorder
Increasing entropy

DISSOLVING affects Entropy

Dissolving a solid also increases its entropy — dissolved particles can **move freely** as they're no longer held in one place.

MORE PARTICLES means More Entropy

It makes sense — the more particles you've got, the **more ways** they and their energy can be **arranged**. — so in a reaction like $N_2O_{4(g)} \rightarrow 2NO_{2(g)}$, entropy increases because the **number of moles** increases.

Entropy Increase May Explain Spontaneous Endothermic Reactions

A **spontaneous** (or **feasible**) change is one that'll **just happen** by itself — you don't need to give it energy.

You might think that only exothermic reactions could be spontaneous because you need to supply energy to endothermic reactions. But the weird thing is, some **endothermic** reactions are **spontaneous**. You normally do have to supply **energy** to make an endothermic reaction happen, but in some reactions the **entropy** increases such a lot that the reaction will happen **by itself**, without you supplying any energy.

For example,

- Water evaporates at room temperature. This change needs **energy** to break the bonds between the molecules (i.e. it's endothermic) — but because it's **changing state** (from a liquid to a gas), the **entropy increases**.
- The reaction of sodium hydrogencarbonate with hydrochloric acid is a **spontaneous endothermic reaction**. Again there's an **increase in entropy**.

$$NaHCO_{3(s)} + H^+_{(aq)} \rightarrow Na^+_{(aq)} + CO_{2(g)} + H_2O_{(l)}$$

1 mole solid   1 mole aqueous ions   1 mole aqueous ions   1 mole gas   1 mole liquid

The product has more particles — and gases and liquids have more entropy than solids too.

Entropy Change

Reactions *Won't* Happen Unless the *Total Entropy Change* is *Positive*

During a reaction, there's an entropy change (ΔS) between the **reactants and products** — the entropy change of the **system**. The entropy of the **surroundings** changes too (because **energy** is transferred to or from the system). The **TOTAL entropy change** is the sum of the entropy changes of the **system** and the **surroundings**.

The units of entropy are JK⁻¹mol⁻¹

$$\Delta S_{total} = \Delta S_{system} + \Delta S_{surroundings}$$

This equation isn't much use unless you know ΔS_{system} and $\Delta S_{surroundings}$. Luckily, there are formulas for them too:

This is just the difference between the entropies of → the reactants and products.

$$\Delta S_{system} = S_{products} - S_{reactants}$$ and $$\Delta S_{surroundings} = -\frac{\Delta H}{T}$$

ΔH = enthalpy change (in Jmol⁻¹)
T = temperature (in K)

Here's an *Example* of How to *Calculate* the *Total Entropy Change*

Example: Calculate the total entropy change for the reaction of ammonia and hydrogen chloride under standard conditions.

$$NH_{3(g)} + HCl_{(g)} \rightarrow NH_4Cl_{(s)} \qquad \Delta H^\ominus = -315 \text{ kJmol}^{-1} \text{ (at 298 K)}$$

$$S^\ominus[NH_{3(g)}] = 192.3 \text{ JK}^{-1}\text{mol}^{-1}, \; S^\ominus[HCl_{(g)}] = 186.8 \text{ JK}^{-1}\text{mol}^{-1}, \; S^\ominus[NH_4Cl_{(s)}] = 94.6 \text{ JK}^{-1}\text{mol}^{-1}$$

First find the entropy change of the **system**:

$$\Delta S^\ominus_{system} = S^\ominus_{products} - S^\ominus_{reactants} = 94.6 - (192.3 + 186.8) = \textbf{-284.5 JK}^{-1}\textbf{mol}^{-1}$$

This shows a negative change in entropy. It's not surprising, as 2 moles of gas have combined to form 1 mole of solid.

Now find the entropy change of the **surroundings**:

$$\Delta H^\ominus = -315 \text{ kJmol}^{-1} = -315 \times 10^3 \text{ Jmol}^{-1}$$ ← Put $\Delta H^\ominus$ in the right units.

$$\Delta S_{surroundings} = -\frac{\Delta H}{T} = \frac{-(-315 \times 10^3)}{298} = \textbf{+1057.0 JK}^{-1}\textbf{mol}^{-1}$$

Finally you can find the **total** entropy change:

$$\Delta S^\ominus_{total} = \Delta S^\ominus_{system} + \Delta S^\ominus_{surroundings} = -284.5 + (+1057.0) = \textbf{+772.5 JK}^{-1}\textbf{mol}^{-1}$$

The total entropy has **increased**. The entropy increase in the surroundings was big enough to make up for the entropy decrease in the system.

Practice Questions

Q1 What does the term 'entropy' mean?

Q2 In each of the following pairs choose the one with the greater entropy value.
   a) 1 mole of $NaCl_{(aq)}$ and 1 mole of $NaCl_{(s)}$ b) 1 mole of $Br_{2(l)}$ and 1 mole of $Br_{2(g)}$
   c) 1 mole of $Br_{2(g)}$ and 2 moles of $Br_{2(g)}$

Q3 Write down the formulas for the following:
   a) total entropy change   b) entropy change of the surroundings   c) entropy change of the system

Exam Question

1   The equation for the formation of magnesium oxide from magnesium and oxygen is shown below.

$$Mg_{(s)} + \tfrac{1}{2}O_{2(g)} \rightarrow MgO_{(s)}$$

a) Based on just the equation, predict whether the reaction is likely to be spontaneous. Give a reason for your answer.   [2 marks]

b) Use the data on the right to calculate the entropy change for the system above.   [3 marks]

Substance	Entropy — standard conditions (JK⁻¹mol⁻¹)
$Mg_{(s)}$	32.7
$\tfrac{1}{2}O_{2(g)}$	102.5
$MgO_{(s)}$	26.9

Being neat and tidy is against the laws of nature...

I know there are three different formulas on this page, but at least they aren't too hard to use. Watch out for your units, though — make sure the temperature's in kelvin. If you're given a temperature in °C, you need to add 273 to it to get it in kelvin. And check that all your enthalpy and entropy values involve joules, not kilojoules (so Jmol⁻¹, not kJmol⁻¹, etc.).

Free-Energy Change

Free energy — I could do with a bit of that. My gas bill is astronomical.

For Spontaneous Reactions ΔG must be **Negative** or **Zero**

Free energy change, ΔG, is a measure used to predict whether a reaction is **feasible**.

If ΔG is **negative or equal to zero**, then the reaction might happen by itself.

Free energy change takes into account the changes in **enthalpy** and **entropy** in the system. And of course, there's a formula for it:

ΔG = free energy change (in Jmol⁻¹)

$$\Delta G = \Delta H - T\Delta S_{system}$$

ΔH = enthalpy change (in Jmol⁻¹)
T = temperature (in K)
ΔS_{system} = entropy of the system (in JK⁻¹mol⁻¹) (see page 151)

Even if ΔG shows that a reaction is theoretically feasible, it might have a really high activation energy and be so slow that you wouldn't notice it happening at all.

Example: Calculate the free energy change for the following reaction at 298 K.

$$MgCO_{3(s)} \rightarrow MgO_{(s)} + CO_{2(g)} \qquad \Delta H^{\ominus} = +117 \text{ kJmol}^{-1}, \ \Delta S^{\ominus}_{system} = +175 \text{ JK}^{-1}\text{mol}^{-1}$$

$$\Delta G = \Delta H - T\Delta S_{system} = +117 \times 10^3 - [(298 \times (+175)] = +64\ 850 \text{ Jmol}^{-1}$$

ΔH must be in J not kJ. So, if you're given a value in kJ, multiply it by 10³.

ΔG is positive — so the reaction isn't feasible at this temperature.

The **Feasiblity** of Some Reactions Depends on **Temperature**

If a reaction is exothermic (**negative ΔH**) and has a **positive entropy** change, then ΔG is **always negative** since $\Delta G = \Delta H - T\Delta S_{system}$. These reactions are feasible at **any** temperature.

If a reaction is endothermic (**positive ΔH**) and has a **negative entropy** change, then ΔG is **always positive**. These reactions are **not** feasible at any temperature.

But for other combinations, temperature has an effect.

1) If ΔH is **positive** (endothermic) and ΔS_{system} is **positive** then the reaction won't be feasible at some temperatures but **will be** at a **higher** temperature.

   For example, the decomposition of **calcium carbonate** is **endothermic** but results in an **increase in entropy** (the number of molecules increases and CO_2 is a gas).

   $$CaCO_{3(s)} \rightarrow CaO_{(s)} + CO_{2(g)}$$

   The reaction will **only** occur when $CaCO_3$ is **heated** — it isn't feasible at 298 K.

After her surgery, Anne found that a reaction wasn't feasible.

Here's an example with numbers: ΔH = +10 kJ mol⁻¹, ΔS_{system} = +10 JK⁻¹mol⁻¹

| at 300 K | $\Delta G = \Delta H - T\Delta S_{system}$ | $\Delta G = +10 \times 10^3 - (300 \times +10) = +7000$ J mol⁻¹ |
| at 1500 K | $\Delta G = \Delta H - T\Delta S_{system}$ | $\Delta G = +10 \times 10^3 - (1500 \times +10) = -5000$ J mol⁻¹ |

So a reaction with these enthalpy and entropy changes is feasible at 1500 K, but not at 300 K...

2) If ΔH is **negative** (exothermic) and ΔS_{system} is **negative** then the reaction will be feasible at some temperatures but won't be **feasible** at a **higher** temperature.

   For example, the process of turning **water** from a **liquid** to a **solid** is **exothermic** but results in a **decrease in entropy** (a solid is more ordered than a liquid), which means it will only occur at **certain temperatures** (i.e. at 0 °C or below).

And another example with numbers: ΔH = −10 kJ mol⁻¹, ΔS_{system} = −10 JK⁻¹mol⁻¹

| at 300 K | $\Delta G = \Delta H - T\Delta S_{system}$ | $\Delta G = -10 \times 10^3 - (300 \times -10) = -7000$ J mol⁻¹ |
| at 1500 K | $\Delta G = \Delta H - T\Delta S_{system}$ | $\Delta G = -10 \times 10^3 - (1500 \times -10) = +5000$ J mol⁻¹ |

...and this one is feasible at 300 K, but not at 1500 K.

Free-Energy Change

You can Calculate the Temperature at which a Reaction Becomes Feasible

When ΔG is **zero**, a reaction is **just feasible**.

You can find the **temperature** when ΔG is zero by rearranging the free energy equation from the previous page:

$\Delta G = \Delta H - T\Delta S_{system}$, so when $\Delta G = 0$, $T\Delta S_{system} = \Delta H$. So: $\boxed{T = \dfrac{\Delta H}{\Delta S_{system}}}$

$T =$ temperature at which a reaction becomes feasible (in K)

$\Delta H =$ enthalpy change (in Jmol^{-1})

$\Delta S_{system} =$ entropy change of the system (in JK^{-1}mol^{-1})

Example: Tungsten, W, can be extracted from its ore, WO_3, by reduction using hydrogen.

$$WO_{3(s)} + 3H_{2(g)} \rightarrow W_{(s)} + 3H_2O_{(g)} \qquad \Delta H^{\ominus} = +117 \text{ kJmol}^{-1}$$

Use the data in the table to find the minimum temperature at which the reaction becomes feasible.

Substance	$S^{\ominus}/ \text{JK}^{-1}\text{mol}^{-1}$
$WO_{3(s)}$	76
$H_{2(g)}$	65
$W_{(s)}$	33
$H_2O_{(g)}$	189

First, convert the **enthalpy change**, ΔH, to joules per mole:

$$\Delta H = 117 \times 10^3 = \textbf{117 000 Jmol}^{-1}$$

Then find the **entropy change**, ΔS_{system}:

$$\Delta S_{system} = S_{products} - S_{reactants} = [33 + (3 \times 189)] - [76 + (3 \times 65)] = \textbf{+329 JK}^{-1}\textbf{mol}^{-1}$$

See page 151 for more on this formula.

Then divide ΔH by ΔS_{system} to find the temperature at which the reaction just becomes feasible:

$$T = \frac{\Delta H}{\Delta S_{system}} = \frac{117\,000}{329} = \textbf{356 K}$$

Practice Questions

Q1 What does ΔG stand for? What is it used for?

Q2 State whether each of the following reactions are feasible, feasible at certain temperatures or not feasible:
    Reaction A: negative ΔH and negative ΔS Reaction B: negative ΔH and positive ΔS
    Reaction C: positive ΔH and positive ΔS Reaction D: positive ΔH and negative ΔS

Q3 Write down the formula for calculating the temperature at which a reaction becomes feasible.

Exam Questions

1   The enthalpy change that occurs when water turns from a liquid to a solid is -6 kJmol^{-1} .

$$H_2O_{(l)} \rightarrow H_2O_{(s)} \qquad S^{\ominus}[H_2O_{(l)}] = 70 \text{ JK}^{-1}\text{mol}^{-1}, \quad S^{\ominus}[H_2O_{(s)}] = 48 \text{ JK}^{-1}\text{mol}^{-1},$$

a)   Calculate the total entropy change at   (i) 250 K   (ii) 300 K   [5 marks]

b)   Will this reaction be spontaneous at 250 K or 300 K? Explain your answer.   [2 marks]

2   Magnesium carbonate decomposes to form magnesium oxide and carbon dioxide:

$$MgCO_{3(s)} \rightarrow MgO_{(s)} + CO_{2(g)} \qquad \Delta H^{\ominus} = +117 \text{ kJmol}^{-1} \text{ and } \Delta S^{\ominus}_{system} = +175 \text{ JK}^{-1}\text{mol}^{-1}.$$

a)   Will this reaction occur spontaneously at a temperature of 550K?
     Explain your answer.   [2 marks]

b)   Calculate the minimum temperature at which the reaction becomes feasible.   [2 marks]

The feasibility of revision depends on what's on the telly...

These pages are a bit confusing if you ask me — so make sure you've properly understood them before you move on. The most important bit to learn is the formula for ΔG. If you know that, then you can always work out whether a reaction is feasible even if you can't remember the rules about positive and negative enthalpy and entropy.

Period 3 Elements and Oxides

Period 3's the third row down on the Periodic Table — the one that starts with sodium and ends with argon.

Sodium is More Reactive Than Magnesium

1) **Sodium** and **magnesium** are the first two elements in **Period 3**. Sodium is in **Group 1**, and magnesium is in **Group 2**. When they react, sodium **loses one electron** to form an **Na^+ ion**, while **magnesium loses two electrons** to form **Mg^{2+}**.

2) Sodium is **more reactive** than **magnesium** because it takes **less energy** to lose **one electron** than it does to lose two. So **more energy** (usually **heat**) is needed for magnesium to react. This is shown in their reactions with **water**.

Sodium Reacts Vigorously With Water

1) Sodium reacts **vigorously** with **cold water**, forming a molten ball on the surface, fizzing and producing H_2 gas.

$$2\,Na_{(s)} + 2\,H_2O_{(l)} \rightarrow 2\,NaOH_{(aq)} + H_{2(g)}$$

2) The reaction produces **sodium hydroxide**, so forms a strongly **alkaline** solution (pH 12 – 14).

3) Magnesium reacts **very slowly** with **cold water**. You can't see any reaction, but it forms a **weakly alkaline** solution (pH 9 – 10), which shows that a reaction has occurred. The solution is only weakly alkaline because magnesium hydroxide is **not very soluble** in water, so relatively **few hydroxide ions** are produced.

$$Mg_{(s)} + 2\,H_2O_{(l)} \rightarrow Mg(OH)_{2(aq)} + H_{2(g)}$$

4) **Magnesium** reacts much faster with **steam** (i.e. when there is **more energy**), to form **magnesium oxide**.

Most Period 3 Elements React Readily With Oxygen

Period 3 elements form **oxides** when they react with **oxygen**. They're usually oxidised to their **highest** oxidation states — the same as their **group numbers**. Sulfur is the exception to this — it forms **SO_2**, in which it's only got a **+4** oxidation state (a **high temperature** and a **catalyst** are needed to make **SO_3**).

The equations are all **really similar** — **element + oxygen → oxide**:

P_4 is a common allotrope (form) of phosphorus.

$2Na_{(s)} + \frac{1}{2}O_{2(g)} \rightarrow Na_2O_{(s)}$	sodium oxide	$Si_{(s)} + O_{2(g)} \rightarrow SiO_{2(s)}$ silicon dioxide
$Mg_{(s)} + \frac{1}{2}O_{2(g)} \rightarrow MgO_{(s)}$	magnesium oxide	$P_{4(s)} + 5O_{2(g)} \rightarrow P_4O_{10(s)}$ phosphorus(V) oxide
$2Al_{(s)} + 1\frac{1}{2}O_{2(g)} \rightarrow Al_2O_{3(s)}$	aluminium oxide	$S_{(s)} + O_{2(g)} \rightarrow SO_{2(g)}$ sulfur dioxide

The **more reactive metals** (Na, Mg) and the **non-metals** (P, S) react **readily** in air, while **Al** and **Si** react **slowly**.

Element	Na	Mg	Al	Si	P	S
Formula of oxide	Na_2O	MgO	Al_2O_3	SiO_2	P_4O_{10}	SO_2
Reaction in air	Vigorous	Vigorous	Slow	Slow	Spontaneously combusts	Burns steadily
Flame	Yellow	Brilliant white			Brilliant white	Blue

You can identify some of the oxides using flame tests.

Bonding and Structure Affect Melting Points

1) **Na_2O, MgO** and **Al_2O_3** — the metal oxides — all have **high melting points** because they form **giant ionic lattices**. The **strong forces of attraction** between each ion mean it takes a lot of heat energy to **break the bonds** and melt them.

2) MgO has a **higher melting point** than Na_2O because it forms **2+ ions**, so bonds more strongly than the 1+ ions in Na_2O.

3) Al_2O_3 has a **lower melting point** than you might expect because the 3+ ions distort the oxygen's electron cloud making the bonds **partially covalent**.

4) **SiO_2** has a **higher melting point** than the other non-metal oxides because it has a **giant macromolecular** structure.

5) **P_4O_{10}** and **SO_2** have relatively **low melting points** because they form **simple molecular** structures. The molecules are bound by **weak intermolecular forces** (dipole-dipole and van der Waals), which take little energy to overcome.

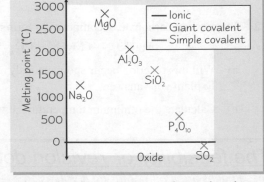

Period 3 Elements and Oxides

Ionic Oxides are Alkaline, Covalent Oxides are Acidic

1) The **ionic oxides** of the **metals** Na and Mg dissolve in water to form **hydroxides**. The solutions are both **alkaline**, but **sodium hydroxide** is more soluble in water, so it forms a **more alkaline** solution than magnesium hydroxide.

$$Na_2O_{(s)} + H_2O_{(l)} \rightarrow 2NaOH_{(aq)} \quad \textbf{pH 12 - 14} \qquad MgO_{(s)} + H_2O_{(l)} \rightarrow Mg(OH)_{2(aq)} \quad \textbf{pH 9 - 10}$$

2) The **simple covalent oxides** of the **non-metals** phosphorus and sulfur form **acidic** solutions. All of the acids are **strong** and so the pH of their solutions is about **0-2**.

$$P_4O_{10(s)} + 6H_2O_{(l)} \rightarrow 4H_3PO_{4(aq)} \qquad \textbf{phosphoric(V) acid}$$
$$SO_{2(g)} + H_2O_{(l)} \rightarrow H_2SO_{3(aq)} \qquad \textbf{sulfurous acid (or sulfuric(IV) acid)}$$
$$SO_{3(l)} + H_2O_{(l)} \rightarrow H_2SO_{4(aq)} \qquad \textbf{sulfuric(VI) acid}$$

An ironic ox-side?

3) The **giant covalent structure** of **silicon dioxide** means that it is **insoluble** in water. However, it will **react with bases** to form salts so it is classed as **acidic**.

4) **Aluminium oxide**, which is partially **ionic** and partially **covalently** bonded, is also **insoluble** in water. But, it will react with **acids and bases** to form salts — i.e. it can act as an acid or a base, so it's classed as **amphoteric**.

Acid + Base → Salt + Water

The equation for **neutralising** an **acid** with a **base** is a classic — **acid + base → salt + water** — and it's no different for reactions of the Period 3 oxides. You may be asked to **write equations** for these reactions, so here are some examples:

1
Basic oxides neutralise acids:
$$Na_2O_{(s)} + 2HCl_{(aq)} \rightarrow 2NaCl_{(aq)} + H_2O_{(l)}$$
$$MgO_{(s)} + H_2SO_{4(aq)} \rightarrow MgSO_{4(aq)} + H_2O_{(l)}$$

2
Acidic oxides neutralise bases:
$$SiO_{2(s)} + 2NaOH_{(aq)} \rightarrow Na_2SiO_{3(aq)} + H_2O_{(l)}$$
$$P_4O_{10(s)} + 12NaOH_{(aq)} \rightarrow 4Na_3PO_{4(aq)} + 6H_2O_{(l)}$$
$$SO_{2(g)} + 2NaOH_{(aq)} \rightarrow Na_2SO_{3(aq)} + H_2O_{(l)}$$
$$SO_{3(g)} + 2NaOH_{(aq)} \rightarrow Na_2SO_{4(aq)} + H_2O_{(l)}$$

3
Amphoteric oxides neutralise acids and bases:
$$Al_2O_{3(s)} + 3H_2SO_{4(aq)} \rightarrow Al_2(SO_4)_{3(aq)} + 3H_2O_{(l)}$$
$$Al_2O_{3(s)} + 2NaOH_{(aq)} + 3H_2O_{(l)} \rightarrow 2NaAl(OH)_{4(aq)}$$

Practice Questions

Q1 Why is Na more reactive than Mg with water?

Q2 What type of oxide is: a) Na_2O,   b) P_4O_{10}.

Q3 Write an equation for the reaction of Na_2O with water.

Q4 Explain why MgO forms a less alkaline solution than Na_2O.

Exam Questions

1 X and Y are oxides of Period 3 elements.
   The Period 3 element in X has an oxidation state of +6 and the oxide forms an acidic solution in water.
   The Period 3 element in Y has an oxidation state of +1 and a high melting point.
   a) Identify oxide X and write an equation for its reaction with water.   [2 marks]
   b) (i) Identify oxide Y and write an equation for its reaction with water.   [2 marks]
       (ii) Explain the reason for its high melting point.   [3 marks]

2 a) Write an equation for the formation of P_4O_{10} from phosphorus and oxygen.   [1 mark]
   b) (i) Write an equation for the reaction between P_4O_{10} and water.   [1 mark]
       (ii) Suggest a value for the pH of the solution obtained from the reaction in part (i). [1 mark]
   c) Write an equation for the reaction between P_4O_{10} and potassium hydroxide.   [1 mark]

These pages have got more trends than a school disco...

Hang on a minute, I hear you cry — what about chlorine and argon? Aren't they in Period 3 too? Well, yes, they are, but you don't need to know about them. Argon's a noble gas, anyway, so it doesn't really react with anything... yawn.

Redox Equations

Redox reactions are all about the movement of electrons.

If Electrons are Transferred, it's a **Redox Reaction**

1) A **loss** of electrons is called **oxidation**. A **gain** of electrons is called **reduction**.
2) Reduction and oxidation happen **simultaneously** — hence the term '**redox**' reaction.
3) An **oxidising agent accepts** electrons and gets reduced.
4) A **reducing agent donates** electrons and gets oxidised.

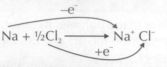

$$Na + \tfrac{1}{2}Cl_2 \xrightarrow{\begin{array}{c}-e^-\\ \\+e^-\end{array}} Na^+ \; Cl^-$$

Na is oxidised
Cl is reduced

I couldn't find a red ox, so you'll have to make do with a multicoloured donkey instead.

Sometimes it's easier to talk about **Oxidation States** ◄ (It's also called oxidation *number*.)

There are lots of rules. Take a deep breath...

1) All atoms are treated as **ions** for this, even if they're covalently bonded.

2) Uncombined **elements** have an oxidation state of **0**.

3) Elements just bonded to **identical atoms**, like O_2 and H_2, also have an oxidation state of **0**.

4) The oxidation state of a simple **monatomic ion**, e.g. Na^+, is the same as its **charge**.

5) In **compounds** or **compound ions**, the **overall oxidation state** is just the ion charge.

SO_4^{2-} — **overall oxidation state = –2**,
oxidation state of **O = –2** (total = –8), ◄
so oxidation state of **S = +6**

Within an ion, the most electronegative element has a negative oxidation state (equal to its ionic charge). Other elements have more positive oxidation states.

6) The sum of the oxidation states for a **neutral compound** is 0.

Fe_2O_3 — **overall oxidation state = 0**, oxidation state of **O = –2**
(total = –6), so oxidation state of **Fe = +3**

7) Combined **oxygen** is nearly always –2, except in peroxides, where it's –1,
(and in the fluorides OF_2, where it's +2, and O_2F_2, where it's +1 (and O_2 where it's 0).

In H_2O, oxidation state of **O = –2**, but in H_2O_2, oxidation state of **H** has to be **+1** (an H atom can only lose one electron), so oxidation state of **O = –1**

8) Combined **hydrogen** is +1, except in metal hydrides where it is –1 (and H_2 where it's 0).

In **HF**, oxidation state of **H = +1**, but in **NaH**, oxidation state of **H = –1**

9) The oxidation state of a **ligand** is equivalent to the charge on the ligand. ◄ See page 166.

So the oxidation state of $CN^- = -1$ and the oxidation state of $NH_3 = 0$.

If you see **Roman numerals** in a chemical name, it's an **oxidation number** — it applies to the atom or group immediately before it. E.g. copper has oxidation number **+2** in **copper(II) sulfate**, and manganese has oxidation number **+7** in a **manganate(VII) ion** (MnO_4^-).

Oxidation States go **Up** or **Down** as Electrons are **Lost** or **Gained**

To work out which atoms are **oxidised** and which are **reduced** in a reaction, you need to look at the **oxidation states**.

1) The oxidation state for an atom will **increase by 1** for each **electron lost**.
2) The oxidation state will **decrease by 1** for each **electron gained**.

Example:

$$V_2O_5 + SO_2 \rightarrow V_2O_4 + SO_3$$

Oxidation state: $(+5)_{\times 2}(-2)_{\times 5}$ $(+4)(-2)_{\times 2}$ $(+4)_{\times 2}(-2)_{\times 4}$ $(+6)(-2)_{\times 3}$

V is reduced from +5 to +4 (it gains 1 electron) and **S is oxidised** from +4 to +6 (it loses two electrons).

Redox Equations

You can separate Redox Reactions into Half-Equations

1) **Ionic half-equations** show oxidation or reduction.
2) An oxidation half-equation can be **combined** with a reduction half-equation to make a **full equation**.

Example: **Zinc metal** displaces **silver ions** from silver nitrate solution to form **zinc nitrate** and a deposit of **silver metal**.

The zinc atoms each lose 2 electrons (oxidation) $Zn_{(s)} \rightarrow Zn^{2+}_{(aq)} + 2e^-$ ← You need to add
The silver ions each gain 1 electron (reduction) $Ag^+_{(aq)} + e^- \rightarrow Ag_{(s)}$ electrons to balance the charges on both sides.

Two silver ions are needed to accept the **two electrons** released by each zinc atom.
So you need to double the silver half-equation before the two half-equations can be combined: $2Ag^+_{(aq)} + 2e^- \rightarrow 2Ag_{(s)}$

Now the number of electrons lost and gained
balance, so the half-equations can be combined: $Zn_{(s)} + 2Ag^+_{(aq)} \rightarrow Zn^{2+}_{(aq)} + 2Ag_{(s)}$ ⎯ Electrons aren't included in the full equation.

You May Need to Add H₂O and H⁺ ions to Balance Half-Equations

Example: Acidified manganate(VII) ions can be reduced to Mn^{2+} by Fe^{2+} ions.
The Fe^{2+} ions are oxidised to Fe^{3+}.

Write the half-equations:

Start with the ion being reduced. $MnO_4^-{}_{(aq)} \rightarrow Mn^{2+}_{(aq)}$
Add four water molecules to balance the oxygen. $MnO_4^-{}_{(aq)} \rightarrow Mn^{2+}_{(aq)} + 4H_2O_{(l)}$
Then add H^+ ions to balance the hydrogen. $MnO_4^-{}_{(aq)} + 8H^+_{(aq)} \rightarrow Mn^{2+}_{(aq)} + 4H_2O_{(l)}$
And finally, add electrons to balance the charge. $MnO_4^-{}_{(aq)} + 8H^+_{(aq)} + 5e^- \rightarrow Mn^{2+}_{(aq)} + 4H_2O_{(l)}$
Do the same thing for the other half-equation. $Fe^{2+}_{(aq)} \rightarrow Fe^{3+}_{(aq)} + e^-$

To balance the electrons you have to multiply the second half-equation by 5: $5Fe^{2+}_{(aq)} \rightarrow 5Fe^{3+}_{(aq)} + 5e^-$

Now you can combine both half-equations: $MnO_4^-{}_{(aq)} + 8H^+_{(aq)} + 5Fe^{2+}_{(aq)} \rightarrow Mn^{2+}_{(aq)} + 4H_2O_{(l)} + 5Fe^{3+}_{(aq)}$
Finish by checking that the charges balance: $(-1) + (+1)_{\times 8} + (+2)_{\times 5} \rightarrow (+2) + 0 + (+3)_{\times 5}$
        $+17 \rightarrow +17$

Practice Questions

Q1 What is an oxidising agent?
Q2 What are you allowed to add when writing half-equations?

Exam Questions

1 What is the oxidation number of the following elements?
a) Ti in $TiCl_4$ b) V in V_2O_5 c) Cr in CrO_4^{2-} d) Cr in $Cr_2O_7^{2-}$ [4 marks]

2 Dichromate ions ($Cr_2O_7^{2-}$) are reduced to Cr^{3+} ions by zinc. Zinc is oxidised during this process.
Write ionic half-equations to show the reduction and oxidation processes.   [2 marks]

3 Acidified manganate(VII) ions will react with aqueous iodide ions to form iodine.
The two half-equations for the changes that occur are:
$MnO_4^-{}_{(aq)} + 8H^+_{(aq)} + 5e^- \rightarrow Mn^{2+}_{(aq)} + 4H_2O_{(l)}$ and $2I^-_{(aq)} \rightarrow I_{2(aq)} + 2e^-$
a) Write the balanced equation to show the reaction taking place.   [2 marks]
b) Use oxidation numbers to explain the redox processes which have occurred.   [4 marks]
c) Suggest why a fairly reactive metal such as zinc will not react with aqueous iodide ions in a similar manner to manganate(VII) ions.   [2 marks]

Oxidise what? You'll have to speak to my agent...

The words oxidation and reduction are tossed about a lot in chemistry — so they're important.
Don't forget, oxidation is really about electrons being lost, **not** oxygen being gained, but you'll
hopefully remember that from AS. Here's that memory aid thingy again just in case...

OIL RIG
- **O**xidation **I**s **L**oss
- **R**eduction **I**s **G**ain
(of electrons)

Electrode Potentials

There's electrons toing and froing in redox reactions. And when electrons move, you get electricity.

Electrochemical Cells Make Electricity

Electrochemical cells can be made from **two different metals** dipped in salt solutions of their **own ions** and connected by a wire (the **external circuit**).

There are always **two** reactions within an electrochemical cell — one's an oxidation and one's a reduction — so it's a **redox process** (see page 156).

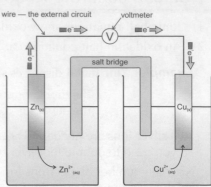

Here's what happens in the **zinc/copper** electrochemical cell on the right:

1) Zinc **loses electrons** more easily than copper. So in the half-cell on the left, zinc (from the zinc electrode) is **OXIDISED** to form $Zn^{2+}_{(aq)}$ ions. This releases electrons into the external circuit.

2) In the other half-cell, the **same number of electrons** are taken from the external circuit, **REDUCING** the Cu^{2+} ions to copper atoms.

*The solutions are connected by a **salt bridge** made from filter paper soaked in $KNO_{3(aq)}$. This allows ions to flow through and balance out the charges.*

So **electrons** flow through the wire from the most reactive metal to the least.

A voltmeter in the external circuit shows the **voltage** between the two half-cells. This is the **cell potential** or **e.m.f.**, E_{cell}.

The boys tested the strength of the bridge, whilst the girls just stood and watched.

You can also have half-cells involving **solutions of two aqueous ions of the same element**, such as $Fe^{2+}_{(aq)}/Fe^{3+}_{(aq)}$.

The conversion from Fe^{2+} to Fe^{3+}, or vice versa, happens on the surface of the **electrode**.

The electrode is made of **platinum** because it is an **inert metal**.

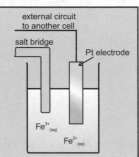

The Reactions at Each Electrode are Reversible

1) The **reactions** that occur at each electrode in the **zinc/copper cell** above are:

$$Zn^{2+}_{(aq)} + 2e^- \rightleftharpoons Zn_{(s)}$$
$$Cu^{2+}_{(aq)} + 2e^- \rightleftharpoons Cu_{(s)}$$

2) The **reversible arrows** show that both reactions can go in **either direction**. **Which direction** each reaction goes in depends on **how easily** each metal **loses electrons** (i.e. how easily it's **oxidised**).

3) How easily a metal is oxidised is measured using **electrode potentials**. A metal that's **easily oxidised** has a very **negative electrode potential**, while one that's harder to oxidise has a less negative or **positive electrode potential**.

Half-cell	Electrode potential (V)
$Zn^{2+}_{(aq)}/Zn_{(s)}$	−0.76
$Cu^{2+}_{(aq)}/Cu_{(s)}$	+0.34

4) The table on the left shows the electrode potentials for the copper and zinc half-cells. The **zinc half-cell** has a **more negative** electrode potential, so **zinc is oxidised** (the reaction goes **backwards**), while **copper is reduced** (the reaction goes **forwards**).

There's a Convention for Drawing Electrochemical Cells

It's a bit of a faff drawing pictures of electrochemical cells. There's a **shorthand** way of representing them though — this is the **Zn/Cu cell**:

There are a couple of important **conventions** when drawing cells:

1) The **half-cell** with the **more negative** potential goes on the **left**.

2) The **oxidised forms** go in the **centre** of the cell diagram.

If you follow the conventions, you can **calculate** the **cell potential** by doing the calculation:

$$E^{\circ}_{cell} = \left(E^{\circ}_{right\ hand\ side} - E^{\circ}_{left\ hand\ side}\right)$$

The symbol for electrode potential is E°.

$$Zn_{(s)} \mid Zn^{2+}_{(aq)} \mid\mid Cu^{2+}_{(aq)} \mid Cu_{(s)}$$

Changes go in this direction

reduced form	oxidised form	oxidised form	reduced form

The cell potential will always be a **positive voltage**, because the more negative E° value is being subtracted from the more positive E° value. For example, the cell potential for the Zn/Cu cell = +0.34 − (−0.76) = **+1.10 V**

Electrode Potentials

Conditions Affect the Value of the Electrode Potential

Half-cell reactions are **reversible**. So just like any other reversible reaction, the **equilibrium position** is affected by changes in **temperature**, **pressure** and **concentration**. Changing the equilibrium position changes the **cell potential**. To get around this, **standard conditions** are used to measure electrode potentials — using these conditions means you always get the **same value** for the electrode potential and you can **compare values** for different cells.

Electrode Potentials are Measured Against Standard Hydrogen Electrodes

You measure the electrode potential of a half-cell against a **standard hydrogen electrode**.

> The **standard electrode potential** of a half-cell is the **voltage measured** under **standard conditions** when the **half-cell** is connected to a **standard hydrogen electrode**.

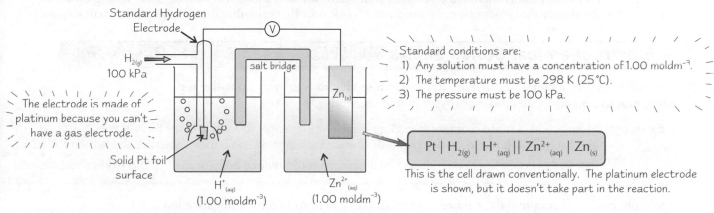

Standard Hydrogen Electrode

$H_{2(g)}$
100 kPa

salt bridge

$Zn_{(s)}$

The electrode is made of platinum because you can't have a gas electrode.

Solid Pt foil surface

$H^+_{(aq)}$
(1.00 moldm^{-3})

$Zn^{2+}_{(aq)}$
(1.00 moldm^{-3})

Standard conditions are:
1) Any solution must have a concentration of 1.00 moldm^{-3}.
2) The temperature must be 298 K (25 °C).
3) The pressure must be 100 kPa.

$$Pt \mid H_{2(g)} \mid H^+_{(aq)} \parallel Zn^{2+}_{(aq)} \mid Zn_{(s)}$$

This is the cell drawn conventionally. The platinum electrode is shown, but it doesn't take part in the reaction.

1) The **standard hydrogen electrode** is always shown on the **left** — it doesn't matter whether or not the other half-cell has a more positive value. The standard hydrogen electrode half-cell has an electrode potential of **0.00 V**.

2) The whole cell potential = $E^\circ_{\text{right-hand side}} - E^\circ_{\text{left-hand side}}$.

$E^\circ_{\text{left hand side}} = 0.00$ V, so the **voltage reading** will be equal to $E^\circ_{\text{right-hand side}}$.
This reading could be **positive** or **negative**, depending which way the **electrons flow**.

Practice Questions

Q1 Draw the cell used for determining the value of E° for the Fe^{3+}/Fe^{2+} system using the conventional representation.

Q2 $Fe^{3+} + e^- \rightleftharpoons Fe^{2+}$, $E^\circ = +0.77$ V $Mn^{3+} + e^- \rightleftharpoons Mn^{2+}$, $E^\circ = +1.48$ V
Show that the e.m.f. of an iron/manganese cell under standard conditions is +0.71 V.

Q3 List the three standard conditions used when measuring standard electrode potentials.

Exam Question

1   An electrochemical cell containing a zinc half-cell and a silver half-cell was set up using a potassium nitrate salt bridge. The cell potential at 25 °C was measured to be 1.40 V.

$$Zn^{2+}_{(aq)} + 2e^- \rightleftharpoons Zn_{(s)} \quad\quad E^\circ = -0.76 \text{ V}$$
$$Ag^+_{(aq)} + e^- \rightleftharpoons Ag_{(s)} \quad\quad E^\circ = +0.80 \text{ V}$$

a) Draw this cell using the conventional representation.   [2 marks]

b) Use the standard electrode potentials given to calculate the standard cell potential for a zinc-silver cell.   [1 mark]

c) Suggest two possible reasons why the actual cell potential was different from the value calculated in part (b).   [2 marks]

d) Write an equation for the overall cell reaction.   [1 mark]

e) Which half-cell released the electrons into the circuit? Explain how you know this.   [1 mark]

Like greased lightning — IT'S ELECTRIFYING...

You've just got to think long and hard about this stuff. The metal on the left-hand electrode disappears off into the solution, leaving its electrons behind. This makes the left-hand electrode the negative one. So the right-hand electrode's got to be the positive one. It makes sense if you think about it. This electrode gives up electrons to turn the positive ions into atoms.

Electrochemical Series

Elements have different standard electrode potentials. So what do chemists do — they write a list of them all in order.

Cells Provide Evidence that Electrons are Transferred

1) **Redox reactions** involve the **transfer of electrons** from one substance to another — or so the **theory** goes. But you might wonder **how** anyone knows this — you **can't see** the electrons move.

2) What you can do is use the **theory** to make a **prediction**, and then **test** the prediction with an **experiment**. For example, you could predict that a **current** will flow between the electrodes of an **electrochemical cell** if an **oxidation** reaction happens at one electrode and a **reduction** at the other electrode.

3) And **a current does flow** — you can even stick an ammeter in the circuit and measure it. So, the electrochemical cell **provides evidence** that **electrons are transferred** in redox reactions.

4) Don't forget that an experiment **doesn't prove** a theory — someone, someday might do an experiment that shows electrons aren't transferred in chemical reactions. That seems **unlikely** at the moment because all the **evidence** we have **fits the theory** that electrons are transferred. So for now that's the theory that chemists are sticking to.

The Electrochemical Series Shows You What's Reactive and What's Not

1) The **more reactive** a **metal** is, the **more** it wants to **lose electrons** to form a **positive ion**. **More reactive metals** have **more negative** standard electrode potentials.

> **Example:** Magnesium is **more reactive** than zinc — so it's more eager to form 2+ ions than zinc is. The list of standard electrode potentials shows that Mg^{2+}/Mg has a **more negative** value than Zn^{2+}/Zn.
>
> In terms of oxidation and reduction, magnesium would **reduce** Zn^{2+} (or Zn^{2+} would **oxidise** magnesium).

2) The more reactive a **non-metal** the **more** it wants to **gain electrons** to form a **negative ion**. **More reactive non-metals** have **more positive** standard electrode potentials.

> **Example:** Chlorine is **more reactive** than bromine — so it's more eager to form a negative ion than bromine is. The list of standard electrode potentials shows that $Cl_2/2Cl^-$ is **more positive** than $Br_2/2Br^-$.
>
> In terms of oxidation and reduction, chlorine would **oxidise** Br^- (or Br^- would **reduce** chlorine).

3) Here's an **electrochemical series** showing some standard electrode potentials:

Half-reaction	E°/V
$Mg^{2+}_{(aq)} + 2e^- \rightleftharpoons Mg_{(s)}$	−2.38
$Al^{3+}_{(aq)} + 3e^- \rightleftharpoons Al_{(s)}$	−1.66
$Zn^{2+}_{(aq)} + 2e^- \rightleftharpoons Zn_{(s)}$	−0.76
$Ni^{2+}_{(aq)} + 2e^- \rightleftharpoons Ni_{(s)}$	−0.25
$2H^+_{(aq)} + 2e^- \rightleftharpoons H_{2(g)}$	0.00
$Sn^{4+}_{(aq)} + 2e^- \rightleftharpoons Sn^{2+}_{(aq)}$	+0.15
$Cu^{2+}_{(aq)} + 2e^- \rightleftharpoons Cu_{(s)}$	+0.34
$Fe^{3+}_{(aq)} + e^- \rightleftharpoons Fe^{2+}_{(aq)}$	+0.77
$Ag^+_{(aq)} + e^- \rightleftharpoons Ag_{(s)}$	+0.80
$Br_{2(aq)} + 2e^- \rightleftharpoons 2Br^-_{(aq)}$	+1.07
$Cr_2O_7^{2-}_{(aq)} + 14H^+_{(aq)} + 6e^- \rightleftharpoons 2Cr^{3+}_{(aq)} + 7H_2O_{(l)}$	+1.33
$Cl_{2(aq)} + 2e^- \rightleftharpoons 2Cl^-_{(aq)}$	+1.36
$MnO_4^-_{(aq)} + 8H^+_{(aq)} + 5e^- \rightleftharpoons Mn^{2+}_{(aq)} + 4H_2O_{(l)}$	+1.52

Chestnut wondered if his load was hindering his pulling potential.

More positive electrode potentials mean that:
1. The left-hand substances are more easily reduced (they gain electrons more easily).
2. The right-hand substances are more stable.

More negative electrode potentials mean that:
1. The right-hand substances are more easily oxidised (they lose electrons more easily).
2. The left-hand substances are more stable.

Electrochemical Series

The **Anticlockwise Rule** Predicts Whether a Reaction Will Happen

You can use the **anticlockwise rule** to **predict** whether a reaction will occur and to show which **direction** it will go in.

For example, will zinc react with aqueous copper ions?

First you write the two half-equations down, putting the one with the **more negative** standard electrode potential on **top**. In each equation, the **oxidised** state of the substance goes on the **left** and the **reduced** state on the **right**.

Then you draw on some **anticlockwise arrows** — these give you the **direction** of each half-reaction.

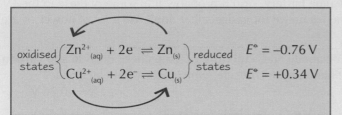

$$\text{oxidised states} \begin{cases} Zn^{2+}_{(aq)} + 2e \rightleftharpoons Zn_{(s)} \\ Cu^{2+}_{(aq)} + 2e^- \rightleftharpoons Cu_{(s)} \end{cases} \text{reduced states} \quad \begin{array}{l} E^\circ = -0.76\,V \\ E^\circ = +0.34\,V \end{array}$$

The **half-equations** are:
$$Zn_{(s)} \rightleftharpoons Zn^{2+}_{(aq)} + 2e^-$$
$$Cu^{2+}_{(aq)} + 2e^- \rightleftharpoons Cu_{(s)}$$

Which combine to give: $\quad Zn_{(s)} + Cu^{2+}_{(aq)} \rightleftharpoons Zn^{2+}_{(aq)} + Cu_{(s)}$

So zinc **does** react with aqueous copper ions.

To find the **cell potential** you do $E^\circ_{bottom} - E^\circ_{top}$, so the cell potential for this reaction is $+0.34 - (-0.76) = +1.10\,V$.

The anticlockwise rule shows that **no reaction** happens if your solution contains Zn^{2+} ions and Cu metal — you can't draw anticlockwise arrows from these substances. You can use the rule to work out whether any two substances will react — just remember to put the half-cell with the **more negative** standard electrode potential on the **top**.

Practice Questions

Q1 Cu is less reactive than Pb.
Which half-reaction has a more negative standard electrode potential, $Pb^{2+} + 2e^- \rightleftharpoons Pb$ or $Cu^{2+} + 2e^- \rightleftharpoons Cu$?

Q2 Use electrode potentials to show that magnesium will reduce Zn^{2+}.

Q3 What is the anticlockwise rule used for? Outline how you use it.

Q4 Use the table on the opposite page to predict whether or not Zn^{2+} ions can oxidise Fe^{2+} ions to Fe^{3+} ions.

Exam Questions

1 Use E° values quoted on the opposite page to determine the outcome of mixing the following solutions.
If there is a reaction, determine the E° value and write the equation. If there isn't a reaction, state this and explain why.
a) Zinc metal and Ni^{2+} ions [2 marks]
b) Acidified MnO_4^- ions and Sn^{2+} ions [2 marks]
c) $Br_{2(aq)}$ and acidified $Cr_2O_7^{2-}$ ions [2 marks]
d) Silver ions and Fe^{2+} ions [2 marks]

2 Use the following data to answer the questions below.
$$O_{2(g)} + 2H_2O_{(l)} + 4e^- \rightleftharpoons 4OH^-_{(aq)} \qquad E^\circ = +0.40\,V$$
$$Fe^{2+}_{(aq)} + 2e^- \rightleftharpoons Fe_{(s)} \qquad E^\circ = -0.44\,V$$

a) (i) Draw a cell diagram for the reaction using the conventional representation. [1 mark]

(ii) Calculate the e.m.f. for the cell. [1 mark]

b) Use electrode potentials to explain why iron is oxidised in the presence of oxygen and water. [1 mark]

The forward reaction that happens is the one with the most positive $E^\ominus$ value...

To see if a reaction will happen, you basically find the two half-equations in the electrochemical series and check whether you can draw anticlockwise arrows on them to get from your reactants to your products. If you can — great. The reaction will have a positive electrode potential, so it should happen. If you can't — well, it ain't gonna work.

Electrochemical Cells

It turns out that electrochemical cells aren't just found in the lab — they're all over your house too... read on.

Electrochemical Cells Are Used as Batteries

The **batteries** that we use to power everything from watches to digital cameras and mobile phones are all types of **electrochemical cell**. Some types of cell are **rechargeable** while others can only be used until they **run out**.

Non-Rechargeable Cells use Irreversible Reactions

A common type of non-rechargeable cell is a **dry cell alkaline battery**. You'll probably find some of these in the TV remote control, torch or smoke alarms in your house. They're useful for gadgets that **don't** use a lot of **power** or are only used for **short periods** of time.

For example, **zinc-carbon dry cell batteries** have a **zinc anode** and a **mixture of manganese dioxide and carbon** for a **cathode**. In between the electrodes is a paste of **ammonium chloride**, which acts as an **electrolyte**.

The **half-equations** are:

$Zn_{(s)} \rightarrow Zn^{2+}_{(aq)} + 2e^-$ $E^\circ = -0.76\,V$

$2MnO_{2(s)} + 2NH_4^+_{(aq)} + 2e^- \rightarrow Mn_2O_{3(s)} + 2NH_{3(aq)} + H_2O_{(l)}$ $E^\circ = +0.75\,V$

So the **cell** is drawn as: $Zn_{(s)} \mid Zn^{2+}_{(aq)} \parallel MnO_{2(s)} \mid Mn_2O_{3(s)}$

The **e.m.f.** of this type of cell is: $E^\circ_{cell} = (E^\circ_{right\,hand\,side} - E^\circ_{left\,hand\,side})$

$= +0.75 - (-0.76) = +1.51\,V$

You draw the cell and calculate the e.m.f. in exactly the same way as for any other electrochemical cell.

The half-equations have **non-reversible arrows** because it is **not practical** to reverse them in a battery. They can be made to run backwards under the **right conditions**, but trying to do this in a **battery** can make it **leak** or **explode**. This is because the **zinc anode** forms the **casing** of the battery, so becomes **thinner** as the zinc is **oxidised**.

Another reason why these batteries **cannot be recharged** is that the ammonium ions produce **hydrogen gas**, which **escapes** from the battery. Without the hydrogen, the ammonium ions **cannot be reformed** by reversing the reactions.

Rechargeable Cells use Reversible Reactions

Rechargeable batteries are found in loads of devices, such as **mobile phones**, **laptops** and **cars**.

For example, **lead-acid cells** are used in car batteries. They normally consist of **6 cells** connected in series. Each cell is made up of a **lead anode** and a **lead(IV) dioxide cathode** immersed in a **sulfuric acid** electrolyte. Both electrodes end up coated in **lead(II) sulfate**.

The **half-equations** are: $Pb_{(s)} + SO_4^{2-}_{(aq)} \rightleftharpoons PbSO_{4(s)} + 2e^-$ $E^\circ = -0.36\,V$

$PbO_{2(s)} + SO_4^{2-}_{(aq)} + 4H^+_{(aq)} + 2e^- \rightleftharpoons PbSO_{4(s)} + 2H_2O_{(l)}$ $E^\circ = +1.69\,V$

So the cell is drawn as: $Pb_{(s)} \mid PbSO_{4(s)} \parallel PbO_{2(s)} \mid PbSO_{4(s)}$

The **e.m.f.** of this type of cell is: $E^\circ_{cell} = (E^\circ_{right\,hand\,side} - E^\circ_{left\,hand\,side}) = +1.69 - (-0.36) = \mathbf{+2.05\,V}$

Two other types of rechargeable battery are **NiCad (nickel-cadmium)** and **L ion (lithium ion)**. To recharge these batteries, a **current** is supplied to force **electrons** to flow in the **opposite direction** around the circuit and **reverse the reactions**. This is possible because **none of the substances** in a rechargeable battery **escape** or are **used up**.

Non-Rechargeable Batteries Have Advantages and Disadvantages

1) **Cost**: non-rechargeable batteries are **cheaper** than rechargeables to buy. However, non-rechargeable batteries have to be **replaced** every time they run out, so **rechargeables are cheaper in the long run**.

2) **Lifetime**: a non-rechargeable battery will **work for longer** than a rechargeable battery. But once a rechargeable battery has run out, you can **recharge it** and use it again whereas non-rechargeables have to be **disposed** of.

3) **Power**: non-rechargeable batteries **can't supply as much power** as rechargeables, so are **no use** in devices that use a **lot of power** — like a mobile phone or a laptop.

4) **Use of resources and waste**: more non-rechargeable batteries are produced because they can only be used once, which **uses more resources** and means they create **more waste** than rechargeables. Both types of battery can be **recycled** and the metals in them recovered to use again, but often we just chuck them in the bin and they end up in **landfill**.

5) **Toxicity**: non-rechargeable batteries are **less likely** to contain the **toxic metals lead** and **cadmium** (although they may contain **mercury**), so they are less hazardous in landfill if the contents **leak** out and **pollute water sources**.

Electrochemical Cells

Fuel Cells can Generate Electricity From Hydrogen and Oxygen

In most cells the **chemicals** that generate the electricity are contained in the **electrodes** and the **electrolyte** that form the cell. In a fuel cell the chemicals are **stored separately** outside the cell and fed in when electricity is required. One example of this is the **hydrogen-oxygen fuel cell**, which can be used to **power electric vehicles**.

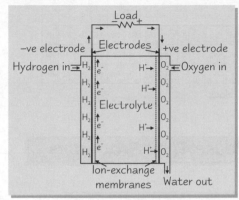

Hydrogen and **oxygen gases** are fed into two separate platinum-containing electrodes. The electrodes are separated by an **ion-exchange membrane** that **allows protons** (H^+ ions) to pass through it, but **stops electrons** going through it.

Hydrogen is fed to the **negative electrode**.
The **reaction** that occurs is:

$$2 H_2 \rightarrow 4 H^+ + 4 e^-$$

The **electrons** flow from the **negative electrode** through an **external circuit** to the **positive electrode**. The H^+ ions pass through the **ion-exchange membrane** towards the positive electrode.

Oxygen is fed to the positive electrode.
The **reaction** here is:

$$O_2 + 4 H^+ + 4 e^- \rightarrow 2 H_2O$$

The **overall effect** is that H_2 and O_2 react to make **water**:

$$2 H_2 + O_2 \rightarrow 2 H_2O$$

Fuel Cells Don't Need Recharging

The **major advantage** of fuel cells over batteries is that they **don't need electrical recharging**. As long as **hydrogen** and **oxygen** are supplied, the cell will continue to **produce electricity**. Another benefit is that the only **waste product** is **water**, so there are **no nasty toxic chemicals** to dispose of and **no CO_2 emissions** from the cell itself.

The **downside** is that you **need energy** to produce a supply of **hydrogen** and **oxygen**. They can be produced from the **electrolysis of water** — i.e. by **reusing the waste product** from the fuel cell, but this requires **electricity** — and this **electricity** is normally generated by **burning fossil fuels**. So the whole process isn't usually carbon neutral. **Hydrogen** is also **highly flammable** so it needs to be handled carefully when it is **stored** or **transported**.

Practice Questions

Q1 Explain the difference between non-rechargeable and rechargeable batteries in terms of the chemical reactions.

Q2 Give one advantage of non-rechargeable batteries.

Q3 Describe a possible environmental problem caused by rechargeable batteries.

Q4 What metal is used in the electrodes in a hydrogen-oxygen fuel cell?

Exam Questions

1 Simple cells can be created from the following half-cells.

Half-cell (1)	$Cu^{2+}_{(aq)} + 2 e^- \rightleftharpoons Cu_{(s)}$	$E^\circ = +0.34$ V
Half-cell (2)	$Mg^{2+}_{(aq)} + 2 e^- \rightleftharpoons Mg_{(s)}$	$E^\circ = -2.38$ V
Half-cell (3)	$Zn^{2+}_{(aq)} + 2 e^- \rightleftharpoons Zn_{(s)}$	$E^\circ = -0.76$ V

Calculate the E° cell value for cells formed between:

a) half-cells (1) and (2) [1 mark]

b) half-cells (1) and (3) [1 mark]

2 This question is about the hydrogen-oxygen fuel cell.

a) (i) Write a half-equation for the reaction of oxygen gas at the positive electrode. [1 mark]

(ii) Where do the electrons come from in this reaction? [1 mark]

b) State one advantage and one disadvantage of hydrogen fuel cells. [2 marks]

Why is chemistry revision like a good battery? It just keeps on going...

You've got to love batteries — they sit there in their shiny metal cases all ready to release the energy stored inside them just when you need it most. So, have some respect for batteries and don't just throw them in the bin when they go flat, recharge them if you can, and if not recycle them. Oh, and you'd probably best learn those half-equations too.

Transition Metals — The Basics

This section's all about transition metals — and there's a lot of it. It's obviously important stuff in the A2 chemistry world.

Transition Elements are Found in the d-Block

The **d-block** is the block of elements in the middle of the periodic table. Most of the elements in the d-block are **transition elements** (or transition metals).

You mainly need to know about the ones in the first row of the d-block. These are the elements from **titanium to copper**.

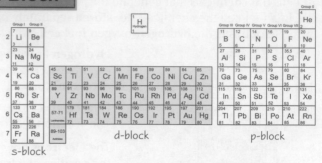

s-block   d-block   p-block

The Electronic Configurations of Transition Metals Cause Their Properties

Here's the definition of a transition metal:

> A **transition metal** is a metal that can form **one or more stable ions** with a **partially filled d-subshell**.

A d-orbital can fit **10** electrons in. So transition metals must form **at least one ion** that has **between 1 and 9 electrons** in the d-orbital. All the Period 4 d-block elements are transition metals apart from **scandium** and **zinc**. Here are their electronic configurations:

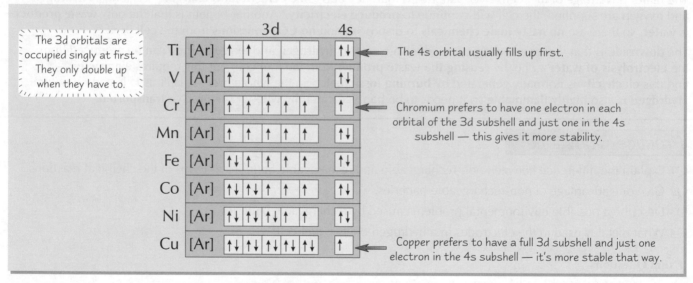

The 3d orbitals are occupied singly at first. They only double up when they have to.

The 4s orbital usually fills up first.

Chromium prefers to have one electron in each orbital of the 3d subshell and just one in the 4s subshell — this gives it more stability.

Copper prefers to have a full 3d subshell and just one electron in the 4s subshell — it's more stable that way.

It's the incomplete d-subshell that causes the **special chemical properties** of transition metals on the next page.

Sc and Zn Aren't Transition Metals

Scandium only forms one ion, Sc^{3+}, which has an **empty d-subshell**. Scandium has the electronic configuration **[Ar]$3d^1 4s^2$**, so when it loses three electrons to form Sc^{3+}, it ends up with the electronic configuration **[Ar]**.

Zinc only forms one ion, Zn^{2+}, which has a **full d-subshell**. Zinc has the electronic configuration **[Ar]$3d^{10} 4s^2$**. When it forms Zn^{2+} it loses 2 electrons, both from the 4s subshell. This means it keeps its full 3d subshell.

When Ions are Formed, the s Electrons are Removed First

Transition metal atoms form **positive** ions. When this happens, the **s electrons** are removed **first**, **then** the d electrons.

For example, iron forms Fe^{2+} ions and Fe^{3+} ions.

When it forms 2+ ions, it loses **both its 4s electrons**. $Fe = [Ar]3d^6 4s^2 \rightarrow Fe^{2+} = [Ar]3d^6$

Only once the 4s electrons are removed can a **3d electron** be removed. E.g. $Fe^{2+} = [Ar]3d^6 \rightarrow Fe^{3+} = [Ar]3d^5$

Transition Metals — The Basics

The Transition Metals All Have Similar Physical Properties

The transition elements don't gradually change across the periodic table like you might expect.
They're all typical metals and have **similar physical properties**:

> 1) They all have a **high density**.
> 2) They all have **high melting** and **high boiling points**.
> 3) Their **ionic radii** are more or less the same.

Transition Metals Have Special Chemical Properties

1) They can form **complex ions** — see pages 166-167. E.g. iron forms a **complex ion with water** — $[Fe(H_2O)_6]^{2+}$.
2) They form **coloured ions** — see pages 168-169. E.g. Fe^{2+} ions are **pale green** and Fe^{3+} ions are **yellow**.
3) They're **good catalysts** — see pages 174-175. E.g. iron is the catalyst used in the **Haber process**.
4) They can exist in **variable oxidation states** — see pages 170-171. E.g. iron can exist in the **+2** oxidation state as Fe^{2+} ions and in the **+3** oxidation state as Fe^{3+} ions.

Some common **coloured** ions and **oxidation states** are shown below. The colours refer to the **aqueous ions**.

oxidation state	+7	+6	+5	+4	+3	+2
			VO_2^+ (yellow)	VO^{2+} (blue)	V^{3+} (green)	V^{2+} (violet)
		$Cr_2O_7^{2-}$ (orange)			Cr^{3+} (green/violet)	
	MnO_4^- (purple)			*See page 170.*		Mn^{2+} (pale pink)
					Fe^{3+} (yellow)	Fe^{2+} (pale green)
						Co^{2+} (pink)
						Ni^{2+} (green)
						Cu^{2+} (blue)

These elements show **variable** oxidation states because the **energy levels** of the 4s and the 3d subshells are **very close** to one another. So different numbers of electrons can be gained or lost using fairly **similar** amounts of energy.

Practice Questions

Q1 Give the electronic arrangement of: (a) a vanadium atom, (b) a V^{2+} ion.

Q2 What is the definition of a transition metal?

Q3 State four chemical properties which are characteristic of transition elements.

Exam Questions

1 When solid copper(I) sulfate is added to water, a blue solution forms with a red-brown precipitate of copper metal.
   a) Give the electron configuration of copper(I) ions. [1 mark]
   b) Does the formation of copper(I) ions show copper acting as a transition metal? Explain your answer. [2 marks]
   c) Identify the blue solution. [1 mark]

2 V^{3+} can be oxidised to VO_2^+ by MnO_4^- in acid solution. The MnO_4^- is reduced to Mn^{2+}.
   a) What is the oxidation state of vanadium in VO_2^+ ? [1 mark]
   b) Write balanced half-equations for:
      (i) the reduction of MnO_4^- to Mn^{2+} (ii) the oxidation of V^{3+} to VO_2^+ [4 marks]
   c) Combine the two half-equations to make a balanced equation for the reaction. [2 marks]

s electrons — like rats leaving a sinking ship...

Definitely have a quick read of the electronic configuration stuff in your AS notes if it's been pushed to a little corner of your mind labelled, "Well, I won't be needing that again in a hurry". It should come flooding back pretty quickly. This page is just an overview of transition metal properties. Don't worry — they're all looked at in lots more detail in the coming pages...

Complex Ions

Transition metals are always forming complex ions. These aren't as complicated as they sound, though. Honest.

Complex Ions are Metal Ions Surrounded by Ligands

> A **complex** is a **metal ion** surrounded by **coordinately bonded ligands**.

1) A **coordinate bond** (or dative covalent bond) is a covalent bond in which **both electrons** in the shared pair come from the **same atom**. In a complex, they come from the **ligands**.

2) So, a **ligand** is an atom, ion or molecule that **donates a pair of electrons** to a central metal ion.

3) The **coordination number** is the **number** of **coordinate bonds** that are formed with the central metal ion.

4) The usual coordination numbers are **6** and **4**. If the ligands are **small**, like H_2O or NH_3, **6** can fit around the central metal ion. But if the ligands are **larger**, like Cl^-, **only 4** can fit around the central metal ion.

6 COORDINATE BONDS MEAN AN OCTAHEDRAL SHAPE

Here are a few examples.

The ligands don't have to be all the same.

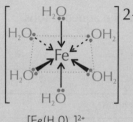

$[Fe(H_2O)_6]^{2+}_{(aq)}$

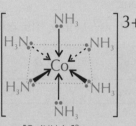

$[Co(NH_3)_6]^{3+}_{(aq)}$

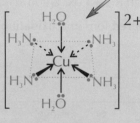

$[Cu(NH_3)_4(H_2O)_2]^{2+}_{(aq)}$

The different types of **bond arrow** show that the complex is **3-D**. The **wedge-shaped arrows** represent bonds coming **towards you** and the **dashed arrows** represent bonds **sticking out behind** the molecule.

4 COORDINATE BONDS USUALLY MEAN A TETRAHEDRAL SHAPE...

E.g. $[CuCl_4]^{2-}$, which is yellow, and $[CoCl_4]^{2-}$, which is blue.

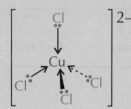

...BUT NOT ALWAYS

In a **few** complexes, e.g. **cisplatin** (shown on the right), **4 coordinate bonds** form a **square planar** shape.

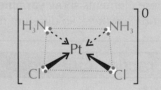

SOME SILVER COMPLEXES HAVE 2 COORDINATE BONDS AND FORM A LINEAR SHAPE

$[Ag(NH_3)_2]^+$ forms a **linear shape**, as shown.

$$\left[H_3N\!:\!\longrightarrow\!Ag\!\longleftarrow\!:\!NH_3 \right]^+$$

See page 176 for the uses of $[Ag(NH_3)_2]^+$ and cisplatin.

Complex Ions Have an Overall Charge or Total Oxidation State

The **overall charge** on the complex ion is its **total oxidation state**. It's put **outside** the **square** brackets. For example:

$[Cu(H_2O)_6]^{2+}_{(aq)}$ ← Overall charge is 2+.

You have to know how to work out the **oxidation state of the metal**:

> **The oxidation state of the metal ion = the total oxidation state − the sum of the oxidation states of the ligands**

E.g. $[CoCl_4]^{2-}_{(aq)}$ The total oxidation state is **−2** and each Cl^- ligand has an oxidation state of **−1**. So in this complex, cobalt's oxidation state = $-2 - (4 \times -1) = +2$.

Complex Ions

A Ligand Must Have at Least One Lone Pair of Electrons

A ligand must have **at least one lone pair of electrons**, or it won't have anything to use to form a **coordinate bond**.

1) Ligands that can only form **one coordinate bond** are called **unidentate** — e.g. $H_2\ddot{O}$, $\ddot{N}H_3$, $\ddot{C}l^-$.

2) Ligands that can form **more than one coordinate bond** are called **multidentate** — e.g. EDTA^{4-} has six lone pairs (it's **hexadentate** to be precise) so it can form **six coordinate bonds** with a metal ion.

3) **Bidentate** ligands are multidentate ligands that can form **two coordinate bonds** — e.g. ethane-1,2-diamine, $\ddot{N}H_2CH_2CH_2\ddot{N}H_2$ has two lone pairs, so it can form **two coordinate bonds** with a metal ion.

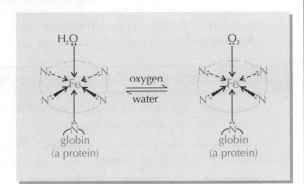

Each **ethane-1,2-diamine** molecule has 2 lone pairs and forms 2 coordinate bonds with the metal ion.

The EDTA^{4-} ion has 6 lone pairs, so it forms **6 coordinate bonds** with the metal ion.

Haem in Haemoglobin Contains a Multidentate Ligand

1) **Haemoglobin** is a protein found in **blood** that helps to **transport oxygen** around the body.

2) **Haemoglobin** contains Fe^{2+} ions, which are **hexa-coordinated** — **six lone pairs** are donated to them to form **six coordinate bonds**.

3) **Four** of the lone pairs come from **nitrogen atoms**, which form a **circle** around the Fe^{2+}. This part of the molecule is called **haem**.

4) The **molecule** that the **four nitrogen atoms** are part of is a **multidentate ligand**.

5) A protein called **globin** and either an **oxygen** or a **water** molecule also bind to the to the Fe^{2+} ion to form an **octahedral structure**. (See page 175 for more on haemoglobin.)

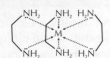

Practice Questions

Q1 What is meant by the term 'complex ion'?

Q2 Describe how a ligand, such as ammonia, bonds to a central metal ion.

Q3 What is meant by the term 'bidentate ligand'? Give an example of one.

Q4 Draw the shape of the complex ion $[Co(NH_3)_6]^{3+}$. Name the shape.

Exam Questions

1   a)   Using $[Ag(NH_3)_2]^+$ as an example, explain what is meant by the following terms:
     (i)   ligand   [2 marks]
     (ii)   coordinate bond   [2 marks]
     (iii)   coordination number   [2 marks]

    b) $S_2O_3^{2-}$ is a unidentate ligand. What shape is $[Ag(S_2O_3)_2]^{3-}$?   [1 mark]

2   When concentrated hydrochloric acid is added to an aqueous solution of $Cu^{2+}_{(aq)}$ a yellow solution is formed.

    a)   State the coordination number and shape of the $Cu^{2+}_{(aq)}$ complex ion in the **initial solution**.   [2 marks]

    b)   State the coordination number, shape and formula of the complex ion responsible for the yellow solution.   [3 marks]

    c)   Explain why the coordination number is different in the yellow solution.   [2 marks]

3   a)   Write an equation for the complete replacement of the water ligands in $[Co(H_2O)_6]^{2+}$ with ethane-1,2-diamine ligands.   [2 marks]

    b)   Draw a diagram to show the bonding of one ethane-1,2-diamine ligand with Co^{2+}.   [2 marks]

Put your hands up — we've got you surrounded...

You never get transition metal ions floating round by themselves in a solution — they'll always be surrounded by other molecules. It's kind of like what'd happen if you put a dish of sweets in a room of eight (or eighteen) year-olds. When you're drawing complex ions, don't forget to include the dashed and wedge-shaped bonds to show that it's 3-D.

Formation of Coloured Ions

Transition metal complex ions have distinctive colours, which is handy when it comes to identifying them.
This page explains why they're so colourful.

Ligands **Split** the 3d Subshell into **Two Energy Levels**

Normally the 3d orbitals of transition element ions **all** have the same energy. But when **ligands** come along and bond to the ions, some of the orbitals are given more energy than others. This splits the 3d orbitals into **two different energy levels**.

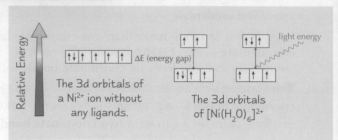

The 3d orbitals of a Ni^{2+} ion without any ligands.

The 3d orbitals of $[Ni(H_2O)_6]^{2+}$

Electrons tend to **occupy the lower orbitals** (the ground state). To jump up to the higher orbitals (excited states) they need **energy** equal to the energy gap, ΔE. They get this energy from **visible light**.

The energy **absorbed** when electrons jump up can be worked out using this formula:

$$\Delta E = h\nu$$

where ν = frequency of light absorbed (hertz/Hz) and h = Planck's constant (6.63×10^{-34} Js)

The amount of energy needed to make electrons jump depends upon the **central metal ion** and its **oxidation state**, the **ligands** and the **coordination number**, as these affect the **size of the energy gap**.

The **Colours** of Compounds are the **Complement** of Those That are **Absorbed**

When **visible light** hits a transition metal ion, some frequencies are **absorbed** when electrons jump up to the higher orbitals. The frequencies absorbed depend on the size of the **energy gap**.

The rest of the frequencies are **reflected**. These **reflected** frequencies combine to make the **complement** of the colour of the absorbed frequencies — this is the **colour** you see.

frequency increases ⟹

For example, **$[Cu(H_2O)_6]^{2+}$** ions absorb **yellow light**. The remaining frequencies **combine** to produce the **complementary colour** — in this case that's blue. So $[Cu(H_2O)_6]^{2+}$ solution appears **blue**.

If there are **no** 3d electrons or the 3d subshell is **full**, then no electrons will jump, so **no energy** will be absorbed. If there's no energy absorbed, the compound will look **white** or **colourless**.

Transition Metal Ions can be Identified by their Colour

It'd be nice if each transition metal formed ions or complexes with just one colour, but sadly it's not that simple. The **colour of a complex** can be altered by any of the factors that can affect the size of the **energy gap**.

1) **Changes in oxidation state** — see page 170-171 for more examples.

Complex:	$[Fe(H_2O)_6]^{2+}_{(aq)}$	$\rightarrow$	$[Fe(H_2O)_6]^{3+}_{(aq)}$		$[V(H_2O)_6]^{2+}_{(aq)}$	$\rightarrow$	$[V(H_2O)_6]^{3+}_{(aq)}$
Oxidation state:	+2	$\rightarrow$	+3	and	+2	$\rightarrow$	+3
Colour:	pale green	$\rightarrow$	yellow		violet	$\rightarrow$	green

2) **Changes in coordination number** — this always involves a change of ligand too.

Complex:	$[Cu(H_2O)_6]^{2+} + 4Cl^-$	$\rightarrow$	$[CuCl_4]^{2-} + 6H_2O$
Coordination number:	6	$\rightarrow$	4
Colour:	blue	$\rightarrow$	yellow

3) **Changes in ligand** — this can cause a colour change even if the oxidation state and coordination number remain the same.

Complex:	$[Co(H_2O)_6]^{2+} + 6NH_3$	$\rightarrow$	$[Co(NH_3)_6]^{2+} + 6H_2O$
Oxidation state:	+2	$\rightarrow$	+2
Colour:	pink	$\rightarrow$	straw coloured

Formation of Coloured Ions

Spectrometry can be used to Find Concentrations of Transition Metal Ions

Spectrometry can be used to determine the **concentration of a solution** by measuring how much **light** it **absorbs**.

1) **White light** is shone through a **filter**, which is chosen to **only** let the **colour of light** through that is **absorbed** by the sample.

2) The light then passes through the sample to a **colorimeter**, which calculates **how much light** was **absorbed** by the sample.

3) The more **concentrated** a coloured solution is, the more light it will absorb. So you can use this measurement to work out the **concentration** of a solution of transition metal ions.

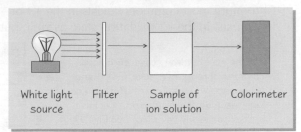

White light source   Filter   Sample of ion solution   Colorimeter

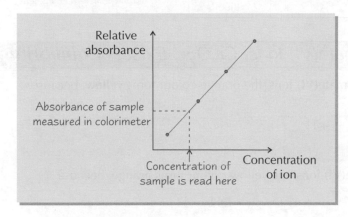

Relative absorbance

Absorbance of sample measured in colorimeter

Concentration of sample is read here

Concentration of ion

Before you can find the unknown concentration of a sample, you have to produce a **calibration graph** — like the lovely one on the left.

This involves measuring the **absorbances** of **known concentrations** of solutions and plotting the results on a graph.

Once you've done this, you can measure the absorbance of your sample and read its **concentration** off the graph.

Stanley's concentration was strong — he'd show this cowboy who's boss.

Practice Questions

Q1 Which subshell is split by the presence of ligands?

Q2 What **two** factors are changing in the following reaction that will cause a change in colour?

$$[Co(H_2O)_6]^{2+} + 4Cl^- \rightarrow [CoCl_4]^{2-} + 6H_2O$$

Q3 What does a colorimeter measure?

Q4 What is the purpose of the filter in spectrometry?

Exam Questions

1 a) Explain why complex transition metal ions such as $[Fe(H_2O)_6]^{2+}$ are coloured. [2 marks]

   b) State three changes to a complex ion that would result in a change in colour. [3 marks]

2 Colorimetry can be used to determine the concentration of a coloured solution. Briefly describe how you would construct a calibration graph, given a coloured solution of known concentration. [3 marks]

3 The frequency of light absorbed by a transition metal complex ion can be determined from the equation $\Delta E = hv$.

   a) State what is meant by ΔE and what change this represents within the complex ion. [2 marks]

   b) Using a noble gas core, [Ar], complete the electron arrangements for the following ions

     (i) Cu^+ [1 mark]

     (ii) Cu^{2+} [1 mark]

   c) Which one of the above ions has coloured compounds? State the feature of its electron arrangement that suggests this. [2 marks]

Blue's not my complementary colour — it clashes with my hair...

Transition metal ions are pretty colours, don't you think? The Romans did — they used iron, copper, cobalt and manganese compounds to add colour to glass objects. I'm not sure that they knew the colours were affected by variable oxidation states, ligands and coordination number, but it's pretty impressive even so.

Variable Oxidation States

One of the reasons why transition metal complexes have such a big range of colours is their variable oxidation states.

Chromium Can Exist in the +2, +3 and +6 Oxidation States

Chromium most commonly exists in the **+3** or **+6** oxidation state.
It can exist in the +2 oxidation state as well, but it's much **less stable**.

In the +6 oxidation state, chromium can form chromate(VI) ions, CrO_4^{2-} **and** dichromate(VI) ions, $Cr_2O_7^{2-}$. Both these ions are good oxidising agents because they can easily be reduced to Cr^{3+}.

Oxidation state	Formula of ion	Colour of ion
+6	$Cr_2O_7^{2-}{}_{(aq)}$	Orange
+6	$CrO_4^{2-}{}_{(aq)}$	Yellow
+3	$Cr^{3+}{}_{(aq)}$	Green (Violet)
+2	$Cr^{2+}{}_{(aq)}$	Blue

When Cr^{3+} ions are surrounded by 6 water ligands they're violet. But the water ligands are often substituted, so this solution usually looks green instead.

Chromate(VI) Ions, CrO_4^{2-}, and Dichromate(VI) Ions, $Cr_2O_7^{2-}$, Exist in Equilibrium

1) When an **alkali** (OH⁻ ions) is added to aqueous **dichromate(VI) ions**, the orange colour turns **yellow**, because aqueous **chromate(VI) ions** form.

$$Cr_2O_7^{2-}{}_{(aq)} + OH^-{}_{(aq)} \rightarrow 2CrO_4^{2-}{}_{(aq)} + H^+{}_{(aq)}$$
orange   **yellow**

2) When an **acid** (H⁺ ions) is added to aqueous **chromate(VI) ions**, the yellow colour turns **orange**, because aqueous **dichromate(VI) ions** form.

$$2CrO_4^{2-}{}_{(aq)} + H^+{}_{(aq)} \rightarrow Cr_2O_7^{2-}{}_{(aq)} + OH^-{}_{(aq)}$$
yellow   **orange**

3) These are **opposite processes** and the two ions exist in **equilibrium**.

$$Cr_2O_7^{2-}{}_{(aq)} + H_2O_{(l)} \rightleftharpoons 2CrO_4^{2-}{}_{(aq)} + 2H^+{}_{(aq)}$$

This isn't a redox process because chromium stays in the +6 oxidation state.

The **position** of equilibrium depends on the **pH** — yep, it's good ol' Le Chatelier's principle again.
If **H⁺ ions** are added, the equilibrium shifts to the **left** so orange $Cr_2O_7^{2-}$ ions are formed.
If **OH⁻ ions** are added, H+ ions are **removed** and the equilibrium shifts to the **right**, forming **yellow CrO_4^{2-} ions**.

Chromium Ions can be Oxidised and Reduced

1) Dichromate(VI) ions can be **reduced** using a good reducing agent, such as **zinc and dilute acid**.

Oxidation states: +6   0   +2   +3
$$Cr_2O_7^{2-}{}_{(aq)} + 14H^+{}_{(aq)} + 3Zn_{(s)} \rightarrow 3Zn^{2+}{}_{(aq)} + 2Cr^{3+}{}_{(aq)} + 7H_2O_{(l)}$$
orange   **green**

2) Zinc will **reduce** Cr^{3+} further to Cr^{2+} —

$$2Cr^{3+}{}_{(aq)} + Zn_{(s)} \rightarrow Zn^{2+}{}_{(aq)} + 2Cr^{2+}{}_{(aq)}$$
green   **blue**

But unless you use an inert atmosphere, you're wasting your time — Cr^{2+} is so **unstable** that it oxidises straight back to Cr^{3+} in air.

3) You can oxidise Cr^{3+} to chromate(VI) ions with **hydrogen peroxide, H_2O_2**, in an **alkaline** solution.

Oxidation states: +3   +6
$$2Cr^{3+}{}_{(aq)} + 10OH^-{}_{(aq)} + 3H_2O_{2(aq)} \rightarrow 2CrO_4^{2-}{}_{(aq)} + 8H_2O_{(l)}$$
green   **yellow**

Here's a summary of all the chromium reactions you need to know:

$Cr_2O_7^{2-}{}_{(aq)}$ $\xrightarrow{H^+{}_{(aq)}}$ $\xleftarrow{OH^-{}_{(aq)}}$ $CrO_4^{2-}{}_{(aq)}$

REDUCTION
$H^+{}_{(aq)}/Zn_{(s)}$

OXIDATION
$OH^-{}_{(aq)}/H_2O_{2(aq)}$

$Cr^{3+}{}_{(aq)}$

REDUCTION
$H^+{}_{(aq)}/Zn_{(s)}$
(inert atmosphere)

OXIDATION
air

$Cr^{2+}{}_{(aq)}$

Variable Oxidation States

Cobalt Can Exist as Co²⁺ and Co³⁺

Cobalt can exist in two oxidation states — **+2** as **Co²⁺**, and **+3** as **Co³⁺**. It much prefers to be in the **+2** state though.

1) Co³⁺ can be made by oxidising Co²⁺$_{(aq)}$ with **hydrogen peroxide** in alkaline conditions.

$$2Co^{2+}_{(aq)} + H_2O_{2(aq)} \rightarrow 2Co^{3+}_{(aq)} + 2OH^-_{(aq)}$$

2) You can also oxidise Co²⁺ with **air in an ammoniacal solution**. You place Co²⁺ ions in an excess of **aqueous ammonia**, which causes **[Co(NH₃)₆]²⁺ ions** to form. If these complex ions are left to stand in **air**, **oxygen** oxidises them to **[Co(NH₃)₆]³⁺**.

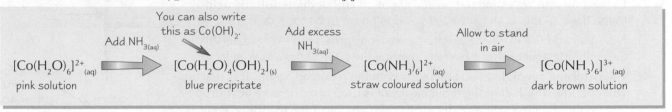

$[Co(H_2O)_6]^{2+}_{(aq)}$ — pink solution — Add NH₃$_{(aq)}$ → (You can also write this as Co(OH)₂.) $[Co(H_2O)_4(OH)_2]_{(s)}$ — blue precipitate — Add excess NH₃$_{(aq)}$ → $[Co(NH_3)_6]^{2+}_{(aq)}$ — straw coloured solution — Allow to stand in air → $[Co(NH_3)_6]^{3+}_{(aq)}$ — dark brown solution

The reaction is done this way because [Co(NH₃)₆]²⁺$_{(aq)}$ ions are far **easier to oxidise** than [Co(H₂O)₆]²⁺$_{(aq)}$ ions.

Practice Questions

Q1 What colour change is seen when acid is added to aqueous chromate(VI) ions? Name the new chromium ion.

Q2 Name the two reagents needed to oxidise chromium(III) to chromium(VI).

Q3 Describe two ways of oxidising Co²⁺$_{(aq)}$ to Co³⁺$_{(aq)}$.

Q4 Show that the oxidation state of cobalt in Co(OH)₂ is +2.

Exam Questions

1   Acidified dichromate ions can act as powerful oxidising agents.

   a)   Write a half-equation for the reduction of dichromate ions to chromium(III) ions in acid solution.   [1 mark]

   b)   Ethanol is oxidised to ethanal by dichromate ions.

      (i) Write a balanced equation for the oxidation of ethanol to ethanal by dichromate ions in acid solution.   [2 marks]

      (ii) Give the colour change seen in this reaction.   [1 mark]

   c)   SO₂ gas is oxidised to SO₄²⁻ ions by acidified dichromate.

      (i) Write a half-equation for the oxidation of SO₂ to SO₄²⁻.   [1 mark]

      (ii) Write an equation for the oxidation of SO₂ by acidified dichromate ions.   [1 mark]

2   The following reaction sequence involves complex ions of a transition element. The ion is initially in its +3 oxidation state.

$$\mathbf{X} \xrightarrow[\text{alkaline solution}]{\text{+ H}_2\text{O}_2 \text{ in}} \mathbf{Y} \xrightarrow{\text{+ acid}} \mathbf{Z}$$

       X green solution   **Y** yellow solution   **Z** solution

   a)   Identify the transition metal ion responsible for the green solution **X**.   [1 mark]

   b)   The half-equation for the hydrogen peroxide reaction is H₂O₂ + 2e⁻ → 2OH⁻.
      Give the oxidation number of oxygen in hydrogen peroxide.   [1 mark]

   c)   Addition of an acid to the yellow solution causes a further colour change that is **not** a redox reaction, forming Z. Write an equation for this reaction and give the colour of the solution formed.   [2 marks]

My girlfriend is usually in the +6 or +8 aggravation states...

Sorry, I'm too tired to tell you anything meaningful right now, so you're on your own I'm afraid... I'm going to have a nap and probably have another dream where I'm being attacked by a wallaby the size of a house... ....

Titrations with Transition Metals

These titrations are redox titrations. They're like acid-base titrations, but different. You don't need an indicator for a start.

Titrations Using **Transition Element Ions** are **Redox** Titrations

Titrations using transition element ions let you find out how much **oxidising agent** is needed to **exactly** react with a quantity of **reducing agent**. If you know the **concentration** of either the oxidising agent or the reducing agent, you can use the titration results to work out the concentration of the other.

1) First you measure out a quantity of **reducing agent**, e.g. aqueous Fe^{2+} ions, using a pipette, and put it in a conical flask.

2) Using a **measuring cylinder**, you add about **20 cm³ of dilute sulfuric acid** to the flask — this is an excess, so you don't have to be too exact.

3) Now you add the **oxidising agent**, e.g. aqueous potassium manganate(VII), to the reducing agent using a **burette**, **swirling** the conical flask as you do so.

4) The **oxidising agent** that you add reacts with the reducing agent. This reaction will continue until **all** of the reducing agent is used up. The **very next drop** you add to the flask will give the mixture the **colour of the oxidising agent**. (You could use a coloured reducing agent and a colourless oxidising agent instead — then you'd be watching for the moment that the colour in the flask disappears.)

5) Stop when the mixture in the flask **just** becomes tainted with the colour of the oxidising agent (the **end point**) and record the volume of the oxidising agent added. This is the **rough titration**.

6) Now you do some **accurate titrations**. You need to do a few until you get **two or more** readings that are **within 0.10 cm³** of each other.

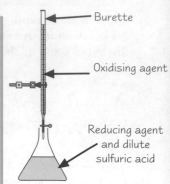

Burette

Oxidising agent

Reducing agent and dilute sulfuric acid

You can also do titrations the other way round — adding the reducing agent to the oxidising agent.

The two main **oxidising agents** used are:

1) **Manganate(VII) ions** (MnO_4^-) in **aqueous potassium manganate(VII)** ($KMnO_4$) — these are **purple**.

2) **Dichromate(VI) ions** ($Cr_2O_7^{2-}$) in **aqueous potassium dichromate(VI)** ($K_2Cr_2O_7$) — these are **orange**.

The **acid** is added to make sure there are plenty of **H⁺ ions** to allow the oxidising agent to be reduced.

Use the **Titration Results** to **Calculate** the **Concentration** of a Reagent...

Example: 27.5 cm³ of 0.0200 moldm⁻³ aqueous potassium manganate(VII) reacted with 25.0 cm³ of acidified iron(II) sulfate solution. Calculate the concentration of Fe^{2+} ions in the solution.

$$MnO_4^-{}_{(aq)} + 8H^+{}_{(aq)} + 5Fe^{2+}{}_{(aq)} \rightarrow Mn^{2+}{}_{(aq)} + 4H_2O_{(l)} + 5Fe^{3+}{}_{(aq)}$$

1) Work out the number of **moles of MnO_4^- ions** added to the flask.

Number of moles MnO_4^- added $= \dfrac{\text{concentration} \times \text{volume}}{1000} = \dfrac{0.0200 \times 27.5}{1000} = 5.50 \times 10^{-4}$ moles

2) Look at the balanced equation to find how many moles of **Fe^{2+}** react with **every mole** of MnO_4^-. Then you can work out the **number of moles of Fe^{2+}** in the flask.

5 moles of Fe^{2+} react with 1 mole of MnO_4^-. So moles of Fe^{2+} = $5.50 \times 10^{-4} \times 5 = 2.75 \times 10^{-3}$ moles.

3) Work out the **number of moles of Fe^{2+}** that would be in 1000 cm³ (1 dm³) of solution — this is the **concentration**.

25.00 cm³ of solution contained 2.75×10^{-3} moles of Fe^{2+}.

1000 cm³ of solution would contain $\dfrac{(2.75 \times 10^{-3}) \times 1000}{25.0} = 0.11$ moles of Fe^{2+}.

So the concentration of Fe^{2+} is **0.11 moldm⁻³**.

Manganate 007, licensed to oxidise.

Titrations with Transition Metals

...Or the *Volume* of a *Reagent*

Example: Aqueous potassium dichromate(VI) with a concentration of 0.008 moldm^{-3} was used to completely oxidise 25.0 cm^3 of 0.06 moldm^{-3} acidified iron(II) sulfate solution. Calculate the volume of potassium dichromate(VI) solution used.

$$Cr_2O_7^{2-}{}_{(aq)} + 14H^+{}_{(aq)} + 6Fe^{2+}{}_{(aq)} \rightarrow 2Cr^{3+}{}_{(aq)} + 7H_2O_{(l)} + 6Fe^{3+}{}_{(aq)}$$

1) Work out the number of **moles of Fe^{2+} ions** in the flask to begin with.

$$\text{Number of moles Fe}^{2+} = \frac{\text{concentration} \times \text{volume}}{1000} = \frac{0.06 \times 25.0}{1000} = 1.5 \times 10^{-3} \text{ moles}$$

2) Look at the balanced equation to find how many moles of **Fe^{2+}** react with **every mole** of $Cr_2O_7^{2-}$. Then you can work out the **number of moles of $Cr_2O_7^{2-}$** needed.

6 moles of Fe^{2+} react with 1 mole of $Cr_2O_7^{2-}$. So moles of $Cr_2O_7^{2-}$ = $1.5 \times 10^{-3} \div 6 = 2.5 \times 10^{-4}$ moles.

3) Rearrange the formula above to find the **volume** of **0.008 moldm^{-3} dichromate solution** that contains **2.5 × 10^{-4} moles of $Cr_2O_7^{2-}$**.

$$\text{Volume of } Cr_2O_7^{2-} = \frac{\text{number of moles} \times 1000}{\text{concentration}} = \frac{(2.5 \times 10^{-4}) \times 1000}{0.008} = 31.25 \text{ cm}^3$$

Practice Questions

Q1 Why is an excess of acid added to potassium manganate(VII) titrations?

Q2 Outline how you would carry out a titration of Fe^{2+} with MnO_4^- in acid solution.

Q3 What colour marks the end-point of a redox titration when potassium manganate(VII) is added from the burette?

Q4 What colour is aqueous potassium dichromate?

Exam Questions

1 0.100 g of a sample of steel containing carbon and iron only was dissolved in sulfuric acid. The resulting solution was titrated using 29.40 cm^3 of 0.0100 moldm^{-3} potassium dichromate(VI). The equation for the reaction is $Cr_2O_7^{2-} + 6Fe^{2+} + 14H^+ \rightarrow 2Cr^{3+} + 6Fe^{3+} + 7H_2O$

   a) Calculate the number of moles of dichromate ions added. [2 marks]

   b) How many moles of Fe^{2+} will have reacted? [1 mark]

   c) Calculate the mass of pure iron in the steel. [2 marks]

   d) Calculate the percentage of iron in the steel. [1 mark]

2 A sample of sodium ethanedioate is dissolved in dilute sulfuric acid. The solution is titrated using 18.30 cm^3 of 0.0200 moldm^{-3} potassium manganate(VII).

   The equation for the reaction is: $2MnO_4^- + 5C_2O_4^{2-} + 16H^+ \rightarrow 2Mn^{2+} + 10CO_2 + 8H_2O$

   Calculate the mass of $Na_2C_2O_4$ in the initial sample. [7 marks]

3 A 0.1 moldm^{-3} solution of iron(II) sulfate has been partly oxidised by the air to an iron(III) compound. Explain how you could use a standard solution of potassium manganate(VII) to find the actual concentration of Fe^{2+} remaining. [4 marks]

And how many moles does it take to change a light bulb....

The example calculations on these two pages might look complicated, but if you break them down into a series of steps, they're really not that bad — especially as you've done titration calculations before. You know, with acids and bases? Back in Unit 4: Section 1? What do you mean you don't remember — go back and have another look.

Uses of Transition Metals

Transition metals aren't just good — they're grrrreat.

Transition Metal Catalysts Work by Changing Oxidation States

Transition metals and their compounds make good catalysts because they can **change oxidation states** by gaining or losing electrons within their **d orbitals**. This means they can **transfer electrons** to **speed up** reactions.

For example, in the **Contact Process**, vanadium(V) oxide is able to **oxidise SO_2 to SO_3** because it can be **reduced** to vanadium(IV) oxide. It's then **oxidised** back to vanadium(V) oxide by oxygen ready to start all over again.

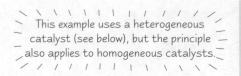 This example uses a heterogeneous catalyst (see below), but the principle also applies to homogeneous catalysts.

Vanadium oxidises SO_2 to SO_3 and is reduced itself.

$$\overset{+5}{V_2O_5} + SO_2 \rightarrow \overset{+4}{V_2O_4} + SO_3$$

vanadium(V) $\rightarrow$ vanadium(IV)

The reduced catalyst is then oxidised by oxygen gas back to its original state.

$$\overset{+4}{V_2O_4} + \tfrac{1}{2}O_2 \rightarrow \overset{+5}{V_2O_5}$$

vanadium(IV) $\rightarrow$ vanadium(V)

Heterogeneous Catalysts are in a Different Phase From the Reactants

A heterogeneous catalyst is one that is in a **different phase** from the reactants — i.e. in a different **physical state**. For example, in the Haber Process (see below) **gases** are passed over a **solid iron catalyst**.

The **reaction** happens on the **surface** of the **heterogeneous catalyst**. So, **increasing** the **surface area** of the catalyst increases the number of molecules that can **react** at the same time, **increasing the rate** of the reaction.

Support mediums are often used to make the **area** of a catalyst **as large as possible**. For example, **catalytic converters** (which 'clean up' emissions from car engines) contain a **ceramic lattice** coated with a thin layer of **rhodium**. The **rhodium** acts as a **catalyst** helping to convert the **really nasty waste gases** to less harmful products.

$$2CO_{(g)} + 2NO_{(g)} \xrightarrow{\;Rh_{(s)}\text{ catalyst}\;} 2CO_{2(g)} + N_{2(g)}$$

The lattice structure **maximises the surface area** of the catalyst, making it more effective. And, it **minimises the cost** of the catalyst because only a **thin coating** is needed.

Examples of **heterogeneous catalysts** include:

You need to know these examples of heterogeneous catalysts.

1) **Iron** in the **Haber Process** for making **ammonia**: $N_{2(g)} + 3H_{2(g)} \xrightarrow{\;Fe_{(s)}\text{ catalyst}\;} 2NH_{3(g)}$

2) **Vanadium(V) oxide** in the **Contact Process** for making **sulfuric acid**: $SO_{2(g)} + \tfrac{1}{2}O_{2(g)} \xrightarrow{\;V_2O_{5(s)}\text{ catalyst}\;} SO_{3(g)}$

3) **Chromium(III) oxide** in the manufacture of **methanol from CO**: $CO_{(g)} + 2H_{2(g)} \xrightarrow{\;Cr_2O_{3(s)}\text{ catalyst}\;} CH_3OH_{(g)}$

Impurities Can Poison Heterogeneous Catalysts

1) During a reaction, reactants are **adsorbed** onto active sites on the **surfaces** of **heterogeneous catalysts**.

2) **Impurities** in the reaction mixture may also **bind** to the catalyst's surface and **block reactants** from being adsorbed. This process is called **catalyst poisoning**.

3) Catalyst poisoning **reduces the surface area** of the catalyst available to the reactants, **slowing down the reaction**.

4) Catalyst poisoning **increases the cost** of a chemical process because **less product** can be made in a certain **time** or with a certain amount of **energy**. The **catalyst** may even need **replacing or regenerating**, which also costs money.

Examples of catalyst poisoning include:

1) **Lead** poisons the catalyst in **catalytic converters**:
Catalytic converters **reduce harmful emissions** from car engines. **Lead** can **coat the surface** of the catalyst in a catalytic converter, so vehicles that have them fitted must only be run on **unleaded petrol**.

2) **Sulfur** poisons the iron catalyst in the **Haber Process**:
The **hydrogen** in the Haber process is produced from **methane**. The methane is obtained from natural gas, which contains impurities, including **sulfur** compounds. Any sulfur that is not removed is **adsorbed** onto the iron, forming iron sulfide, and stopping the iron from catalysing the reaction efficiently.

Uses of Transition Metals

Homogeneous Catalysts are in the Same Phase as the Reactants

Homogeneous catalysts are in the **same physical state** as the reactants.
Usually a **homogeneous** catalyst is an **aqueous catalyst** for a reaction between two **aqueous solutions**.

A homogeneous catalyst works by forming an intermediate species.
The reactants combine with the catalyst to make an intermediate species,
which then reacts to form the products and reform the catalyst.

This causes the enthalpy profile for a homogeneously catalysed
reaction to have two humps in it, corresponding to the two reactions.

The activation energy needed to form the intermediates (and to form the
products from the intermediates) is lower than that needed to make the
products directly from the reactants.

The catalyst is always reformed so it can carry on catalysing the reaction.

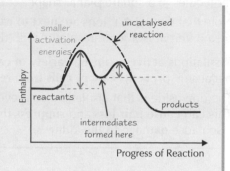

Fe²⁺ Catalyses the Reaction Between $S_2O_8^{2-}$ and I⁻

The redox reaction between iodide ions and peroxodisulfate ($S_2O_8^{2-}$) ions takes place annoyingly slowly
because both ions are negatively charged. The ions repel each other, so it's unlikely they'll collide and react.

$$S_2O_8{}^{2-}{}_{(aq)} + 2I^-{}_{(aq)} \rightarrow I_{2(aq)} + 2SO_4{}^{2-}{}_{(aq)}$$

But if **Fe²⁺ ions** are added, things are really **speeded up** because each stage of
the reaction involves a **positive and a negative ion**, so there's **no repulsion**.
First, the Fe²⁺ ions are **oxidised** to Fe³⁺ ions by the $S_2O_8^{2-}$ ions.

The newly formed intermediate Fe³⁺ ions
now **easily oxidise** the I⁻ ions to iodine,
and the **catalyst is regenerated**.

$$2Fe^{3+}{}_{(aq)} + 2I^-{}_{(aq)} \rightarrow I_{2(aq)} + 2Fe^{2+}{}_{(aq)}$$

The negative charges of the two
ions is one reason why this reaction
has high activation energy.

$$S_2O_8{}^{2-}{}_{(aq)} + 2Fe^{2+}{}_{(aq)} \rightarrow 2Fe^{3+}{}_{(aq)} + 2SO_4{}^{2-}{}_{(aq)}$$

Fe²⁺ is in the same phase as the
reactants, so it's a homogeneous catalyst.

You can test for **iodine** by adding **starch solution** — it'll turn **blue-black** if iodine is present.

Autocatalysis is when a Product Catalyses the Reaction

Another example of a **homogeneous catalyst** is Mn²⁺ in the reaction between $C_2O_4^{2-}$ and MnO_4^-.
It's an **autocatalysis reaction** because Mn²⁺ is a **product** of the reaction and **acts as a catalyst** for the reaction.
This means that as the reaction progresses and the **amount** of the **product increases**, the reaction **speeds up**.

$$2MnO_4{}^-{}_{(aq)} + 16H^+{}_{(aq)} + 5C_2O_4{}^{2-}{}_{(aq)} \rightarrow 2Mn^{2+}{}_{(aq)} + 8H_2O_{(l)} + 10CO_{2(g)}$$

Transition metals make good catalysts, but that isn't their only use — it's time to look at some other places they turn up.

Fe²⁺ in Haemoglobin Allows Oxygen to be Carried in the Blood

The blood product, **haemoglobin**, is a complex formed from **Fe²⁺ ions** (see page 167 for the structure). Both **water**
and **oxygen** will bind to the Fe²⁺ ion as **ligands**, so the complex can **transport oxygen** to where it's needed, and then
swap it for a water molecule — here's **how it works**.

1) In the lungs, where the oxygen concentration is high, the water ligand is **substituted** for an **oxygen molecule**
   to form **oxyhaemoglobin**, which is carried **around the body** in the blood.

2) When the **oxyhaemoglobin** gets to a place where oxygen is needed, the **oxygen molecule** is **exchanged** for a
   water molecule. The haemoglobin then **returns to the lungs** and the whole process starts again.

This process can be disrupted if **carbon monoxide** is inhaled. The **haemoglobin** swaps its **water** ligand for a **carbon
monoxide** ligand, forming **carboxyhaemoglobin**. This is bad news because carbon monoxide is a **strong** ligand and
doesn't readily exchange with oxygen or water ligands, meaning the haemoglobin **can't transport oxygen** any more.
Carbon monoxide poisoning starves the organs of oxygen — it can cause **headaches**, **dizziness**, **unconsciousness** and
even **death** if it's not treated.

Uses of Transition Metals

Cisplatin is an Anti-Cancer Drug

Cisplatin is a complex of platinum(II) with two chloride ions and two ammonia molecules in a square planar shape.

Note that the two Cl⁻ ions are next to each other, never opposite, as this would be a different isomer (transplatin) with different biological properties.

Cisplatin is active against a variety of cancers, including lung and bladder cancer because it prevents cancer cells from reproducing.

The downside is that cisplatin also prevents normal cells from reproducing, including blood and hair cells. This can cause hair loss and suppress the immune system, increasing the risk of infection. Cisplatin may also cause damage to the kidneys.

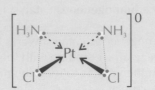

Tollens' Reagent Contains a Silver Complex

Tollens' reagent is prepared by adding just enough ammonia solution to silver nitrate solution to form a colourless solution containing the complex ion $[Ag(NH_3)_2]^+$.

Tollens' reagent is used to distinguish between aldehydes and ketones — aldehydes react to give a silver mirror on the inside of the test tube.

$$RCHO + 2[Ag(NH_3)_2]^+ + 3OH^- \rightarrow RCOO^- + 2Ag + 4NH_3 + 2H_2O$$

Ketones don't react with Tollens' reagent.

Steve was seeking professional help about his silver complex.

Practice Questions

Q1 What property of transition elements makes them good catalysts?

Q2 Which catalyst is used in the manufacture of methanol from carbon monoxide?

Q3 What term describes the process when a product catalyses a reaction?

Q4 State a medical use of the Pt(II) complex cisplatin.

Exam Questions

1 a) Using equations, explain how vanadium(V) oxide acts as a catalyst in the Contact Process. [4 marks]
  b) (i) Describe how heterogeneous catalysts can become poisoned and give an example. [3 marks]
    (ii) What are the consequences of catalytic poisoning? [2 marks]

2 a) Explain the difference between homogeneous and heterogeneous catalysts. [2 marks]
  b) H⁺ ions are used to catalyse the reaction between $C_2H_5OH_{(aq)}$ and $CH_3COOH_{(aq)}$.
    (i) What type of catalysis is this? [1 mark]
    (ii) Sketch an enthalpy profile diagram for this reaction to show the catalysed and uncatalysed routes for this exothermic reaction. [3 marks]

3 Tollens' reagent contains a transition metal complex ion.
  a) Which two types of organic compound is Tollens' reagent used to distinguish between? [1 mark]
  b) What is seen when a positive reaction occurs with Tollens' reagent? [1 mark]
  c) Give the:
    (i) formula (ii) shape (iii) colour
    of the transition metal complex ion in Tollens' reagent. [3 marks]

Bagpuss, Sylvester, Tom — the top three on my catalyst...

Hopefully you'll have realised from these last three pages that transition metals are really quite useful. You should also know the difference between heterogeneous and homogeneous catalysts and why the variable oxidation states of transition metals make them good catalysts. If you do, then go grab yourself a cup of tea— it speeds up the rate of revision, you know.

Metal-Aqua Ions

Remember all the stuff on pages 166-169 about complex ions and their pretty coloured solutions?
Well this section's all about their reactions. So prepare yourself for more colour changes than a chameleon playing twister...

Metal Ions Become **Hydrated** in Water

When **transition metal compounds** dissolve in water, the water molecules form **coordinate bonds** with the **metal ions**.
This forms **metal-aqua complex ions**. In general, **six water molecules** form coordinate bonds with each metal ion.

The water molecules do this by donating a
non-bonding pair of electrons from their oxygen.

The diagrams show the metal-aqua ions formed by
iron — $Fe(H_2O)_6^{2+}$ and by **chromium — $Cr(H_2O)_6^{3+}$**.

This is the charge on
the metal ion (water
molecules are neutral).

A Lewis Acid Accepts Electrons — A Lewis Base Donates Electrons

The Brønsted-Lowry theory of acids and bases says that an **acid** is a **proton donor**, and a **base** is a **proton acceptor**.

$$NH_{3(aq)} + HCl_{(aq)} \rightarrow NH_{4\ (aq)}^+ + Cl_{(aq)}^-$$

HCl is donating a proton, so it's acting as a **Brønsted-Lowry acid**.
NH_3 is accepting a proton, so it's acting as a **Brønsted-Lowry base**.

Now look at the reaction that happens between **ammonia** and **boron trifluoride**:

$$NH_{3(aq)} + BF_{3(aq)} \rightarrow NH_3BF_{3(aq)}$$

This **doesn't** involve transferring protons, so it doesn't fit in with the Brønsted-Lowry acid-base theory. You'd think this
would mean that it's not an acid-base reaction, but a chemist (Mr Lewis) came up with an alternative theory that
broadens the definition of acids and bases. The Lewis theory is based on the transfer of **electrons**, rather than protons:

> A **Lewis acid** is an **electron pair acceptor**.
> A **Lewis base** is an **electron pair donor**.

In the reaction above, the **ammonia** molecule **donates** a pair of electrons to the boron atom, forming a **coordinate bond**.

$$H_3N\!:\!\rightarrow BF_3$$

So **NH_3** is acting as a **Lewis base**, and **BF_3** is acting as a **Lewis acid**.

Some substances fit both the Brønsted-Lowry and the Lewis definitions of an acid or base. Others only fit one.
So, if you're describing something as an acid or base, you should say whether it's a **Brønsted-Lowry** or a **Lewis** acid or base.

You Need to be Able to Spot Lewis Acids and Bases

Have a look at these examples:

$$H_2O + H^+ \rightarrow H_3O^+$$

Coordinate bond

The water molecule's donating an electron pair to the hydrogen ion, and
the hydrogen ion is accepting an electron pair from the water. So the
water molecule is the Lewis base and the hydrogen ion's the Lewis acid.

$$AlCl_3 + Cl^- \rightarrow AlCl_4^-$$

Coordinate
bond

The aluminium chloride is accepting an electron pair — so it's the Lewis acid.
The chloride ion is donating an electron pair — so it's the Lewis base.

Coordinate bonds always involve one substance donating an electron pair to another. So if there's a coordinate bond,
there are no two ways about it — it **must** have been formed in a **Lewis acid-base reaction**.

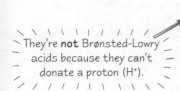

They're **not** Brønsted-Lowry
acids because they can't
donate a proton (H^+).

Metal ions act as **Lewis acids** in aqueous solution because they **accept electron pairs**
from the water molecules that surround them. And the water molecules, like any **ligands**,
are **electron pair donors** so they must be **Lewis bases**.

Metal-Aqua Ions

Solutions Containing Metal-Aqua Ions are Acidic

Aqua-ironing —
it keeps those
flat fish smooth.

In a solution containing metal-aqua **2+** ions, there's a reaction between the metal-aqua ion and the water — this is a **hydrolysis** or **acidity reaction**.

E.g. $$Fe(H_2O)_6^{2+}{}_{(aq)} + H_2O_{(l)} \rightleftharpoons [Fe(H_2O)_5(OH)]^+{}_{(aq)} + H_3O^+{}_{(aq)}$$

The metal-aqua **2+** ions release H^+ ions, so an **acidic** solution is formed.
There's only **slight** dissociation though, so the solution is only **weakly acidic**.

Metal-aqua **3+** ions react in the same way. They form **more acidic** solutions though.

E.g. $$Al(H_2O)_6^{3+}{}_{(aq)} + H_2O_{(l)} \rightleftharpoons [Al(H_2O)_5(OH)]^{2+}{}_{(aq)} + H_3O^+{}_{(aq)}$$

> In these forward reactions, the **metal-aqua ion** is acting as a **Brønsted-Lowry acid**. It **donates a proton** from one of its water ligands to a free water molecule.

Here's why 3+ metal-aqua ions form more acidic solutions than 2+ metal-aqua ions:

Metal 3+ ions are pretty **small** but have a **big charge** — so they've got a **high charge density** (otherwise known as **charge/size ratio**). The metal 2+ ions have a **much lower** charge density.

This makes the 3+ ions much more **polarising** than the 2+ ions. More polarising power means that they attract **electrons** from the oxygen atoms of the coordinated water molecules more strongly, weakening the O–H bond.

So it's more likely that a **hydrogen ion** will be released. And more hydrogen ions means a **more acidic** solution.

You Can Hydrolyse Metal-Aqua Ions Further to Form Precipitates

Adding **OH⁻ ions** to solutions of **metal-aqua ions** produces **insoluble metal hydroxides**. Here's why:

M might be Fe, Al or Cr.

1) In water, **metal-aqua 3+ ions** form the equilibrium: $M(H_2O)_6^{3+}{}_{(aq)} + H_2O_{(l)} \rightleftharpoons [M(H_2O)_5(OH)]^{2+}{}_{(aq)} + H_3O^+{}_{(aq)}$.
   If you add **OH⁻ ions** to the equilibrium H_3O^+ ions are removed — this shifts the equilibrium to the **right**.

2) Now another equilibrium is set up in the solution: $[M(H_2O)_5(OH)]^{2+}{}_{(aq)} + H_2O_{(l)} \rightleftharpoons [M(H_2O)_4(OH)_2]^+{}_{(aq)} + H_3O^+{}_{(aq)}$.
   Again the OH⁻ ions remove H_3O^+ ions from the solution, pulling the equilibrium to the right.

3) This happens one last time — now you're left with an **insoluble uncharged metal hydroxide**:
   $[M(H_2O)_4(OH)_2]^+{}_{(aq)} + H_2O_{(l)} \rightleftharpoons M(H_2O)_3(OH)_{3(s)} + H_3O^+{}_{(aq)}$.

The same thing happens with **metal-aqua 2+ ions** (e.g. Fe, Co or Cu), except this time there are only **two steps**:

$$M(H_2O)_6^{2+}{}_{(aq)} + H_2O_{(l)} \rightleftharpoons [M(H_2O)_5(OH)]^+{}_{(aq)} + H_3O^+{}_{(aq)} \longrightarrow [M(H_2O)_5(OH)]^+{}_{(aq)} + H_2O_{(l)} \rightleftharpoons M(H_2O)_4(OH)_{2(s)} + H_3O^+{}_{(aq)}$$

All the metal hydroxide precipitates **will dissolve in acid**. They act as Brønsted-Lowry bases and **accept H⁺ ions**. This **reverses** the hydrolysis reactions above.

Some metal hydroxides are **amphoteric** — they can act as both acids and bases. This means they'll **dissolve in an excess of base** as well as in **acids**. **Aluminium hydroxide** and **chromium(III) hydroxide** are **amphoteric**. They act as **Brønsted-Lowry acids** and **donate H⁺ ions** to the OH⁻ ions, forming **soluble compounds**.

$$Al(H_2O)_3(OH)_{3(s)} + OH^-{}_{(aq)} \rightleftharpoons [Al(H_2O)_2(OH)_4]^-{}_{(aq)} + H_2O_{(l)} \qquad Cr(OH)_3(H_2O)_{3(s)} + 3OH^-{}_{(aq)} \rightleftharpoons [Cr(OH)_6]^{3-}{}_{(aq)} + 3H_2O_{(l)}$$

Precipitates Form with Ammonia Solution...

The obvious way of adding hydroxide ions is to use a strong alkali, like **sodium hydroxide solution** — but you can use **ammonia solution** too. When ammonia dissolves in water this equilibrium occurs: $NH_3 + H_2O \rightleftharpoons NH_4^+ + OH^-$ Because hydroxide ions are formed, adding a **small** amount of ammonia solution gives the same results as sodium hydroxide.

In some cases, a further reaction happens if you add an **excess** of ammonia solution — the H_2O and OH⁻ ligands are displaced by NH_3 ligands. Have a look at the table on the next page to see which complexes this applies to.

...and Sodium Carbonate Too

Metal 2+ ions react with **sodium carbonate** to form **insoluble metal carbonates**, like this: $$M(H_2O)_6^{2+}{}_{(aq)} + CO_3^{2-}{}_{(aq)} \rightleftharpoons MCO_{3(s)} + 6H_2O_{(l)}$$

But, **metal 3+ ions** are stronger acids so they always form **hydroxide precipitates** when you add sodium carbonate. The **carbonate ions** react with the H_3O^+ ions, removing them from the solution just like OH⁻ ions do.

> So you only get MCO_3 formed, not $M_2(CO_3)_3$.

Metal-Aqua Ions

Learn the *Colours* of All the *Complex Ion Solutions* and *Precipitates*

This handy table summarises all the compounds that are formed in the reactions on these pages.
You need to know the formulas of all the complex ions, and their colours.

Metal-aqua ion	With OH$^-_{(aq)}$ or NH$_{3(aq)}$	With excess OH$^-_{(aq)}$	With excess NH$_{3(aq)}$	With Na$_2$CO$_{3(aq)}$
$[Co(H_2O)_6]^{2+}$ pink solution	$Co(H_2O)_4(OH)_2$ blue-green precipitate	no change	$[Co(NH_3)_6]^{2+}$ straw coloured solution	$CoCO_3$ pink precipitate
$[Cu(H_2O)_6]^{2+}$ blue solution	$Cu(H_2O)_4(OH)_2$ blue precipitate	no change	$[Cu(NH_3)_4(H_2O)_2]^{2+}$ deep blue solution	$CuCO_3$ green-blue precipitate
$[Fe(H_2O)_6]^{2+}$ green solution	$Fe(H_2O)_4(OH)_2$ green precipitate	no change	no change	$FeCO_3$ green precipitate
$[Al(H_2O)_6]^{3+}$ colourless solution	$Al(H_2O)_3(OH)_3$ white precipitate	$[Al(H_2O)_2(OH)_4]^-$ colourless solution	no change	$Al(H_2O)_3(OH)_3$ white precipitate
$[Cr(H_2O)_6]^{3+}$ violet solution	$Cr(H_2O)_3(OH)_3$ green precipitate	$[Cr(OH)_6]^{3-}$ green solution	$[Cr(NH_3)_6]^{3+}$ purple solution	$Cr(H_2O)_3(OH)_3$ green precipitate
$[Fe(H_2O)_6]^{3+}$ yellow solution	$Fe(H_2O)_3(OH)_3$ brown precipitate	no change	no change	$Fe(H_2O)_3(OH)_3$ brown precipitate

Practice Questions

Q1 Explain why AlCl$_3$ can act as a Lewis acid.

Q2 Show by equations how Al(OH)$_3$ can act as both a Brønsted-Lowry acid and a Brønsted-Lowry base.

Q3 Explain why 3+ metal-aqua ions form more acidic solutions than 2+ metal-aqua ions.

Q4 What colour solution is formed when you add excess ammonia to a solution containing [Co(H$_2$O)$_6$]$^{2+}$ ions?

Exam Questions

1   Explain why separate solutions of iron(II) sulfate and iron(III) sulfate with equal concentrations
    have different pH values.   [4 marks]

2   Describe what you would see when ammonia solution is added slowly to a solution containing
    copper(II) sulfate until it is in excess. Write equations for each reaction that occurs.   [8 marks]

3   Aqueous ammonia was added to an aqueous solution of chromium(III) sulfate.
    a)   Identify the chromium complex ion present in:
         (i)   the aqueous chromium(III) sulfate.   [1 mark]
         (ii)   the green precipitate initially formed when aqueous ammonia is added.   [1 mark]
         (iii) the purple solution when an excess of aqueous ammonia is added.   [1 mark]
    b)   Write an equation for the reaction in which the purple solution is formed from the green precipitate.   [1 mark]

4   a)   Describe what you would observe when aqueous sodium carbonate is added to:
         (i)   aqueous iron(III) chloride.   [1 mark]
         (ii)   freshly-prepared aqueous iron(II) sulfate.   [1 mark]
    b)   Write an equation for the reaction of the iron(II)-aqua ion with the carbonate ion.   [1 mark]
    c)   If iron(II) sulfate solution is left to stand overnight in an open beaker before the aqueous sodium carbonate
         is added, then a different reaction is observed.
         (i)   Describe the new observation.   [1 mark]
         (ii)   Explain this change.   [1 mark]

Test-tube reactions — proper chemistry at last...

So many pretty colours. The only downside is that you have to remember them. But examiners do love to ask questions about colours of solutions and precipitates. So learn them all, or come exam day you'll end up feeling blue. Or possibly blue-green...

Substitution Reactions

There are more equations on this page than the number of elephants you can fit in a Mini.

Ligands can Change Places with One Another

One ligand can be **swapped** for another ligand — this is **ligand exchange**. It pretty much always causes a **colour change**.

1) If the ligands are of **similar size**, e.g. H_2O and NH_3, then the **coordination number** of the complex ion doesn't change, and neither does the **shape**.

$$[Co(H_2O)_6]^{2+}_{(aq)} + 6NH_{3(aq)} \rightarrow [Co(NH_3)_6]^{2+}_{(aq)} + 6H_2O_{(l)}$$
octahedral   octahedral
pink   straw coloured

$$[Cr(H_2O)_6]^{3+}_{(aq)} + 6OH^-_{(aq)} \rightarrow [Cr(OH)_6]^{3-}_{(aq)} + 6H_2O_{(l)}$$
octahedral   octahedral
violet   green

2) If the ligands are **different sizes**, e.g. H_2O and Cl^-, there's a **change of coordination number** and a **change of shape**.

$$[Cu(H_2O)_6]^{2+}_{(aq)} + 4Cl^-_{(aq)} \rightleftharpoons [CuCl_4]^{2-}_{(aq)} + 6H_2O_{(l)}$$
octahedral   tetrahedral
pale blue   yellow

$$[Co(H_2O)_6]^{2+}_{(aq)} + 4Cl^-_{(aq)} \rightleftharpoons [CoCl_4]^{2-}_{(aq)} + 6H_2O_{(l)}$$
octahedral   tetrahedral
pink   blue

The forward reaction is endothermic, so the equilibrium can be shifted to the right-hand side by heating. The equilibrium will also shift to the right if you add more concentrated hydrochloric acid. Adding water to this equilibrium shifts it back to the left.

3) Sometimes the substitution is only **partial**.

$$[Cu(H_2O)_6]^{2+}_{(aq)} + 4NH_{3(aq)} \rightarrow [Cu(NH_3)_4(H_2O)_2]^{2+}_{(aq)} + 4H_2O_{(l)}$$
octahedral   elongated octahedral
pale blue   deep blue

$$[Fe(H_2O)_6]^{3+}_{(aq)} + SCN^-_{(aq)} \rightarrow [Fe(H_2O)_5SCN]^{2+}_{(aq)} + H_2O_{(l)}$$

but this usually looks yellow ⟶   octahedral   distorted octahedral
pale violet when pure   blood red

Have a quick peek back at the reactions on page 178 — these were **ligand exchange reactions** too. In these, **hydroxide precipitates** were formed when a little bit of **sodium hydroxide** or **ammonia solution** was added to metal-aqua ions. The hydroxide precipitates sometimes **dissolved** when excess sodium hydroxide or ammonia solution was added.

Some Ligands Can Attach to the Metal Atom in More Than One Place

So far all the ligands in this section have been **unidentate**, but you'll remember from page 167 that ligands can also be **bidentate** or **multidentate**. Here's a quick recap of what that means:

1) **Unidentate** ligands have **one lone pair** of electrons to donate to the metal ion to form a **dative covalent bond**.

2) **Bidentate** ligands have **two lone pairs** of electrons to donate.

3) **Multidentate** ligands donate **more than one lone pair** of electrons to form **more than one coordinate bond** with the metal ion.

Polydent ate my hamster.

Different Ligands Form Different Strength Bonds

Ligand exchange reactions can be easily **reversed**, **UNLESS** the new complex ion is much **more stable** than the old one.

1) If the new ligands form **stronger** bonds with the central metal ion than the old ligands did, the change is **less easy** to reverse. E.g. **CN⁻ ions** form stronger coordinate bonds with Fe^{3+} ions than H_2O molecules did, so it's hard to reverse this reaction:

$$[Fe(H_2O)_6]^{3+}_{(aq)} + 6CN^-_{(aq)} \rightarrow [Fe(CN)_6]^{3-}_{(aq)} + 6H_2O_{(l)}$$

2) **Multidentate** ligands form more stable complexes than unidentate ligands, so a change like the one below is hard to reverse:

$$[Cu(H_2O)_6]^{2+}_{(aq)} + 3NH_2CH_2CH_2NH_{2(aq)} \rightarrow [Cu(NH_2CH_2CH_2NH_2)_3]^{2+}_{(aq)} + 6H_2O_{(l)}$$

This is explained on the next page...

Substitution Reactions

A *Positive Entropy Change* Makes a *More Stable* Complex

When a **ligand exchange reaction** occurs, bonds are **broken** and **formed**.
The **strength** of the bonds being broken is often very similar to the strength of the new bonds being made.
So the **enthalpy change** for a ligand exchange reaction is usually very **small**.

Example: Substituting ammonia with ethane-1,2-diamine in a nickel complex:

$$[Ni(NH_3)_6]^{2+} + 3NH_2CH_2CH_2NH_2 \rightarrow [Ni(NH_2CH_2CH_2NH_2)_3]^{2+} + 6NH_3 \quad \Delta H = -13 \text{ kJmol}^{-1}$$

Break 6 coordinate bonds between Ni and N

Form 6 coordinate bonds between Ni and N

This is actually a **reversible** reaction, but the equilibrium lies so **far to the right** that it is thought of as being irreversible. $[Ni(NH_2CH_2CH_2NH_2)_3]^{2+}$ is **much more stable** than $[Ni(NH_3)_6]^{2+}$. This isn't accounted for by an enthalpy change, but an **increase in entropy** explains it:

When unidentate ligands are substituted with bidentate or multidentate ligands, the number of particles increases — the **more particles**, the **greater the entropy**. Reactions that result in an increase in entropy are **more likely** to occur. This is known as the **chelate effect**.

In the reaction above the number of particles increased from 4 to 7.

So that's why multidentate ligands always form much more stable complexes than unidentate ligands.

When the **hexadentate ligand EDTA⁴⁻** replaces unidentate or bidentate ligands, the complex formed is **loads more stable**.

$[Cr(NH_3)_6]^{3+} + EDTA^{4-} \rightarrow [Cr(EDTA)]^- + 6NH_3$ **2 particles → 7 particles**

$[Cr(NH_2CH_2CH_2NH_2)_3]^{3+} + EDTA^{4-} \rightarrow [Cr(EDTA)]^- + 3NH_2CH_2CH_2NH_2$ **2 particles → 4 particles**

It's difficult to reverse these reactions, because reversing them would cause a **decrease in entropy**.

Practice Questions

Q1 What colour is a solution of the complex ion $[Cu(NH_3)_4(H_2O)_2]^{2+}$?

Q2 What is the chelate effect?

Q3 When you add hydroxide ions to $[Cu(H_2O)_6]^{2+}$ you get a precipitate of $Cu(H_2O)_4(OH)_2$.
Adding extra H_2O does not reverse the reaction. Which ligand bonds more strongly to Cu^{2+}, OH^- or H_2O?

Exam Questions

1 An aqueous solution of copper sulfate contains the complex ion $[Cu(H_2O)_6]^{2+}$.
a) Write an equation for a ligand substitution reaction between $[Cu(H_2O)_6]^{2+}$ and chloride ions, in which all of the water ligands are replaced. Give the colour of the new complex formed. [3 marks]
b) The ethanedioate ion, $C_2O_4^{2-}$, is a bidentate ligand. Write an equation for a ligand substitution reaction between $[Cu(H_2O)_6]^{2+}$ and ethanedioate ions in which all of the water ligands are substituted. [2 marks]
c) $[Cu(H_2O)_6]^{2+}$ ions react with sodium hydroxide to form a precipitate.
State the formula and colour of the complex ion contained in the precipitate. [2 marks]

You might need to check back to page 179 for some of these questions.

2 When a solution of EDTA⁴⁻ ions is added to an aqueous solution of $[Fe(H_2O)_6]^{3+}$ ions, a ligand substitution reaction occurs.
a) Write an equation for the reaction that takes place. [2 marks]
b) The new complex that is formed is more stable than $[Fe(H_2O)_6]^{3+}$. Explain why. [2 marks]
c) Another sample of $[Fe(H_2O)_6]^{3+}$ solution is mixed with a weak solution of sodium hydroxide.
A brown precipitate forms. Give the formula of the complex ion contained in the precipitate. [1 mark]

3 A scientist takes three samples of a solution that contains the complex ion $[Co(H_2O)_6]^{2+}$. She mixes the first sample with a solution of chloride ions, the second with an excess of ammonia solution, and the third with a solution containing ethane-1,2-diamine. She observes the reactions that occur. Write an equation for a reaction the scientist observed where:
a) there was no change in the number of ligands surrounding the cobalt ion. [2 marks]
b) the number of ligands surrounding the cobalt ion changed, but the overall charge on the complex ion did not. [2 marks]
c) both the number of ligands surrounding the cobalt ion and the overall charge on the complex ion changed. [2 marks]

Ligand exchange — the musical chairs of the molecular world...

Ligands generally don't mind swapping with other ligands, so long as they're not too tightly attached to the central metal ion. They also won't fancy changing if it means forming fewer molecules and having less entropy. It's kind of like you wouldn't want to swap a cow for a handful of beans. Unless, of course, they're magic beans. But that almost never happens...

AS Answers

Unit 1: Section 1 — Atomic Structure

Page 5 — The Atom

1) a) *Similarity — They've all got the same number of protons/ electrons.* **[1 mark]**
Difference — They all have different numbers of neutrons. **[1 mark]**

b) *1 proton* **[1 mark]**, *1 neutron (2 – 1)* **[1 mark]**, *1 electron* **[1 mark]**.

c) ^3H. **[1 mark]**
Since tritium has 2 neutrons in the nucleus and also 1 proton, it has a mass number of 3. You could also write ^3_1H but you don't really need the atomic number.

2) a) (i) *Same number of electrons.* **[1 mark]**
$^{32}S^{2-}$ *has 16 + 2 = 18 electrons. ^{40}Ar has 18 electrons too.* **[1 mark]**

(ii) *Same number of protons.* **[1 mark]**. *Each has 16 protons (the atomic number of S must always be the same)* **[1 mark]**.

(iii) *Same number of neutrons.* **[1 mark]**
^{40}Ar *has 40 – 18 = 22 neutrons. ^{42}Ca has 42 – 20 = 22 neutrons.* **[1 mark]**

b) **A** *and* **C**. **[1 mark]** *They have the same number of protons but different numbers of neutrons.* **[1 mark]**.
It doesn't matter that they have a different number of electrons because they are still the same element.

Page 7 — Atomic Models

1 a) *In order to explain the observations made during his students' experiments* **[1 mark]**.

b) *The Bohr model gives a better explanation of the observations* **[1 mark]** *of the frequencies of radiation emitted by atoms* **[1 mark]**.

c) *The more accurate models are very complicated / The Bohr model is still useful for explaining most observations* **[1 mark]**.

2 a) Ca^{2+} *is the more stable* **[1 mark]**.

b) *Atomic models predict that noble gas electronic configurations/full electron shells are most stable* **[1 mark]**. *Since Ca^{2+} has a noble gas electronic configuration / full electron shells it should be more stable than Ca^+ which does not* **[1 mark]**.

Page 9 — Relative Mass

1) a) *First multiply each relative abundance by the relative mass —*
$120.8 \times 63 = 7610.4$, $54.0 \times 65 = 3510.0$
Next add up the products —
$7610.4 + 3510.0 = 11\,120.4$ **[1 mark]**
Now divide by the total abundance ($120.8 + 54.0 = 174.8$)

$$A_r(Cu) = \frac{11120.4}{174.8} \approx \mathbf{63.6}$$ **[1 mark]**

You can check your answer by seeing if $A_r(Cu)$ is in between 63 and 65 (the lowest and highest relative isotopic masses).

b) *A sample of copper is a mixture of 2 isotopes in different abundances* **[1 mark]**. *The weighted average mass of these isotopes isn't a whole number* **[1 mark]**.

2) a) *Mass spectrometry.* **[1 mark]**

b) *You use almost the same method here as for question 1) a).*
$93.11 \times 39 = 3631.29$, $(0.12 \times 40) = 4.8$,
$(6.77 \times 41) = 277.57$
$3631.29 + 4.8 + 277.57 = 3913.66$ **[1 mark]**
This time you divide by 100 because they're percentages.

$$A_r(K) = \frac{3913.66}{100} \approx \mathbf{39.14}$$ **[1 mark]**

Again check your answer's between the lowest and highest relative isotopic masses, 39 and 41. $A_r(K)$ is closer to 39 because most of the sample (93.11 %) is made up of this isotope.

Page 11 — Electronic Structure

1) a) *K atom: $1s^2\,2s^2\,2p^6\,3s^2\,3p^6\,4s^1$* **[1 mark]**
K^+ ion: $1s^2\,2s^2\,2p^6\,3s^2\,3p^6$ **[1 mark]**

b)
Oxygen electron Configuration

1s	2s	2p
↑↓	↑↓	↑↓ ↑ ↑

[1 mark for the correct number of electrons in each sub-shell. 1 mark for having spin-pairing in one of the p orbitals and parallel spins in the other two p orbitals. A box filled with 2 arrows is spin pairing — 1 up and 1 down. If you've put the four p electrons into just 2 orbitals, it's wrong.]

c) *The outer shell electrons in potassium and oxygen can get close to the outer shells of other atoms so they can be transferred or shared* **[1 mark]**. *The inner shell electrons are tightly held and shielded from the electrons in other atoms/ molecules* **[1 mark]**.

2) a) $1s^2\,2s^2\,2p^6\,3s^2\,3p^6\,3d^5\,4s^2$. **[1 mark]**

b)
Al^{3+} electron Configuration

1s	2s	2p
↑↓	↑↓	↑↓ ↑↓ ↑↓

[1 mark for the correct number of electrons in each sub-shell. 1 mark for one arrow in each box pointing up, and one pointing down.]

c) *Germanium ($1s^2\,2s^2\,2p^6\,3s^2\,3p^6\,3d^{10}\,4s^2\,4p^2$).* **[1 mark]**
(The 4p sub-shell is partly filled so it must be a p block element.)

d) *Ar (atom)* **[1 mark]**, *K^+ (positive ion)* **[1 mark]**, *Cl^- (negative ion)* **[1 mark]**. *You also could have suggested Ca^{2+}, S^{2-} or P^{3-}.*

Page 13 — Ionisation Energies

1) a) $Li_{(g)} \rightarrow Li^+_{(g)} + e^-$
[1 mark for the correct equation with wrong or missing state symbols. 1 mark for the correct state symbols.]

b) *Increasing number of protons means a stronger pull from the positively charged nucleus* **[1 mark]** *making it harder to remove an electron from the outer shell* **[1 mark]**. *There are no extra inner electrons to add to the shielding effect* **[1 mark]**.

c) (i) *Boron has the configuration $1s^2\,2s^2\,2p^1$ compared to $1s^2\,2s^2$ for beryllium* **[1 mark]**. *The 2p shell is at a slightly higher energy level than the 2s shell. As a result, the extra distance and partial shielding of the 2s orbital make it easier to remove the outer electron* **[1 mark]**.

(ii) *Oxygen has the configuration $1s^2\,2s^2\,2p^4$ compared to $1s^2\,2s^2\,2p^3$ for nitrogen* **[1 mark]**. *Electron repulsion in the shared 2p sub-shell in oxygen makes it easier to remove an electron* **[1 mark]**.

2) *As you go down Group 2, it takes less energy to remove an electron* **[1 mark]**. *This is evidence that the outer electrons are increasingly distant from the nucleus* **[1 mark]** *and additional inner shells of electrons exist to shield the outer shell* **[1 mark]**.

AS Answers

Unit 1: Section 2 — Amount of Substance

Page 15 — The Mole

1) M of CH_3COOH = $(2 \times 12) + (4 \times 1) + (2 \times 16)$
= 60 g mol^{-1} **[1 mark]**
so mass of 0.36 moles = 60×0.36 = **21.6 g [1 mark]**

2) No. of moles = $\frac{0.25 \times 60}{1000}$ = 0.015 moles H_2SO_4 **[1 mark]**
M of H_2SO_4 = $(2 \times 1) + (1 \times 32) + (4 \times 16)$ = 98 g mol^{-1}
Mass of 0.015 mol H_2SO_4 = 98×0.015 = **1.47 g [1 mark]**

3) a) M of C_3H_8 = $(3 \times 12) + (8 \times 1)$ = 44 g mol^{-1}

No. of moles of C_3H_8 = $\frac{88}{44}$ = 2 moles **[1 mark]**

At r.t.p. 1 mole of gas occupies 24 dm^3
so 2 moles of gas occupies 2×24 = **48 dm^3 [1 mark]**

b) $pV = nRT$ **[1 mark]**
so $V = nRT \div p = [2 \times 8.31 \times 308] \div [100 \times 10^3]$
= 51.1896×10^{-3} m^3 ≈ **51.2 dm^3 [1 mark]**

Page 17 — Equations and Calculations

1) M of C_2H_5Cl = $(2 \times 12) + (5 \times 1) + (1 \times 35.5)$
= 64.5 g mol^{-1} **[1 mark]**

Number of moles of C_2H_5Cl = $\frac{258}{64.5}$ = 4 moles **[1 mark]**

From the equation, 1 mole C_2H_5Cl is made from 1 mole C_2H_4
so, 4 moles C_2H_5Cl is made from 4 moles C_2H_4 **[1 mark]**.
M of C_2H_4 = $(2 \times 12) + (4 \times 1)$ = 28 g mol^{-1}
so, the mass of 4 moles C_2H_4 = 4×28 = **112 g [1 mark]**

2) a) M of $CaCO_3$ = $40 + 12 + (3 \times 16)$ = 100 g mol^{-1}

Number of moles of $CaCO_3$ = $\frac{15}{100}$ = 0.15 moles **[1 mark]**

From the equation, 1 mole $CaCO_3$ produces 1 mole CaO
so, 0.15 moles of $CaCO_3$ produces 0.15 moles of CaO
[1 mark]
M of CaO = $40 + 16$ = 56 g mol^{-1}
so, mass of 0.15 moles of CaO = 56×0.15 = **8.4 g [1 mark]**

b) From the equation, 1 mole $CaCO_3$ produces 1 mole CO_2
so, 0.15 moles of $CaCO_3$ produces 0.15 moles of CO_2
[1 mark]
1 mole gas occupies 24 dm^3 **[1 mark]**,
so, 0.15 moles occupies 24×0.15 = **3.6 dm^3 [1 mark]**

3) On the LHS, you need 2 each of K and I, so use 2KI
This makes the final equation:
2KI + Pb(NO$_3$)$_2$ → PbI$_2$ + 2KNO$_3$ [1 mark]
In this equation, the NO_3 group remains unchanged, so it makes
balancing much easier if you treat it as one indivisible lump.

Page 19 — Titrations

1) First write down what you know —
$CH_3COOH + NaOH \rightarrow CH_3COONa + H_2O$
25.4 cm^3 14.6 cm^3
?   0.5 M

Number of moles of NaOH = $\frac{0.5 \times 14.6}{1000}$ = 0.0073 moles

[1 mark]

From the equation, you know 1 mole NaOH neutralises 1
mole of CH_3COOH, so if you've used 0.0073 moles NaOH
you must have neutralised 0.0073 moles CH_3COOH.
[1 mark]

Concentration of CH_3COOH = $\frac{0.0073 \times 1000}{25.4}$ = **0.287 M**

[1 mark]

2) First write down what you know again —
$CaCO_3 + H_2SO_4 \rightarrow CaSO_4 + H_2O + CO_2$
0.75 g   0.25 M

M of $CaCO_3$ = $40 + 12 + (3 \times 16)$ = 100 g mol^{-1} **[1 mark]**
Number of moles of $CaCO_3$ = $\frac{0.75}{100}$ = 7.5 x 10^{-3} moles

[1 mark]
From the equation, 1 mole $CaCO_3$ reacts with 1 mole H_2SO_4
so, 7.5 × 10^{-3} moles $CaCO_3$ reacts with 7.5 × 10^{-3} moles
H_2SO_4. **[1 mark]**

The volume needed is = $\frac{(7.5 \times 10^{-3}) \times 1000}{0.25}$ = 30 cm^3

[1 mark]

If the question mentions concentration or molarities, you can
bet anything you like that you'll need to use the formula:

no. of moles = $\frac{\text{conc. } \times \text{ volume (cm}^3)}{1000}$

(or no. moles = conc. × volume (dm^3)).

Page 21 — Formulas, Yield and Atom Economy

1) Assume you've got 100 g of the compound so you can turn
the % straight into mass.

No. of moles of C = $\frac{92.3}{12}$ = 7.69 moles

No. of moles of H = $\frac{7.7}{1}$ = 7.7 moles **[1 mark]**

Divide both by the smallest number, in this case 7.69.
So ratio C : H = 1 : 1
So, the empirical formula = CH **[1 mark]**
The empirical mass = $12 + 1$ = 13

No. of empirical units in molecule = $\frac{78}{13}$ = 6

So the molecular formula = **C$_6$H$_6$ [1 mark]**

2) a) There is only one product, so the theoretical yield can be
calculated by adding the masses of the reactants **[1 mark]**.
So theoretical yield = $0.275 + 0.142$ = **0.417 g [1 mark]**

b) percentage yield = $(0.198 \div 0.417) \times 100$ = **47.5% [1 mark]**

c) Changing reaction conditions will have no effect on atom
economy **[1 mark]**. Since the equation shows that there is
only one product, the atom economy will always be 100%
[1 mark].
Atom economy is related to the type of reaction — addition,
substitution, etc. — not to the quantities of products and
reactants.

AS Answers

Unit 1: Section 3 — Bonding and Periodicity

Page 23 — Ionic Bonding

1) a)

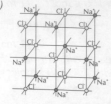

Your diagram should show the following —
- cubic structure with ions at corners **[1 mark]**
- sodium ions and chloride ions labelled **[1 mark]**
- alternating sodium ions and chloride ions **[1 mark]**

b) giant ionic/crystal (lattice) **[1 mark]**

c) You'd expect it to have a high melting point **[1 mark]** because there are strong bonds between the ions **[1 mark]** due to the electrostatic forces **[1 mark]**. A lot of energy is required to overcome these bonds **[1 mark]**.

2) a) Electrons move from one atom to another **[1 mark]**. Any correct examples of ions, one positive (e.g. Na^+) **[1 mark]**, one negative (e.g. Cl^-) **[1 mark]**.

b) In a solid, ions are held in place by strong ionic bonds **[1 mark]**. When the solid is heated to melting point, the ions gain enough energy to overcome the forces of attraction enough to become mobile **[1 mark]** and so carry charge (and hence electricity) through the substance **[1 mark]**.

Page 25 — Covalent Bonding

1) a) Covalent **[1 mark]**

b)

Your diagram should show the following —
- a completely correct electron arrangement in carbon **[1 mark]**
- all 4 overlaps correct (one dot and one cross in each) **[1 mark]**

2) a) Dative covalent/coordinate bonding **[1 mark]**

b) One atom donates **[1 mark]** a pair of electrons to the bond **[1 mark]**.

3) a) Giant molecular/macromolecular/giant covalent **[1 mark]**

b)

Diamond   Graphite

[1 mark for each correctly drawn diagram]
Diamond's a bit awkward to draw without it looking like a bunch of ballet dancing spiders — just make sure each central carbon is connected to four others.

c) Diamond has electrons in localised covalent bonds **[1 mark]**, so is a poor electrical conductor **[1 mark]**. Graphite has delocalised electrons which can flow within the sheets **[1 mark]**, making it a good electrical conductor **[1 mark]**.

Page 27 — Shapes of Molecules

1) a) NCl₃ **[1 mark]**   BCl₃ **[1 mark]**

b) NCl_3 **[1 mark]**

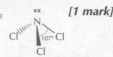

shape: (trigonal) pyramidal **[1 mark]**,
bond angle: 107° (accept between 105° and 109°) **[1 mark]**

BCl_3 **[1 mark]**

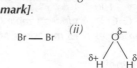

(It must be a reasonable "Y" shaped molecule.)
shape: trigonal planar **[1 mark]**, bond angle: 120° exactly **[1 mark]**

c) BCl_3 has three electron pairs only around B **[1 mark]**. NCl_3 has four electron pairs around N **[1 mark]**, including one lone pair **[1 mark]**.

Page 30 — Polarisation and Intermolecular Forces

1) a) The power of an atom to withdraw electron density **[1 mark]** from a covalent bond **[1 mark]** OR the ability of an atom to attract the bonding electrons **[1 mark]** in a covalent bond **[1 mark]**.

b) (i) Br — Br   (ii) [O with δ−, H δ+, H δ+]   (iii) [N with δ−, H δ+, H δ+, H δ+]

[1 mark for each correct shape, 1 mark for bond polarities correctly marked on H₂O, 1 mark for bond polarities correctly marked on NH₃.]

2) a) Van der Waals OR instantaneous/temporary dipole-induced dipole OR dispersion forces.
Permanent dipole-dipole interactions/forces.
Hydrogen bonding.
(Permanent dipole-induced dipole interactions.)
[1 mark each for any three]

b)

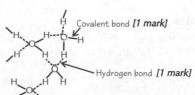

Covalent bond **[1 mark]**

Hydrogen bond **[1 mark]**

You could have shown the H₂O molecules in either of these two ways.

[O with H H or O δ− with H H δ+] **[1 mark]**

Van der Waals OR instantaneous/temporary dipole-induced dipole OR dispersion forces of attraction between water molecules. **[1 mark]**

c) More energy **[1 mark]** is needed to break the hydrogen bonds between water molecules **[1 mark]**.

AS Answers

Page 33 — Metallic Bonding and Properties of Structures

1)

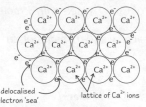

delocalised electron 'sea'

lattice of Ca^{2+} ions

[1 mark for showing closely packed Ca^{2+} ions and 1 mark for showing a sea of delocalised electrons.]
Metallic bonding results from the attraction between positive metal ions *[1 mark]* and a sea of delocalised electrons between them *[1 mark]*.

2) a) A — Ionic   B — (Simple) molecular
C — Metallic   D — Giant molecular (macromolecular)
[1 mark for each]
b) (i) Diamond — D
(ii) Aluminium — C
(iii) Sodium chloride — A
(iv) Iodine — B
[2 marks if all correct. 1 mark for only two correct.]

3) **Magnesium** has a metallic crystal lattice (it has metallic bonding) *[1 mark]*. It has a sea of electrons/delocalised electrons/freely moving electrons *[1 mark]*, which allow it to conduct electricity in the solid or liquid state *[1 mark]*.
Sodium chloride has a (giant) ionic lattice *[1 mark]*. It doesn't conduct electricity when it's solid *[1 mark]* because its ions don't move freely, but vibrate about a fixed point *[1 mark]*. Sodium chloride conducts electricity when liquid/molten *[1 mark]* or in aqueous solution *[1 mark]* because it has freely moving ions (not electrons) *[1 mark]*.
Graphite is giant covalent/macromolecular *[1 mark]*. It has delocalised/freely moving electrons between the layers *[1 mark]*. It conducts electricity along the layers in the solid state *[1 mark]*.

Page 35 — Periodicity

1) Mg has more delocalised electrons per atom *[1 mark]* and the ion has a greater charge density (due to its smaller ionic radius) *[1 mark]*. This gives Mg a stronger metal-metal bond, resulting in a higher boiling point *[1 mark]*.

2) a) Si has a macromolecular (or giant molecular) structure *[1 mark]* consisting of very strong covalent bonds *[1 mark]*.
b) Sulfur (S_8) has a larger molecule than phosphorus (P_4) *[1 mark]* which results in larger van der Waals forces of attraction between molecules *[1 mark]*.

3) The atomic radius decreases across the period from left to right *[1 mark]*. The number of protons increases, so nuclear charge increases *[1 mark]*. Electrons are pulled closer to the nucleus *[1 mark]*. The electrons are all added to the same outer shell, so there's little effect on shielding *[1 mark]*.

4) Neon has the configuration $1s^2 2s^2 2p^6$ and sodium $1s^2 2s^2 2p^6 3s^1$. *[1 mark]* The extra distance of sodium's outer electron from the nucleus and electron shielding make it easier to remove an electron from the 3s sub-shell *[1 mark]*.

Unit 1: Section 4 — Alkanes and Organic Chemistry

Page 37 — Basic Stuff

1) a)

H-C-C-C-C-Br *[1 mark]*
1-bromobutane

b) Haloalkanes / halogenoalkanes *[1 mark]*
c) but-1-ene *[2 marks, or 1 mark for just butene]*

2) a) 1-chloro-2-methylpropane
[2 marks available, lose 1 mark for each mistake]
Remember to put the substituents in alphabetical order.
b) 3-methylbut-1-ene *[2 marks available, lose 1 mark for each mistake]*
c) 2,4-dibromo-but-1-ene
[2 marks available, lose 1 mark for each mistake]
In parts b) and c), the double bond is the most important functional group, so it's given the lowest number.

Page 39 — Formulas and Structural Isomerism

1) a) 1-chlorobutane, 2-chlorobutane, 1-chloro-2-methylpropane, 2-chloro-2-methylpropane *[1 mark for each correct isomer]*
b) 1-chloro-2-methylpropane and 2-chloro-2-methylpropane OR 1-chlorobutane and 2-chlorobutane *[1 mark]*
c) 1-chlorobutane and 1-chloro-2-methylpropane OR 2-chlorobutane and 2-chloro-2-methylpropane *[1 mark]*

2) a)

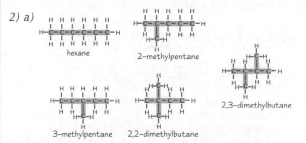

hexane   2-methylpentane

3-methylpentane   2,2-dimethylbutane   2,3-dimethylbutane

[1 mark for each correctly drawn isomer, 1 mark for each correct name]
b) A group of compounds represented by the same general formula OR having the same functional group OR with similar chemical properties *[1 mark]*. Each successive member differs by $-CH_2-$ *[1 mark]*.
c) (i) C_8H_{18} *[1 mark]*
(ii) $CH_3CH_2CH_2CH_2CH_2CH_2CH_2CH_3$ *[1 mark]*

3) a) pentane, (2-)methylbutane, (2,2-)dimethylpropane *[1 mark for each]*
There's only actually one type of methylbutane. You can't have 1-methylbutane — it'd be exactly the same as pentane.
b) Chain isomers have the same molecular formula/are made up of the same atoms *[1 mark]*, but they have different arrangements of the carbon skeleton *[1 mark]*.

AS Answers

Page 41 — Alkanes and Petroleum

1) a) As a mixture, crude oil is not very useful — the different alkanes it's made up of have different uses *[1 mark]*.
   b) Boiling point *[1 mark]*.
   c) (i) C_8H_{18} *[1 mark]*
      (ii) Near the top *[1 mark]*. This is because the molecules in petrol have a relatively low boiling point *[1 mark]* and the fractionating column is cooler at the top than the bottom *[1 mark]*.
      (iii) C_8H_{16} *[1 mark]*
2) a) There's greater demand for smaller fractions *[1 mark]* for motor fuels *[1 mark]* OR for alkenes *[1 mark]* to make petrochemicals/polymers *[1 mark]*.
   b) E.g. $C_{12}H_{26} \rightarrow C_2H_4 + C_{10}H_{22}$ *[1 mark]*.
      There are loads of possible answers — just make sure the C's and H's balance and there's an alkane and an alkene.

Page 43 — Alkanes as Fuels

1) a) $C_7H_{16} + 11O_2 \rightarrow 7CO_2 + 8H_2O$
      [2 marks available, lose 1 mark for each error]
   b) (i) Carbon monoxide *[1 mark]*
      (ii) By fitting a catalytic converter *[1 mark]*
2 a) Nitrogen and oxygen from air *[1 mark]* react together because of the conditions (high pressure and temperature) in the engine *[1 mark]*.
   b) Some fossil fuels contain sulfur that forms sulfur dioxide when burned *[1 mark]*. Sulfur dioxide dissolves in water in the atmosphere to form an acid (sulfuric acid) *[1 mark]*. Power stations remove the sulfur dioxide from their flue gases using calcium oxide *[1 mark]*.

Unit 2: Section 1 — Energetics

Page 45 — Enthalpy Changes

1) a) Total energy required to break bonds
      $= (4 \times 435) + (2 \times 498) = 2736$ kJ mol^{-1} *[1 mark]*
      Energy released when bonds form
      $= (2 \times 805) + (4 \times 464) = 3466$ kJ mol^{-1} *[1 mark]*
      Net energy change $= +2736 + (-3466) = -730$ kJ mol^{-1}
      [1 mark for correct numerical value, 1 mark for correct unit]
   b) The reaction is exothermic, because the enthalpy change is negative / more energy is given out than is taken in *[1 mark]*.
2) a) $CH_3OH_{(l)} + 1\frac{1}{2}O_{2(g)} \rightarrow CO_{2(g)} + 2H_2O_{(l)}$
      Correct balanced equation *[1 mark]*. Correct state symbols for reactants *[1 mark]*.
      It is perfectly OK to use halves to balance equations. Make sure that only 1 mole of CH$_3$OH is combusted, as it says in the definition for $\Delta H_c^\ominus$.
   b) $C_{(s)} + 2H_{2(g)} + \frac{1}{2}O_{2(g)} \rightarrow CH_3OH_{(l)}$
      Correct balanced equation *[1 mark]*. Correct state symbols for reactants *[1 mark]*.
   c) Only 1 mole of C_3H_8 should be shown according to the definition of $\Delta H_c^\ominus$ *[1 mark]*.
      You really need to know the definitions of the standard enthalpy changes off by heart. There's loads of nit-picky little details they could ask you questions about.

Page 47 — Calculating Enthalply Changes

1) $\Delta H_r^\ominus$ = sum of $\Delta H_f^\ominus$(products) — sum of $\Delta H_f^\ominus$(reactants)
      [1 mark]
      $= [0 + (3 \times -602)] - [-1676 + (3 \times 0)]$ *[1 mark]*
      $= -130$ kJ mol^{-1} *[1 mark]*
   Don't forget the units. It's a daft way to lose marks.

2) No. of moles of $CuSO_4 = \dfrac{0.200 \times 50}{1000}$ *[1 mark]*
      $= 0.01$ moles *[1 mark]*
   From the equation, 1 mole of $CuSO_4$ reacts with 1 mole of Zn.
   So, 0.01 moles of $CuSO_4$ reacts with 0.01 moles of Zn *[1 mark]*.
   Heat produced by reaction
      $= mc\Delta T$ *[1 mark]*
      $= 50 \times 4.18 \times 2.6 = 543.4$ J *[1 mark]*
   0.01 moles of zinc produces 543.4 J of heat, therefore 1 mole of zinc produces:
   $\dfrac{543.4}{0.01}$ *[1 mark]* $= 54\,340$ J ≈ 54.3 kJ
   So the enthalpy change is **−54.3 kJ mol^{-1}** (you need the minus sign because it's exothermic) *[1 mark for correct number, 1 mark for minus sign]*.
   It'd be dead easy to work out the heat produced by the reactions, breathe a sigh of relief and sail on to the next question. But you need to find out the enthalpy change when 1 mole of zinc reacts. It's always a good idea to reread the question and check you've actually answered it.

Unit 2: Section 2 — Kinetics and Equilibria

Page 50 — Reaction Rates and Catalysts

1) The molecules don't always have enough energy *[1 mark]*.
2) The particles in a liquid move freely and all of them are able to collide with the solid particles *[1 mark]*. Particles in solids just vibrate about fixed positions, so only those on the touching surfaces between the two solids will be able to react. *[1 mark]*
3) a) $2H_2O_{2(l)} \rightarrow 2H_2O_{(l)} + O_{2(g)}$
      Correct symbols *[1 mark]* and balancing equation *[1 mark]*. You get the marks even if you forgot the state symbols.
   b)

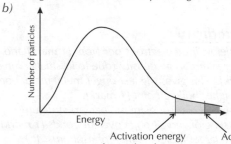

   Correct general shape of the curve *[1 mark]*. Correctly labelling the axes *[1 mark]*. Activation energies marked on the horizontal axis — the catalysed activation energy must be lower than the uncatalysed activation energy *[1 mark]*.
   c) Manganese(IV) oxide lowers the activation energy by providing an alternative reaction pathway *[1 mark]*. So, more reactant molecules have at least the activation energy *[1 mark]*, meaning there are more successful collisions in a given period of time, and so the rate increases *[1 mark]*.
   d) Raising the temperature will increase the rate of reaction *[1 mark]*. The particles will have more kinetic energy on average *[1 mark]*, so a greater proportion of particles will have enough energy to react *[1 mark]*.

AS Answers

Page 53 — Reversible Reactions

1) a) If a reaction at equilibrium is subjected to a change in concentration, pressure or temperature, the equilibrium will shift to try to oppose (counteract) the change. *[1 mark]*. Examiners are always asking for definitions so learn them — they're easy marks.

b) (i) There's no change *[1 mark]*. There's the same number of molecules/moles on each side of the equation *[1 mark]*.

(ii) Reducing temperature removes heat. So the equilibrium shifts in the exothermic direction to release heat *[1 mark]*. The reverse reaction is exothermic (since the forward reaction is endothermic). So, the position of equilibrium shifts left *[1 mark]*.

(iii) Removing nitrogen monoxide reduces its concentration. The equilibrium position shifts right to try and increase the nitrogen monoxide concentration again *[1 mark]*.

c) No effect *[1 mark]*.
Catalysts don't affect the equilibrium position.
They just help the reaction to get there sooner.

2) a) At low temperature the particles move more slowly / have less energy *[1 mark]*. This means fewer successful collisions *[1 mark]* and a slower rate of reaction *[1 mark]*.

b) High pressure is expensive. *[1 mark]* The cost of the extra pressure is greater than the value of the extra yield. *[1 mark]*

Unit 2: Section 3 — Reactions and Elements

Page 55 — Redox Reactions

1) a) Oxidation is the loss of electrons, reduction is the gain of electrons *[1 mark]*.

b) (i) 0 *[1 mark]*
(ii) +1 *[1 mark]*

c) $Li \rightarrow Li^+ + e^-$
[2 marks — 1 mark for electrons on RHS, 1 mark for correct equation, allow different balancing]
$O_2 + 4e^- \rightarrow 2O^{2-}$
[2 marks — 1 mark for electrons on LHS, 1 mark for correct equation, allow different balancing]
Lithium is being oxidised and oxygen is being reduced *[1 mark]*.

2) a) An oxidising agent accepts electrons from another species *[1 mark]*.

b) $In \rightarrow In^{3+} + 3e^-$ *[2 marks — 1 mark for electrons on RHS, 1 mark for correct equation]*

c) $2In + 3Cl_2 \rightarrow 2InCl_3$ *[2 marks — 1 mark for correct reactants and product, 1 mark for correct balancing]*

Page 57 — Group 7 — The Halogens

1) a) $I_2 + 2At^- \rightarrow 2I^- + At_2$ *[1 mark]*

b) The (sodium) astatide *[1 mark]*

2) a) (i) Boiling point increases down the group *[1 mark]* because the size and relative mass of the atoms increases *[1 mark]*, so the Van der Waals forces holding the molecules together get stronger *[1 mark]*.

(ii) Electronegativity decreases down the group *[1 mark]* because the atoms get larger *[1 mark]*, and larger atoms don't attract electrons as strongly as smaller ones *[1 mark]*.

b) Fluorine *[1 mark]*

3 a) $Cl_2 + H_2O \rightarrow HCl$ *[1 mark]* + $HClO$ *[1 mark]*

b) Chlorine (or the chlorate(I) ions) kill bacteria *[1 mark]*.
Too much chlorine would be dangerous because it is toxic *[1 mark]*.

Page 59 — Halide Ions

1) **Aqueous** solutions of both halides are tested. *[1 mark]*

a) **Sodium chloride** — silver nitrate gives white precipitate which dissolves in dilute ammonia solution *[1 mark]*.
$Ag^+ + Cl^- \rightarrow AgCl$ *[1 mark]*
Sodium bromide — silver nitrate gives cream precipitate which is only soluble in concentrated ammonia solution *[1 mark]*.
$Ag^+ + Br^- \rightarrow AgBr$ *[1 mark]*

b) **Sodium chloride** — Misty fumes *[1 mark]*
$NaCl + H_2SO_4 \rightarrow NaHSO_4 + HCl$ *[1 mark]*
Sodium bromide — Misty fumes *[1 mark]*
$NaBr + H_2SO_4 \rightarrow NaHSO_4 + HBr$ *[1 mark]*
$2HBr + H_2SO_4 \rightarrow Br_2 + SO_2 + 2H_2O$ *[1 mark]*
Orange / brown vapour *[1 mark]*

2) a) NaI (via HI) reduces H_2SO_4 to H_2S *[1 mark]*. The reducing power of halide ions increases down the group *[1 mark]* and At is below I in the group *[1 mark]*, so H_2S will be produced *[1 mark]*.

b) AgI is insoluble in concentrated ammonia solution *[1 mark]*. The solubility of halides in ammonia solution decreases down the group *[1 mark]*, so AgAt will **NOT** dissolve. *[1 mark]*

Page 62 — Group 2 — The Alkaline Earth Metals

1) Mg $1s^2 2s^2 2p^6 3s^2$ Ca $1s^2 2s^2 2p^6 3s^2 3p^6 4s^2$ *[1 mark]*
First ionisation energy of Ca is smaller *[1 mark]* because Ca has (one) more electron shell(s) *[1 mark]*. This reduces the attraction between the nucleus and the outer electrons because it increases shielding the effect *[1 mark]* and because the outer shell of Ca is further from the nucleus *[1 mark]*.

2) a) Y *[1 mark]*

b) Y has the largest radius *[1 mark]* so it will have the smallest ionisation energy/lose its outer electrons more easily *[1 mark]*.

3) Add barium chloride (or nitrate) solution to both *[1 mark]*
Zinc chloride would not change/no reaction *[1 mark]*
Zinc sulfate solution would give a white precipitate *[1 mark]*
$BaCl_{2(aq)} + ZnSO_{4(aq)} \rightarrow BaSO_{4(s)} + ZnCl_{2(aq)}$ *[1 mark]*
(or suitable equation using $Ba(NO_3)_{2(aq)}$)
OR $Ba^{2+}_{(aq)} + SO_4^{2-}_{(aq)} \rightarrow BaSO_{4(s)}$ *[1 mark]*
(or a test with silver nitrate for the chloride ions could be done.)

4) a) $NaHCO_3 + HCl \rightarrow NaCl + H_2O + CO_2$ *[1 mark]*

b) Wind/burping etc. *[1 mark]*

c) Magnesium hydroxide *[1 mark]* – NB many other of the compounds would be either toxic or otherwise harmful.
$Mg(OH)_2 + 2HCl \rightarrow MgCl_2 + 2H_2O$ *[1 mark]*

Page 65 — Extraction of Metals

1) a) $Fe_3O_4 + 4CO \rightarrow 3Fe + 4CO_2$ *[1 mark]*

b) $Fe_3O_4 + 2C \rightarrow 3Fe + 2CO_2$
OR $Fe_3O_4 + 4C \rightarrow 3Fe + 4CO$ *[1 mark]*

2) Advantage: the tungsten produced is purer *[1 mark]*
Disadvantage: hydrogen is more expensive OR hydrogen is highly explosive *[1 mark]*

3) a) Aluminium oxide dissolved *[1 mark]* in molten cryolite *[1 mark]*

b) Cathode: $Al^{3+} + 3e^- \rightarrow Al$ *[1 mark]*
Anode: $2O^{2-} \rightarrow O_2 + 4e^-$ *[1 mark]*

c) High energy costs of extracting Al *[1 mark]*.

AS Answers

Unit 2: Section 4 — More Organic Chemistry

Page 67 — Synthesis of Chloroalkanes

1) a) One with no double bonds OR the maximum number of hydrogens OR single bonds only *[1 mark]*. It contains only hydrogen and carbon atoms *[1 mark]*.
   b) It has non-polar bonds/it's a non-polar molecule *[1 mark]*, so it does not attract/react with polar reagents *[1 mark]*.
   c) $CH_3CH_3 + Cl_2 \xrightarrow{U.V.} CH_3CH_2Cl + HCl$ *[1 mark]*
   Initiation: $Cl_2 \xrightarrow{U.V.} 2Cl\cdot$ *[1 mark]*
   Propagation: $CH_3CH_3 + Cl\cdot \rightarrow CH_3CH_2\cdot + HCl$ *[1 mark]*
   $CH_3CH_2\cdot + Cl_2 \rightarrow CH_3CH_2Cl + Cl\cdot$ *[1 mark]*
   Termination: $CH_3CH_2\cdot + Cl\cdot \rightarrow CH_3CH_2Cl$ OR
   $CH_3CH_2\cdot + CH_3CH_2\cdot \rightarrow CH_3CH_2CH_2CH_3$ *[1 mark]*
   [1 mark for mentioning U.V.] It's a free-radical *[1 mark]*
   substitution *[1 mark]* reaction.

Page 70 — Nucleophilic Substitution and Elimination

1) a) **Reaction 1**
   Reagent — NaOH/KOH/OH⁻ *[1 mark]*
   Solvent — Water/aqueous solution *[1 mark]*
   Reaction 2
   Reagent — Ammonia/NH_3 *[1 mark]*
   Solvent — Ethanol/alcohol *[1 mark]*
   Reaction 3
   Reagent — NaOH/KOH *[1 mark]*.
   Solvent — Ethanol/alcohol *[1 mark]*
   b) There'd be a faster reaction *[1 mark]*. The C–I bond is weaker than C–Br, or C–I bond enthalpy is lower *[1 mark]*.

Page 73 — Reactions of Alkenes

1) a) Shake the alkene with bromine water *[1 mark]*, and the solution goes colourless if a double bond is present *[1 mark]*.
   b) Electrophilic *[1 mark]* addition *[1 mark]*.

   c) (i)
   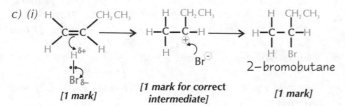
   [1 mark]   *[1 mark for correct intermediate]*   *[1 mark]*
   2-bromobutane

   Check your curly arrows go from exactly where the electrons are from, to where they're going to.
   (ii) The secondary carbocation OR the carbocation with the most attached alkyl groups *[1 mark]* is the most stable intermediate and so is the most likely to form *[1 mark]*.

Page 75 — E/Z Isomers and Polymers

1) a)
   [1 mark]   *[1 mark]*
   E-pent-2-ene *[1 mark]*   Z-pent-2-ene *[1 mark]*

   b) E/Z isomers occur because atoms can't rotate about C=C double bonds *[1 mark]*. Alkenes contain C=C double bonds and alkanes don't, so alkenes can form E/Z isomers and alkanes can't *[1 mark]*.

2) a) (i) *[1 mark]*   (ii) *[1 mark]*

b)

[1 mark]

Page 77 — Alcohols

1) a) Butan-1-ol *[1 mark]*, primary *[1 mark]*
   b) 2-methylpropan-2-ol *[1 mark]*, tertiary *[1 mark]*
   c) Butan-2-ol *[1 mark]*, secondary *[1 mark]*
   d) 2-methylpropan-1-ol *[1 mark]*, primary *[1 mark]*
2) a) Primary *[1 mark]*. The -OH group is bonded to a carbon with one alkyl group/other carbon atom attached *[1 mark]*.
   b) (i) $C_6H_{12}O_{6(aq)} \rightarrow 2C_2H_5OH_{(aq)} + 2CO_{2(g)}$ *[1 mark]*
   (ii) Yeast *[1 mark]*, temperature between 30 and 40 °C *[1 mark]*, anaerobic conditions OR air/oxygen excluded *[1 mark]*
   c) Ethene is cheap and abundantly available / It's a low-cost process / it's a high-yield process / Very pure ethanol is produced / Fast reaction *[1 mark each for up to two of these reasons]*. This might change in the future as crude oil reserves run out / become more expensive *[1 mark]*.

Page 79 — Oxidising Alcohols

1) a) (i) Acidified potassium dichromate(VI) *[1 mark]*
   (ii)

   propanal
   CH_3CH_2CHO *[1 mark]*

   b) (i) Warm with Fehling's/Benedict's solution: turns from blue to brick-red OR warm with Tollen's reagent: a silver mirror is produced *[1 mark for test, 1 mark for result]*
   (ii) Propanoic acid *[1 mark]*
   (iii) $CH_3CH_2CH_2OH + [O] \rightarrow CH_3CH_2CHO + H_2O$ *[1 mark]*
   $CH_3CH_2CHO + [O] \rightarrow CH_3CH_2COOH$ *[1 mark]*
   (iv) Distillation *[1 mark]*. This is so aldehyde is removed immediately as it forms *[1 mark]*.
   If you don't get the aldehyde out quick-smart, it'll be a carboxylic acid before you know it.

   c) (i)
   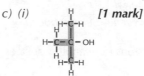
   [1 mark]

   (ii) 2-methylpropan-2-ol is a tertiary alcohol (which is more stable) *[1 mark]*.

Page 82 — Analytical Techniques

1) a) 44 *[1 mark]*
   b) X has a mass of 15. It is probably a methyl group/CH_3. *[1 mark]*
   Y has a mass of 29. It is probably an ethyl group/C_2H_5. *[1 mark]*
   c) *[1 mark]*

   d) If the compound was an alcohol, you would expect a peak with m/z ratio of 17 *[1 mark]*, caused by the OH fragment *[1 mark]*.
2 a) A's due to an O–H group in a carboxylic acid *[1 mark]*. B's due to a C=O in an aldehyde, ketone, acid or ester *[1 mark]*.
   b) The spectrum suggests it's a carboxylic acid — it's got a COOH group *[1 mark]*. This group has a mass of 45, so the rest of the molecule has a mass of 29 (74 – 45), which is likely to be C_2H_5 *[1 mark]*. So the molecule could be C_2H_5COOH — propanoic acid *[1 mark]*.

A2 Answers

Unit 4: Section 1 — Kinetics and Equilibria

Page 89 — Rate Equations

1 a) Experiments 1 and 2: [E] doubles and the initial rate doubles(with [D] remaining constant) **[1 mark]**. So it's 1st order with respect to [E] **[1 mark]**.
Experiments 1 and 3: [E] and [D] both halve, and the initial rate halves **[1 mark]**. Halving [E] alone would halve the rate, so changing [D] cannot affect the rate **[1 mark]**.
The reaction is 0 order with respect to [D] **[1 mark]**.
Always explain your reasoning carefully — state which concentrations are constant and which are changing.
b) rate = k[E] **[1 mark]**
c) The value of the rate constant, k, will increase **[1 mark]**.

Page 91 — More on Rates and Rate Equations

1 H^+ is acting as a catalyst **[1 mark]**. You know this because it is not one of the reactants in the chemical equation **[1 mark]**, but it does affect the rate of reaction/appear in the rate equation **[1 mark]**.

2 a) rate = k[H_2][ICl] **[1 mark]**
b) (i) One molecule of H_2 and one molecule of ICl (or something derived from these molecules) **[1 mark]**. If the molecule is in the rate equation, it must be in the rate determining step **[1 mark]**. The orders of the reaction tell you how many molecules of each reactant are in the rate-determining step **[1 mark]**.
(ii) Incorrect **[1 mark]**. H_2 and ICl are both in the rate equation, so they must both be in the rate-determining step OR the order of the reaction with respect to ICl is 1, so there must be only one molecule of ICl in the rate-determining step **[1 mark]**.

3 a) Rate = k[NO]^2[H_2] **[1 mark]**
b) (i) $0.00267 = k \times 0.004^2 \times 0.002$ **[1 mark]** $\Rightarrow k = 8.34 \times 10^4$
Units: $k = moldm^{-3}s^{-1}/[(moldm^{-3})^2 \times (moldm^{-3})]$
$= dm^6mol^{-2}s^{-1}$.
$k = 8.34 \times 10^4 \, dm^6mol^{-2}s^{-1}$
[1 mark for correct value of k, 1 mark for units].
(ii) It would decrease **[1 mark]**.
If the temperature decreases, the rate decreases too. A lower rate means a lower rate constant.

Page 94 — The Equilibrium Constant

1 $K_c = \dfrac{[H_2][I_2]}{[HI]^2}$ **[1 mark]**

At equilibrium, $[H_2] = [I_2]$ **[1 mark]**

$\Rightarrow [HI]^2 = \dfrac{[H_2][I_2]}{K_c} = \dfrac{0.77 \times 0.77}{0.02} = 29.6$ **[1 mark]**

$\Rightarrow [HI] = \sqrt{29.6} = 5.4 \, moldm^{-3}$ **[1 mark]**

2 a) (i) mass/M_r = 42.5/46 = 0.92 **[1 mark]**
(ii) moles of O_2 = mass/M_r = 14.1/32 = 0.44 **[1 mark]**
moles of NO = 2 × moles of O_2 = 0.88 **[1 mark]**
moles of NO_2 = 0.92 − 0.88 = 0.04 **[1 mark]**

b) Concentration of O_2 = 0.44 ÷ 22.8 = 0.019 $moldm^{-3}$
Concentration of NO = 0.88 ÷ 22.8 = 0.039 $moldm^{-3}$
Concentration of NO_2 = 0.04 ÷ 22.8 = 1.75 × 10^{-3} $moldm^{-3}$
[1 mark]

$K_c = \dfrac{[NO]^2[O_2]}{[NO_2]^2}$ **[1 mark]**

$\Rightarrow K_c = \dfrac{0.039^2 \times 0.019}{(1.75 \times 10^{-3})^2}$ **[1 mark]** = 9.4 **[1 mark]** $moldm^{-3}$ **[1 mark]**

(Units = (moldm^{-3})^2 × (moldm^{-3}) /(moldm^{-3})^2 = moldm^{-3})

3 a) $K_c = \dfrac{[CH_3COO^-][H^+]}{[CH_3COOH]}$ **[1 mark]**

b) Increasing the concentration of CH_3COOH will have no effect on the value of K_c **[1 mark]**.
K_c is fixed at a given temperature. Changing the concentration of a reactant will change the concentrations of the products, but it won't change K_c.
c) The equilibrium position will move to the right OR to the products OR forward reaction will increase **[1 mark]**.
The forward reaction is endothermic, so the equilibrium moves to the right to absorb the extra heat **[1 mark]**.
d) The value of K_c will increase **[1 mark]**.

Page 97 — Acids, Bases and pH

1 a) (i) A proton donor **[1 mark]**
(ii) A proton acceptor **[1 mark]**
b) (i) $HSO_4^- \rightarrow H^+ + SO_4^{2-}$ **[1 mark]**
(ii) $HSO_4^- + H^+ \rightarrow H_2SO_4$ **[1 mark]**
2 Weak acids dissociate (or ionise) a small amount **[1 mark]** to produce hydrogen ions (or protons) **[1 mark]**.
$HCN \rightleftharpoons H^+ + CN^-$ or $HCN + H_2O \rightleftharpoons H_3O^+ + CN^-$
[1 mark for correct formulas, 1 mark for equilibrium sign.]
3 a) A monoprotic acid means that each molecule of acid will release one proton when it dissociates OR each mole of acid will produce one mole of protons when it dissociates **[1 mark]**.
b) $[H^+] = 10^{-pH} = 10^{-0.55}$ **[1 mark]** = 0.28 $moldm^{-3}$ **[1 mark]**.
A strong acid will ionise fully in solution — so $[H^+] = [Acid]$. That's why the question tells you that HNO_3 is a strong acid.
4 $[NaOH] = [OH^-] = 0.125 \, moldm^{-3}$ **[1 mark]**.

$[H^+] = \dfrac{K_w}{[OH^-]} = \dfrac{1 \times 10^{-14}}{0.125} = 8 \times 10^{-14} \, moldm^{-3}$ **[1 mark]**.

$pH = -\log_{10}[H^+] = -\log_{10}(8 \times 10^{-14}) = 13.1$ **[1 mark]**.
A strong base ionises fully in solution too.
5 a) In pure water $[H^+] = [OH^-]$
So $K_w = [H^+][H^+] = [H^+]^2$ **[1 mark]**.
$[H^+] = \sqrt{K_w} = \sqrt{1.47 \times 10^{-14}} = 1.21 \times 10^{-7} \, moldm^{-3}$
[1 mark].
b) $pH = -\log_{10}[H^+] = -\log_{10}(1.21 \times 10^{-7}) = 6.92$ **[1 mark]**.

A2 Answers

Page 99 — More pH Calculations

1 a) $K_a = \frac{[H^+][A^-]}{[HA]}$ *[1 mark]*

b) $K_a = \frac{[H^+]^2}{[HA]}$ $\Rightarrow$ [HA] is 0.280 moldm^{-3} because very few HA will dissociate *[1 mark]*.

$[H^+] = \sqrt{(5.60 \times 10^{-4}) \times 0.280} = 0.0125$ moldm^{-3} *[1 mark]*

$pH = -\log_{10}[H^+] = -\log_{10}(0.0125) = 1.90$ *[1 mark]*

2 a) $[H^+] = 10^{-2.65}$ *[1 mark]* $= 2.24 \times 10^{-3}$ moldm^{-3}

$K_a = \frac{[H^+]^2}{[HX]}$ *[1 mark]* $= \frac{(2.24 \times 10^{-3})^2}{0.15}$

$= 3.34 \times 10^{-5}$ *[1 mark]* moldm^{-3} *[1 mark]*

b) $pK_a = -\log_{10}K_a = -\log_{10}(3.34 \times 10^{-5}) = 4.48$ *[1 mark]*

3 a) $K_a = \frac{[H^+]^2}{[HCN]}$ or $\frac{[H^+][CN^-]}{[HCN]}$ *[1 mark]*

b) $[H^+] = 10^{-4.15}$ *[1 mark]* $= 7.08 \times 10^{-5}$ moldm^{-3}

$K_a = \frac{[H^+]^2}{[HCN]} \Rightarrow [HCN] = \frac{[H^+]^2}{K_a}$ *[1 mark]*

$[HCN] = \frac{[H^+]^2}{K_a} = \frac{(7.08 \times 10^{-5})^2}{4.9 \times 10^{-10}} = 10.23$ moldm^{-3} *[1 mark]*

c) $pK_a = -\log_{10}K_a = -\log_{10}(6.17 \times 10^{-10}) = 9.31$ *[1 mark]*

4 $K_a = 10^{-pKa} = 10^{-3.16} = 6.92 \times 10^{-4}$ moldm^{-3} *[1 mark]*

$K_a = \frac{[H^+]^2}{[HF]} \Rightarrow [H^+]^2 = K_a [HF]$ *[1 mark]*

$K_a [HF] = (6.92 \times 10^{-4})(0.5) = 3.46 \times 10^{-4}$ moldm^{-3} *[1 mark]*

$[H^+] = \sqrt{[H^+]^2} = \sqrt{3.46 \times 10^{-4}} = 1.86 \times 10^{-2}$ moldm^{-3} *[1 mark]*

$pH = -\log_{10}[H^+] = -\log_{10}(1.86 \times 10^{-2}) = 1.73$ *[1 mark]*

Page 101 — pH Curves, Titrations and Indicators

1 a) Any strong monoprotic acid, e.g. HCl, HNO$_3$ *[1 mark]*
   Any strong alkali, e.g. NaOH, KOH *[1 mark]*
   The shape tells you that this is a strong acid/strong base titration curve.

b) Any indicator that changes colour between pH 3 and pH 11, e.g. methyl orange, phenolphthalein *[1 mark]*.

2 a)

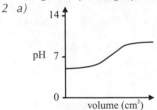

*[1 mark for general shape correct,
1 mark for approximate start and finish pHs correct]*

b) pH 7 *[1 mark]*

3 a) Y *[1 mark]*

b) $NaOH_{(aq)} + CH_3COOH_{(aq)} \rightarrow$
   $CH_3COONa_{(aq)}$ *[1 mark]* $+ H_2O_{(l)}$
   [1 mark for correctly balanced equation]

c) Any indicator that changes colour between pH 7 and pH 12, e.g. phenolphthalein, cresolphthalein, thymolphthalein *[1 mark]*. This indicator is suitable because it changes colour over the vertical part of the pH curve OR it will change colour at the equivalence point/end point of the titration *[1 mark]*.

Page 103 — Titration Calculations

1 a) $HCl_{(aq)} + NaOH_{(aq)} \rightarrow NaCl_{(aq)} + H_2O_{(l)}$ *[1 mark]*

b) i) 25.6 cm^3 *[1 mark]*

   ii) Moles NaOH $= \frac{0.1 \times 25.6}{1000}$ *[1 mark]*

   $= 2.56 \times 10^{-3}$ *[1 mark]*

c) Concentration HCl $= \frac{(2.56 \times 10^{-3}) \times 1000}{25}$ *[1 mark]*

   $= 0.1024$ moldm^{-3} *[1 mark]*

2 a) $H_2SO_{4(aq)} + 2NaOH_{(aq)} \rightarrow Na_2SO_{4(aq)} + 2H_2O_{(l)}$ *[1 mark]*

b) i) Moles NaOH $= \frac{0.1 \times 35.65}{1000}$ *[1 mark]*

   $= 3.565 \times 10^{-3}$ *[1 mark]*

   ii) Moles H_2SO_4 = Moles NaOH $\div 2 = 1.78 \times 10^{-3}$ *[1 mark]*

   iii) Concentration $H_2SO_4 = \frac{(1.78 \times 10^{-3}) \times 1000}{25}$ *[1 mark]*

   $= 0.07$ moldm^{-3} *[1 mark]*

3 Equation: $HCl_{(aq)} + NaOH_{(aq)} \rightarrow NaCl_{(aq)} + H_2O_{(l)}$ *[1 mark]*

   Moles NaOH $= \frac{0.25 \times 10}{1000}$ *[1 mark]* $= 2.5 \times 10^{-3}$ *[1 mark]*

   Moles HCl = Moles NaOH $= 2.5 \times 10^{-3}$ *[1 mark]*

   Volume HCl $= \frac{(2.5 \times 10^{-3}) \times 1000}{0.1}$ *[1 mark]* $= 25$ cm^3 *[1 mark]*.

Page 105 — Buffer Action

1 a) $K_a = \frac{[C_6H_5COO^-][H^+]}{[C_6H_5COOH]}$ *[1 mark]*

   $\Rightarrow [H^+] = 6.4 \times 10^{-5} \times \frac{0.40}{0.20} = 1.28 \times 10^{-4}$ moldm^{-3} *[1 mark]*

   $pH = -\log_{10}1.28 \times 10^{-4} = 3.9$ *[1 mark]*

b) $C_6H_5COOH \rightleftharpoons H^+ + C_6H_5COO^-$ *[1 mark]*
   Adding H_2SO_4 increases the concentration of H$^+$ *[1 mark]*. The equilibrium shifts left to reduce the concentration of H$^+$, so the pH will only change very slightly *[1 mark]*.

2 a) $CH_3(CH_2)_2COOH \rightleftharpoons H^+ + CH_3(CH_2)_2COO^-$ *[1 mark]*

b) $[CH_3(CH_2)_2COOH] = [CH_3(CH_2)_2COO^-]$,
   so $[CH_3(CH_2)_2COOH] \div [CH_3(CH_2)_2COO^-] = 1$ *[1 mark]*
   and $K_a = [H^+]$. $pH = -\log_{10}1.5 \times 10^{-5}$ *[1 mark]* $= 4.8$ *[1 mark]*
   If the concentrations of the weak acid and the salt of the weak acid are equal, they cancel from the K_a expression and the buffer $pH = pK_a$.

Unit 4: Section 2 — Basic Organic Chemistry & Isomerism

Page 107 — Naming Organic Compounds

1 a) 1,4-dibromobutane *[1 mark]*.

b) 2,3-dibromobutane *[1 mark]*.

c) $CH_2=CH-CH=CH_2$ OR

   Buta-1,3-diene OR 1,3-butadiene *[1 mark]*.

2 a)

   Butan-2-ol *[1 mark]*.
   As long as you've drawn one OH group on the second carbon along the chain, counting from either end, you've got butan-2-ol.

b) But-1-ene *[1 mark]*. But-2-ene *[1 mark]*.

A2 Answers

Page 109 — Formulas and Isomers

1 a) Any two from: pentan-1-ol, pentan-2-ol, pentan-3-ol *[1 mark for each]*.
 b) E-pent-2-ene *[1 mark]*.
 Z-pent-2-ene *[1 mark]*.
 c) propene *[1 mark]*.
 cyclopropane *[1 mark]*.

2 a) $C_4H_{10}O$ *[1 mark]*.
 b) $C_nH_{2n+2}O$ *[1 mark]*.
 c) $C_{10}H_{22}O$ *[1 mark]*.

 d) 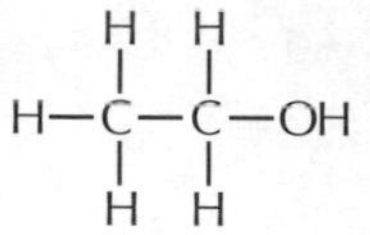 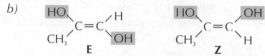 *[1 mark]*.

Page 111 — Optical Isomerism

1 a) The property of having stereoisomers, which are molecules with the same molecular formula and with their atoms arranged in the same way *[1 mark]*, but with a different orientation of the bonds in space *[1 mark]*.

 b)

 [1 mark for each correct structure, 1 mark for correct E/Z labels.]

 c) (i)

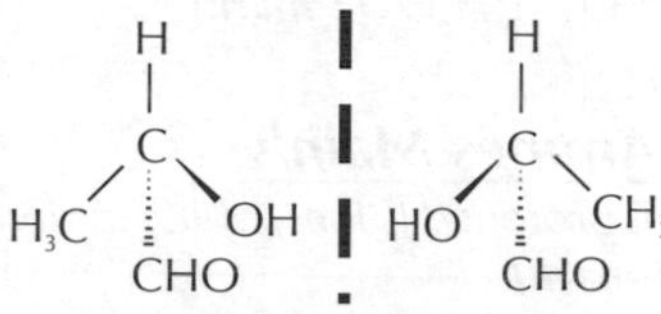

 [1 mark for each correctly drawn structure — they don't have to be orientated in the same way as in the diagram above, as long as the molecules are mirror images of each other.]
 (ii) An asymmetric carbon/a chiral carbon/a carbon with four different groups attached *[1 mark]*.
 (iii) Shine (monochromatic) plane-polarised light through a solution of the molecule *[1 mark]*. The enantiomers will rotate the light in opposite directions *[1 mark]*.

2 a)

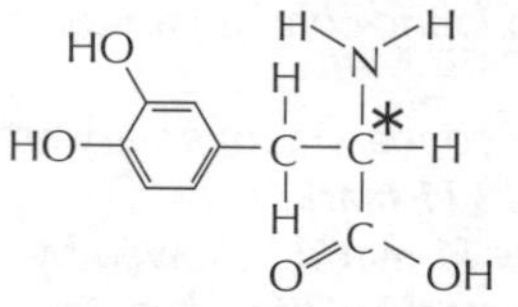

 [1 mark for chiral carbon clearly marked]
 The chiral carbon is the one with 4 different groups attached.
 b) (i) A mixture of equal quantities of each enantiomer of an optically active compound *[1 mark]*.
 (ii) Smaller doses needed/drug is more effective *[1 mark]*, fewer side-effects (because there's no D-DOPA enantiomer) *[1 mark]*.

Unit 4: Module 3 — Carbonyl Compounds

Page 113 — Aldehydes and Ketones

1 a) Propanal *[1 mark]* *[1 mark]*

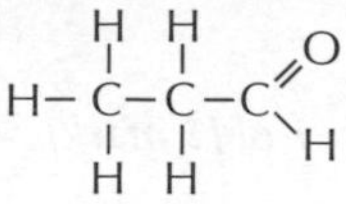

 Propanone *[1 mark]* *[1 mark]*

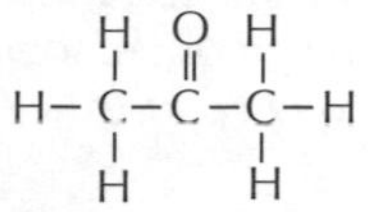

 b) (i) Nucleophilic addition *[1 mark]*
 (ii) From propanal, *[1 mark]*

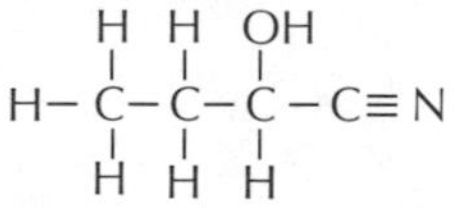

 This product *[1 mark]* will be produced in a racemic mixture, because it's asymmetrical/contains a chiral carbon *[1 mark]*.

 From propanone, *[1 mark]*

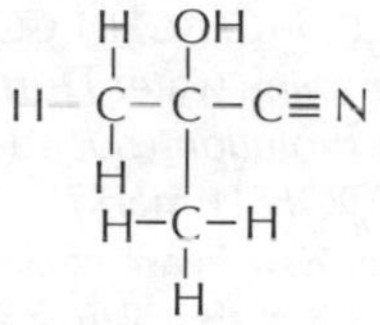

 c) (i) $CH_3CH_2CHO + 2[H] \rightarrow CH_3CH_2CH_2OH$ *[1 mark]*
 (ii) e.g. $NaBH_4$ *[1 mark]* dissolved in water with methanol *[1 mark]*.

2 a) Butanal *[1 mark]* and butanone (or butan-2-one) *[1 mark]*
 b) EITHER
 Tollens' reagent (or silver nitrate dissolved in aqueous ammonia) *[1 mark]*
 silver mirror with butanal *[1 mark]*
 no reaction with butanone *[1 mark]*
 OR
 Fehling's or Benedict's solution (or copper II ions dissolved in NaOH or Na_2CO_3) *[1 mark]*
 brick-red precipitate with butanal *[1 mark]*
 no reaction with butanone *[1 mark]*
 c) butanoic acid *[1 mark]*

Page 115 — Carboxylic Acids and Esters

1 a) $2CH_3COOH_{(aq)} + Na_2CO_{3(s)} \rightarrow$
 $2CH_3COONa_{(aq)} + H_2O_{(l)} + CO_{2(g)}$
 [1 mark for CH_3COONa, 1 mark for CO_2, and 1 mark for correctly balancing the equation.]
 b) methanol *[1 mark]*, esterification (or condensation) *[1 mark]*.
 Substance X is a carboxylic acid (ethanoic acid) and substance Y is an ester (methyl ethanoate).

2 a)

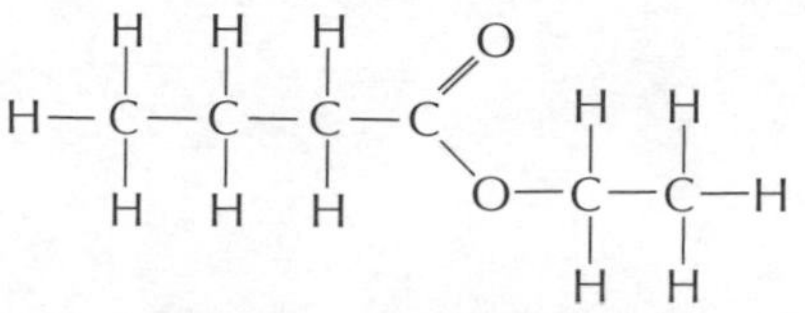

 [1 mark]

 b) Flavouring OR a perfume OR a plasticiser *[1 mark]*

3 a) $CH_3COOH + CH_3CH(CH_3)CH_2CH_2OH \rightarrow$
 $CH_3COOCH_2CH_2CH(CH_3)CH_3 + H_2O$ *[1 mark]*
 b) ethanoic acid *[1 mark]*
 c) Heat OR warm OR reflux *[1 mark]* and (concentrated sulfuric) acid catalyst *[1 mark]*.

A2 Answers

Page 117 — More on Esters

1 a) 2-methylpropyl ethanoate *[1 mark]*

b) Ethanoic acid *[1 mark]*

[1 mark]

2-methylpropan-1-ol *[1 mark]* *[1 mark]*

This is acid hydrolysis *[1 mark]*

c) With sodium hydroxide, sodium ethanoate is produced, but in the reaction in part (b), ethanoic acid is produced *[1 mark]*.

2 a)

[1 mark]

b) $CH_3(CH_2)_7CH=CH(CH_2)_7COONa + H^+ \rightarrow$ $CH_3(CH_2)_7CH=CH(CH_2)_7COOH + Na^+$ *[1 mark]*

c) Shake with bromine water *[1 mark]*. It will turn colourless with oleic acid solution (due to C=C), but not with $CH_3(CH_2)_{16}COOH$ *[1 mark]*.

You might never have heard of oleic acid before, never mind learned a test for it. But you're told it's got a double bond, and you should know how to test for one of them.

Page 119 — Acyl Chlorides

1 a) Ethanoyl chloride:
$CH_3COCl + CH_3OH \rightarrow CH_3COOCH_3 + HCl$ *[1 mark]*
Ethanoic anhydride:
$(CH_3CO)_2O + CH_3OH \rightarrow CH_3COOCH_3 + CH_3COOH$ *[1 mark]*

ethyl methanoate *[1 mark]*

b) Vigorous reaction OR (HCl) gas/fumes produced *[1 mark]*

c) Irreversible reaction OR faster reaction *[1 mark]*

2 a) $CH_3COCl + CH_3CH_2NH_2 \rightarrow CH_3CONHCH_2CH_3 + HCl$
[1 mark for correct reactants, 1 mark for correct products.]
N-ethylethanamide *[1 mark]*

b)

[1 mark for each curly arrow on first diagram.]
[1 mark for all 3 curly arrows on second and third diagrams.]
[1 mark for correct structures.]

Unit 4: Section 4 — More Organic Chemistry

Page 122 — Aromatic Compounds

1 The Kekulé model suggests that benzene has 3 C=C bonds. If this was true, the enthalpy change for benzene + $3H_2$ should be approximately $3 \times -120 = -360$ kJ mol^{-1} *[1 mark]*. The actual value of -208 kJ mol^{-1} means that the reaction is much less exothermic than this, so the Kekulé model does not fit the data *[1 mark]*.

2 a) Conditions: dry ether, reflux *[1 mark]*

b) $H_3C-C^+=O$ *[1 mark]*

c) *[1 mark]*

d) For example, the aluminium only has 6 electrons in its outer shell, or the aluminium has an incomplete outer shell *[1 mark]*.

3 a) A: nitrobenzene *[1 mark]*
B + C: concentrated nitric acid *[1 mark]* and concentrated sulfuric acid *[1 mark]*
D: warm, not more than 55 °C *[1 mark]*
When you're asked to name a compound, write the name, not the formula.

b) See page 122 for the mechanism.
[1 mark for each of the three steps.]

c) $HNO_3 + H_2SO_4 \rightarrow H_2NO_3^+ + HSO_4^-$ *[1 mark]*
$H_2NO_3^+ \rightarrow NO_2^+ + H_2O$ *[1 mark]*

Page 125 — Amines Mainly

1 a) It can accept protons/H^+ ions, or it can donate a lone pair of electrons *[1 mark]*.

b) Methylamine is stronger, as the nitrogen lone pair is more available *[1 mark]* — the methyl group/CH_3 pushes electrons onto/increases electron density on the nitrogen *[1 mark]*. Phenylamine is weaker, as the nitrogen lone pair is less available *[1 mark]* — nitrogen's electron density is decreased as it's partially delocalised around the benzene ring *[1 mark]*.

2 a) $CH_3CH_2NH_2 + 3 CH_3Br \rightarrow CH_3CH_2(CH_3)_3N^+Br^- + 2 HBr$
[1 mark] for correct structure/formula of the quaternary salt
[1 mark] for correct balanced equation

b) An excess of CH_3Br *[1 mark]*.

c) For example, as a surfactant, in fabric conditioners or in hair products *[1 mark]*.

3 a) You get a mixture of primary, secondary and tertiary amines, and quaternary ammonium salts *[1 mark]*.

b) (i) $LiAlH_4$ and dry diethyl ether *[1 mark]*, followed by dilute acid *[1 mark]* OR reflux *[1 mark]* with sodium metal and ethanol *[1 mark]*.

(ii) It's too expensive *[1 mark]*.

(iii) Metal catalyst such as platinum or nickel *[1 mark]* and high temperature and pressure *[1 mark]*.

A2 Answers

4 a) $CH_3CH_2NH_2 + CH_3CH_2Br \rightarrow (CH_3CH_2)_2NH + HBr$ **[1 mark]**
   The product is diethylamine **[1 mark]**
   b) nucleophilic substitution **[1 mark]**

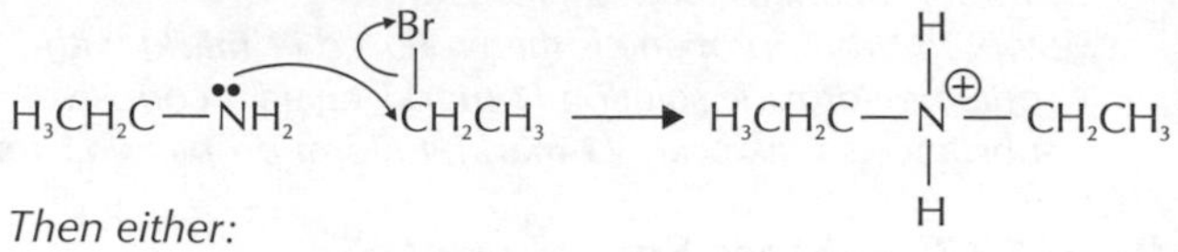

Then either:

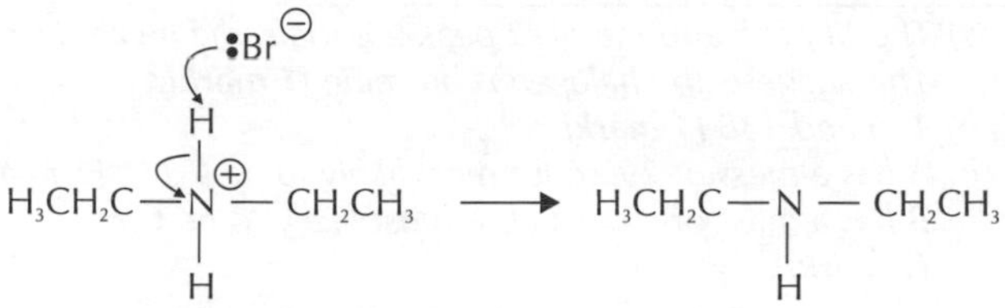

OR

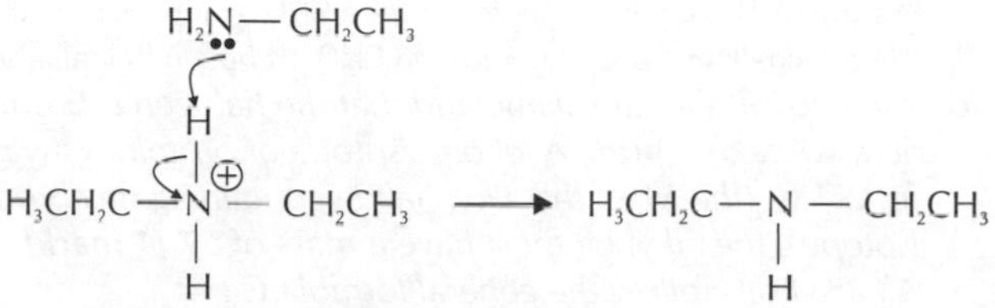

[1 mark for both curly arrows on first diagram.]
[1 mark for both curly arrows on second diagram.]
[1 mark for correct structures.]

Page 127 — Amino Acids and Proteins

1 a)

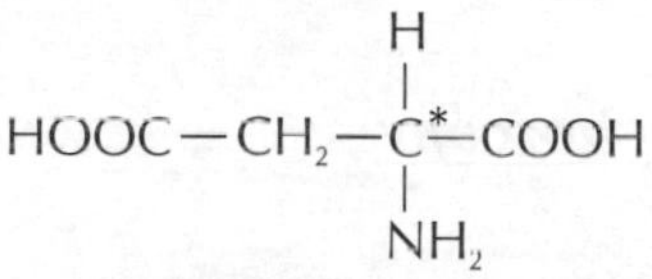

[1 mark for the correct structure, and 1 mark for labelling the chiral carbon]

   b) 2-aminobutanedioic acid **[1 mark]**
   (accept 2-amino-1,4-butanedioic acid)

2 a)

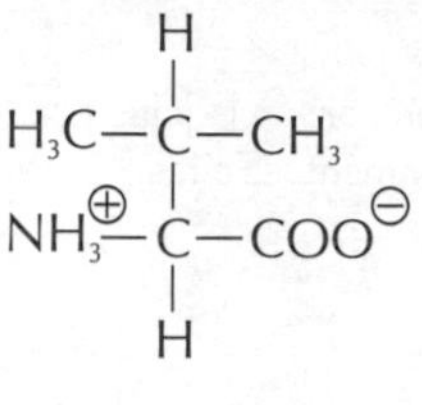

[1 mark]

   b)

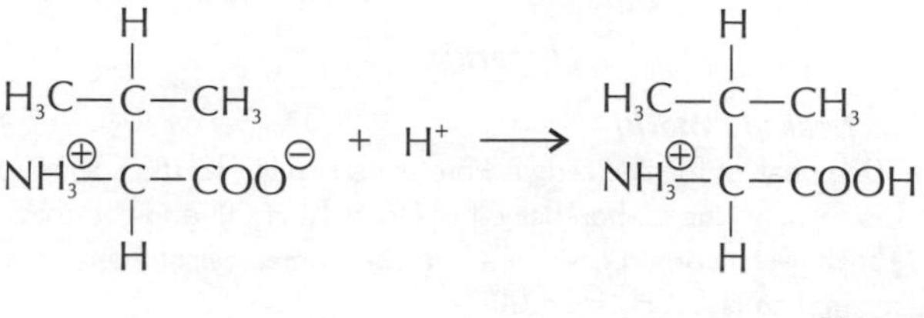

[1 mark]

3 a)

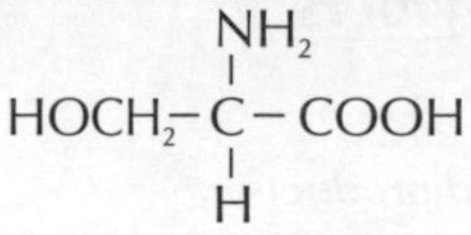

[1 mark]

You'll need a cool head for this one. Start with a 3-carbon chain and then add the groups — you start numbering the carbons from the one with the carboxyl group.

   b) (i) Two amino acids joined together **[1 mark]** by a peptide/amide/–CONH– link **[1 mark]**.
      ii)

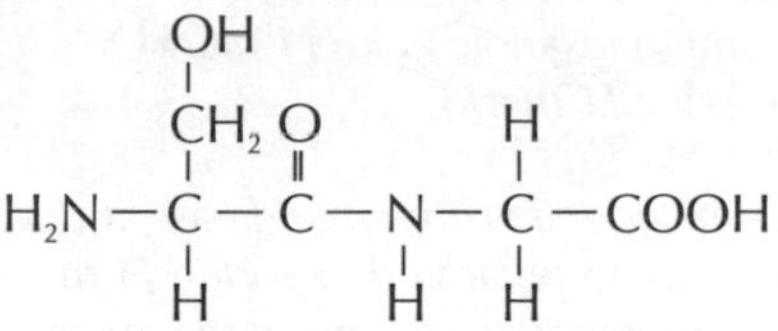

[1 mark]

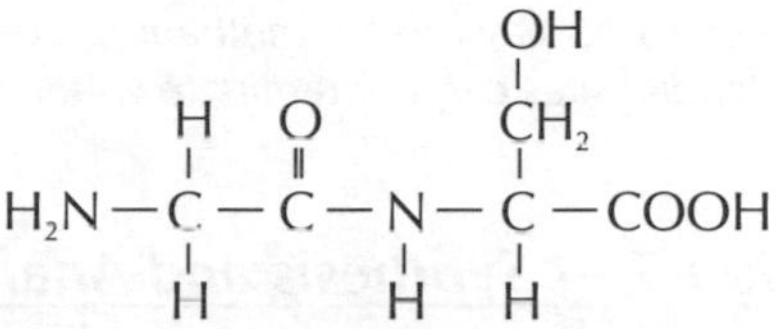

[1 mark]

The amino acids can join together in either order — that's why there are two dipeptides.

   c) Hot aqueous 6 M hydrochloric acid/HCl **[1 mark]**, reflux for 24 hours **[1 mark]**.

Page 129 — Polymers

1 a) Polypropene **[1 mark]**
   b) Addition polymerisation **[1 mark]**
   c)

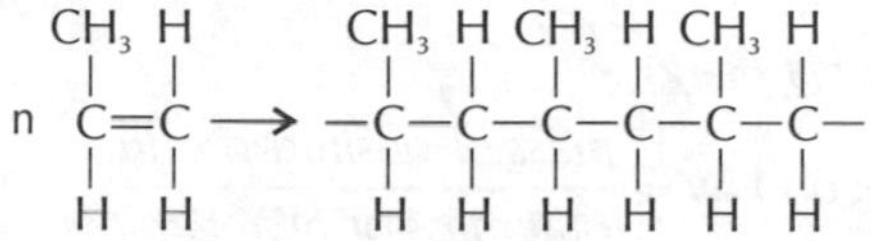

[1 mark]

2 a) (i)

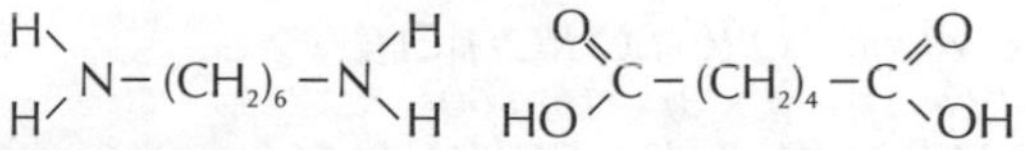

[1 mark each]
      (ii) There are six carbon atoms in each monomer/reagent **[1 mark]**.
         Don't forget to count the carbons in the carboxyl groups too — that's the only way the name will make sense.
      (iii) Amide/peptide link **[1 mark]**
   b) (i)

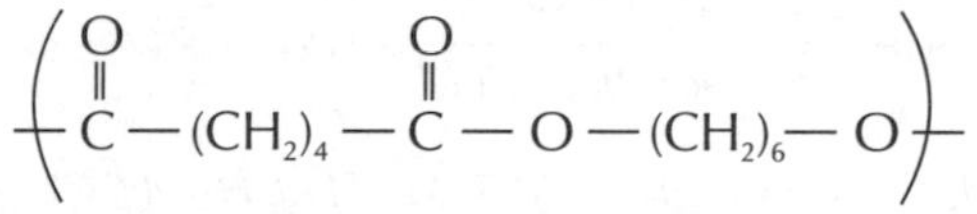

or

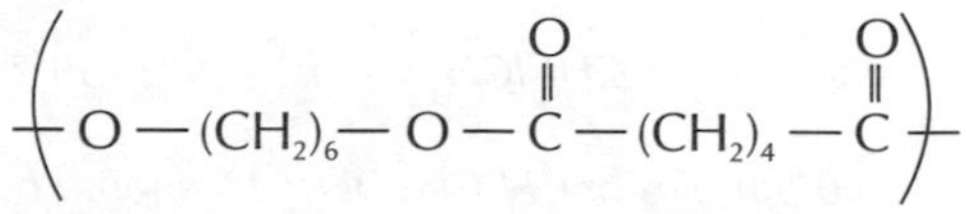

[1 mark]

      (ii) For each link formed, one small molecule (water) is eliminated **[1 mark]**.

A2 Answers

Page 131 — More About Polymers

1 a) $CaCO_3$ *[1 mark]*
   $CaCO_3 + 2HCl \rightarrow CO_2 + H_2O + CaCl_2$
   [1 mark for correct reactants and products, 1 mark for balancing]
   b) CO_2 may contribute to global warming/greenhouse effect *[1 mark]*

2 a) (i) A — condensation polymer/polypeptide/protein/polyamide *[1 mark]*
      B — addition polymer *[1 mark]*
      (ii) A — 6-aminohexanoic acid *[1 mark]*
      B — propene *[1 mark]*
   b) A *[1 mark]*
   c) Advantages include: it is cheap and easy, it doesn't require waste plastics to be separated or sorted *[1 mark]*.
   Disadvantages include: it requires large areas of land, decomposing waste may release methane/greenhouse gases, leaks from landfill sites can contaminate water supplies *[1 mark]*.

Unit 4: Section 5 — Synthesis and Analysis

Page 134 — Organic Synthesis and Analysis

1 a) Shake with bromine water *[1 mark]*.
   No reaction with A *[1 mark]*.
   Turns colourless with B *[1 mark]*.
   b) Tollens' reagent OR Fehling's/Benedict's solution *[1 mark]*
   No reaction with C *[1 mark]*.
   Tollens' reagent gives a silver mirror with D OR Fehling's/Benedict's solution gives a brick red precipitate with D *[1 mark]*.

2 a) propan-2-ol *[1 mark]*

   b) $\%\ atom\ economy = \dfrac{mass\ of\ desired\ product}{total\ mass\ of\ all\ reactants} \times 100$

   M_r of Alcohol Q ($CH_3CH(OH)CH_3$)
   $= (12 \times 3) + (1 \times 8) + 16 = 60$
   OR Mass of Alcohol Q ($CH_3CH(OH)CH_3$) = 60 g *[1 mark]*
   Total M_r of all reactants = $M_r(CH_3CHClCH_3) + M_r(NaOH)$
   $= 78.5 + 40 = 118.5$
   OR Total mass of all reactants = 78.5 g + 40 g = 118.5 g *[1 mark]*
   % atom economy = $(60 \div 118.5) \times 100 = 50.6$ *[1 mark]*

   c) $\%\ yield = \dfrac{actual\ yield}{theoretical\ yield} \times 100$

   $M_r(CH_3CHClCH_3) = 78.5$, $M_r(CH_3CH(OH)CH_3) = 60$ *[1 mark]*
   moles of $CH_3CHClCH_3$ = mass ÷ M_r = 39.25 ÷ 78.5 = 0.5 *[1 mark]*
   So 0.5 moles of $CH_3CH(OH)CH_3$ should be produced.
   Mass of 0.5 moles of $CH_3CH(OH)CH_3$ = 0.5 × 60 = 30 g *[1 mark]*
   Percentage yield = (24.9 ÷ 30) × 100 = 83% *[1 mark]*

   d) $K_2Cr_2O_7$/potassium dichromate and H_2SO_4/sulfuric acid *[1 mark]*
   Heat and reflux *[1 mark]*

3 Step 1: The methanol is refluxed *[1 mark]* with $K_2Cr_2O_7$ *[1 mark]* and sulfuric acid *[1 mark]* to form methanoic acid *[1 mark]*.
   Step 2: The methanoic acid is reacted under reflux *[1 mark]* with ethanol *[1 mark]* using an acid catalyst *[1 mark]*.

4 Step 1: React propane with bromine *[1 mark]* in the presence of UV light *[1 mark]*. Bromine is toxic and corrosive *[1 mark]* so great care should be taken. Bromopropane is formed *[1 mark]*.
   Step 2: Bromopropane is then refluxed *[1 mark]* with sodium hydroxide solution *[1 mark]*, again a corrosive substance so take care *[1 mark]*, to form propanol *[1 mark]*.

Page 137 — Mass Spectrometry

1 a) The M peak and the M+2 peak are the same height *[1 mark]*. This suggests the halogen is bromine *[1 mark]*.
   b) 136 and 138 *[1 mark]*
   c) A has a mass of 29, so it's most likely to be $C_2H_5^+$ *[1 mark]*.
   B has a mass of 43, which is most likely to be $C_3H_7^+$ *[1 mark]*.
   C has a mass of 57, which is most likely to be $C_4H_9^+$ *[1 mark]*.
   Fragment B has the same mass as a CHO group — but you know it's a haloalkane, and there are no CHO groups in haloalkanes.
   d) You know it's a haloalkane and that the halogen is bromine. So if we take 79 (the A_r of one isotope of bromine) away from 136 (the M_r of the alkyl halide containing this bromine isotope), the alkyl bit must have a mass of 57 *[1 mark]*. All alkyl bits follow the general formula C_nH_{2n+1}. n must be 4 to make this add up to 57. So the molecular formula is C_4H_9Br *[1 mark]*. To give fragment C, the bromine atom must be on an end carbon *[1 mark]*.

   [1 mark]

Page 139 — NMR Spectroscopy

1 a)
   [1 mark]

   4 peaks *[1 mark]*
   This molecule has no symmetry — each carbon is joined to different groups and in a unique environment. So its ^{13}C NMR spectrum has four peaks.

   b)
   [1 mark]

   3 peaks *[1 mark]*
   One peak is for the red carbon (joined to $H_2ClCH(CH_3)_2$), another is for the blue carbon (joined to $H(CH_3)_2CH_2Cl$), and the third is for both green carbons, which are in the same environment (both joined to $H_3C(CH_3)HCH_2Cl$).

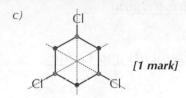

A2 Answers

c)

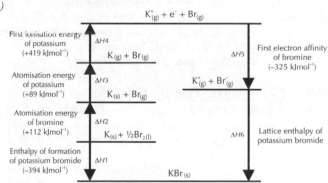

[1 mark]

2 peaks [1 mark]
This molecule has three lines of symmetry — the three (blue) carbons with Cl atoms attached to them are all in the same environment, and the three (red) carbons that don't are all in the same environment. So its ^{13}C NMR spectrum has two peaks.

2 a)

H OH H   H H H
H–C–C–C–H   H–C–C–C–OH
H H H   H H H

[1 mark for each]

b)

H OH H
H–C–C–C–H   OR propan-2-ol *[1 mark]*
H H H

There are only two peaks on the spectrum. So there must be exactly 2 different carbon environments in the isomer [1 mark].

c) *Tetramethylsilane/TMS/$Si(CH_3)_4$ [1 mark]*

Page 141 — More About NMR

1 a) *A CH_2 group adjacent to a halogen OR $R–CH_2–Cl$ [1 mark].*
You gotta read the question carefully — it tells you it's a haloalkane. So the group at 3.6 ppm can't have oxygen in it. It can't be halogen-CH_3 either, as this has 3 hydrogens in it.

b) *A CH_3 group OR $R–CH_3$ [1 mark].*

c) *CH_2 added to CH_3 gives a mass of 29, so the halogen must be chlorine with a mass of 35.5 [1 mark]. So a likely structure is CH_3CH_2Cl [1 mark].*

d) *The quartet at 3.6 ppm is caused by 3 protons on the adjacent carbon [1 mark]. The n +1 rule tells you that 3 protons give 3 + 1 = 4 peaks [1 mark].*
Similarly the triplet at 1.3 ppm is due to 2 adjacent protons [1 mark] giving 2 + 1 = 3 peaks [1 mark].

2 a) *3 [1 mark]*
The three different hydrogen environments come from the H atoms that are attached to three different carbons in this molecule:

H
H–C–C=O
H   H H
    O–C–C–H
        H H

b) *3:2:3 [1 mark]*

c) *singlet, quartet, triplet [1 mark]*

Page 144 — IR Spectroscopy and Chromatography

1 a) *Gas liquid chromatography [1 mark]*

b) *Different substances have different retention times [1 mark]. The retention time of substances in the sample is compared against that for ethanol [1 mark].*

c) *It is unreactive/does not react with the sample [1 mark].*

2 a) *A's due to an O–H group in a carboxylic acid [1 mark]. B's due to a C=O as in an aldehyde, ketone, acid or ester [1 mark]. C's due to a C–O as in an alcohol, ester or acid [1 mark]. D's also due to a C–O as in an alcohol, ester or acid [1 mark].*

b) *The spectrum suggests it's a carboxylic acid — it's got a COOH group [1 mark]. This group has a mass of 45, so the rest of the molecule has a mass of 29 (74 – 45), which is likely to be C_2H_5 [1 mark]. So the molecule could be C_2H_5COOH — propanoic acid [1 mark].*

3 a) i) *$CH_3COOCH_2CH_3$ [1 mark]*
ii) *CH_3COOH [1 mark]*
You can work out which carboxylic was used by looking at the number of carbons — the ester has 4 and ethanol has 2, so the carboxylic acid must also have 2 — i.e. it's ethanoic acid.

b) *Ethanoic acid [1 mark]. The peak is caused by C=O [1 mark]. This is present in carboxylic acids but not in alcohols [1 mark].*

c) i) *The broad peak at around 2500 – 3750 cm^{-1} that is caused by the O–H bond in both ethanol [1 mark] and ethanoic acid [1 mark] would not be present in the spectrum of the ester [1 mark].*
ii) *The peak in the region 1680 – 1800 cm^{-1} which is caused by the ethanoic acid would still be present in the ester [1 mark] because it also has a C=O group [1 mark]. The peak in the region 2800 – 3100 cm^{-1} would still be present [1 mark] because esters, alcohols and carboxylic acids all have C–H bonds [1 mark].*

Unit 5: Section 1 — Thermodynamics

Page 147 — Lattice Enthalpies

1 a)

[1 mark for left of cycle. 1 mark for right of cycle. 1 mark for formulas/state symbols. 1 mark for correct directions of arrows.]

b) *Lattice enthalpy, $\Delta H6 = -\Delta H5 - \Delta H4 - \Delta H3 - \Delta H2 + \Delta H1$*
$= -(-325) - (+419) - (+89) - (+112) + (-394)$ [1 mark]
$= -689$ [1 mark] $kJmol^{-1}$ [1 mark]
Award marks if calculation method matches cycle in part (a).

2 a) *X: 2 × the atomisation enthalpy of chlorine OR the bond dissociation enthalpy of chlorine [1 mark]*
Y: the 2nd ionisation enthalpy of copper [1 mark]
Z: 2 × the 1st electron affinity of chlorine [1 mark]

b) *The enthalpy of formation of copper(II) chloride [1 mark]. The lattice enthalpy of copper(II) chloride [1 mark]. The atomisation enthalpy of copper [1 mark].*

A2 Answers

Page 149 — Enthalpies of Solution

1 a)

$SrF_{2(s)}$ $\xrightarrow{\text{Enthalpy change of solution}}$ $Sr^{2+}_{(aq)} + 2F^-_{(aq)}$

lattice enthalpy of dissociation

Enthalpy of hydration of $Sr^{2+}_{(g)}$
Enthalpy of hydration of $2F^-_{(g)}$

$Sr^{2+}_{(g)} + 2F^-_{(g)}$

[1 mark for each of the 4 enthalpy changes labelled, 1 mark for a complete, correct cycle.]

Don't forget — you have to double the enthalpy of hydration for F^- because there are two in SrF_2.

b) $+2492 + (-1480) + (2 \times -506)$ **[1 mark]** $= 0$ $kJmol^{-1}$ **[1 mark]**

2 By Hess's law:

Enthalpy change of solution $(MgCl_{2(s)})$
= $-$lattice enthalpy of formation $(MgCl_{2(s)})$
+ enthalpy of hydration $(Mg^{2+}_{(g)})$
+ $[2 \times$ enthalpy of hydration $(Cl^-_{(g)})]$ **[1 mark]**
= $-(-2526) + (-1920) + [2 \times (-364)]$ **[1 mark]**
= $2526 - 1920 - 728 = -122$ $kJmol^{-1}$ **[1 mark]**

3 a)

Bonds broken	Bonds formed
$4 \times C\text{–}H = 4 \times 412 = 1648$	$2 \times C=O = 2 \times 743 = 1486$
$2 \times O=O = 2 \times 496 = 992$	$4 \times O\text{–}H = 4 \times 463 = 1852$

[1 mark]

$(1648 + 992) - (1486 + 1852) = -698$ $kJmol^{-1}$ **[1 mark]**
[1 mark for correct (negative) sign on final answer.]

b) Some of the bond enthalpies used are averages **[1 mark]**. These may be different from the actual ones in the substances **[1 mark]**.
OR enthalpy cycles use data for the actual compounds shown **[1 mark]**. **Maximum 2 marks.**

Page 151 — Entropy Change

1 a) Reaction is not likely to be spontaneous **[1 mark]** because there is a decrease in entropy **[1 mark]**.
Remember — more particles means more entropy.
There's 1½ moles of reactants and only 1 mole of products (and this mole is a solid, which has a lower entropy than a gas).

b) $\Delta S_{system} = 26.9 - (32.7 + 102.5)$ **[1 mark]**
$= -108.3$ **[1 mark]** $JK^{-1}mol^{-1}$ **[1 mark]**

Page 153 — Free-Energy Change

1 a) (i) $\Delta S_{system} = 48 - 70 = -22$ $JK^{-1}mol^{-1}$ **[1 mark]**
$\Delta S_{surroundings} = -(-6000)/250 = +24$ $JK^{-1}mol^{-1}$ **[1 mark]**
$\Delta S_{total} = \Delta S_{system} + \Delta S_{surroundings} = -22 + 24 = +2$ $JK^{-1}mol^{-1}$ **[1 mark]**

(ii) $\Delta S_{surroundings} = -(-6000)/300 = +20$ $JK^{-1}mol^{-1}$ **[1 mark]**
$\Delta S_{total} = \Delta S_{system} + \Delta S_{surroundings} = -22 + 20 = -2$ $JK^{-1}mol^{-1}$ **[1 mark]**

b) 250 K: $\Delta G = -6 \times 10^3 - 250 \times -22 = -500$.
ΔG is negative so the reaction is feasible at 250 K **[1 mark]**.
At 300 K, $\Delta G = -6 \times 10^3 - 300 \times -22 = 600$
ΔG is positive so the reaction is not feasible at 300 K **[1 mark]**.

2 a) $\Delta G = 117 \times 10^3 - (550 \times 175) = +20\,750$ $Jmol^{-1}$ **[1 mark]**.
The value of ΔG is positive at $550K$, so the reaction will not occur spontaneously at this temperature **[1 mark]**.

b) $T = \Delta H \div \Delta S_{system} = 117 \times 10^3 \div 175$ **[1 mark]** $= 668.6$ K **[1 mark]**.

Unit 5: Section 2 — Period 3 and Redox Equilibria

Page 155 — Period 3 Elements and Oxides

1 a) SO_3 **[1 mark]**, $SO_3 + H_2O \rightarrow H_2SO_4$ **[1 mark]**
b) (i) Na_2O **[1 mark]**, $Na_2O + H_2O \rightarrow 2NaOH$ **[1 mark]**
(ii) Na_2O has a giant lattice structure **[1 mark]** with strong ionic bonds **[1 mark]**, that take a lot of energy to break **[1 mark]**.

2 a) $P_4 + 5O_2 \rightarrow P_4O_{10}$ or $4P + 5O_2 \rightarrow P_4O_{10}$ **[1 mark]**
b) (i) $P_4O_{10} + 6H_2O \rightarrow 4H_3PO_4$ **[1 mark]**
(ii) The solution is acidic, with a pH of 0-2 **[1 mark]**.
c) $P_4O_{10} + 12KOH \rightarrow 4K_3PO_4 + 6H_2O$ **[1 mark]**

Page 157 — Redox Equations

1 a) $Ti + (4 \times -1) = 0$, $Ti = +4$ **[1 mark]**
b) $(2 \times V) + (5 \times -2) = 0$, $V = +5$ **[1 mark]**
c) $Cr + (4 \times -2) = -2$, $Cr = +6$ **[1 mark]**
d) $(2 \times Cr) + (2 \times -7) = -2$, $Cr = +6$ **[1 mark]**

2 $Cr_2O_7^{2-}{}_{(aq)} + 14H^+_{(aq)} + 6e^- \rightarrow 2Cr^{3+}_{(aq)} + 7H_2O_{(l)}$ **[1 mark]**
$Zn_{(aq)} \rightarrow Zn^{2+}_{(aq)} + 2e^-$ **[1 mark]**

3 a) $2MnO_4^-{}_{(aq)} + 16H^+_{(aq)} + 10I^-_{(aq)} \rightarrow 2Mn^{2+}_{(aq)} + 8H_2O_{(l)} + 5I_{2(aq)}$
[1 mark for correct reactants and products, 1 mark for correct balancing]
You have to balance the number of electrons before you can combine the half-equations. And always double-check that your equation definitely balances. It's easy to slip up and throw away marks.
b) Mn has been reduced **[1 mark]** from +7 to +2 **[1 mark]**.
I^- has been oxidised **[1 mark]** from −1 to 0 **[1 mark]**.
c) Reactive metals have a tendency to lose electrons, so are good reducing agents **[1 mark]**. I^- is already in its reduced form **[1 mark]**.

Page 159 — Electrode Potentials

1 a) $Zn_{(s)} | Zn^{2+}_{(aq)} || Ag^+_{(aq)} | Ag_{(s)}$
[1 mark for zinc on the left and silver on the right, 1 mark for the oxidised products in the middle.]
b) $+0.80$ $V - (-0.76$ $V) = 1.56$ V **[1 mark]**
c) The concentration of Zn^{2+} ions or Ag^+ ions was not 1.00 $moldm^{-3}$ **[1 mark]**. The pressure wasn't 100 kPa **[1 mark]**.
The difference can't be due to temperature, because it was standard during the experiment (25 °C is the same as 298 K).
d) $Zn_{(s)} + 2Ag^+_{(aq)} \rightarrow Zn^{2+}_{(aq)} + 2Ag_{(s)}$ **[1 mark]**
e) The zinc half-cell. It has a more negative standard electrode potential/it's less electronegative **[1 mark]**.

Page 161 — Electrochemical Series

1 a) $Zn_{(s)} + Ni^{2+}_{(aq)} \rightleftharpoons Zn^{2+}_{(aq)} + Ni_{(s)}$ **[1 mark]**
$E^\circ = (-0.25) - (-0.76) = +0.51$ V **[1 mark]**
b) $2MnO_4^-{}_{(aq)} + 16H^+_{(aq)} + 5Sn^{2+}_{(aq)} \rightleftharpoons$
$2Mn^{2+}_{(aq)} + 8H_2O_{(l)} + 5Sn^{4+}_{(aq)}$ **[1 mark]**
$E^\circ = (+1.52) - (+0.15) = +1.37$ V **[1 mark]**
c) No reaction **[1 mark]**. Both reactants are in their oxidised form **[1 mark]**.
d) $Ag^+_{(aq)} + Fe^{2+}_{(aq)} \rightleftharpoons Ag_{(s)} + Fe^{3+}_{(aq)}$ **[1 mark]**
$E^\circ = (+0.80) - (+0.77) = +0.03$ V **[1 mark]**

2 a) (i) $Fe_{(s)} | Fe^{2+}_{(aq)} || O_{2(g)} | OH^-_{(aq)}$ **[1 mark]**
(ii) $0.40 - (-0.44) = +0.84$ V **[1 mark]**
b) The iron half-cell has a more negative electrode potential than the oxygen/water half-cell, so iron is oxidised **[1 mark]**.

A2 Answers

Page 163 — Electrochemical Cells

1 a) $0.34 - (-2.38) = +2.72$ V *[1 mark]*
   b) $0.34 - (-0.76) = +1.10$ V *[1 mark]*
2 a) (i) $O_2 + 4H^+ + 4e^- \rightarrow 2H_2O$ *[1 mark]*
      (ii) From the reaction of hydrogen at the negative electrode *[1 mark]*.
   b) Advantages include: fuel cells don't need electrically recharging, the only waste product is water *[1 mark]*. Disadvantages include: hydrogen is highly flammable so is difficult to store or transport safely, electricity is needed to produce the raw materials *[1 mark]*.

Unit 5: Section 3 — Transition Metals

Page 165 — Transition Metals — The Basics

1 a) $1s^2 2s^2 2p^6 3s^2 3p^6 3d^{10}$ or $[Ar]3d^{10}$ *[1 mark]*
   b) No, it doesn't *[1 mark]*. Cu^+ ions have a full 3d subshell *[1 mark]*.
   c) copper(II) sulfate ($CuSO_{4(aq)}$) *[1 mark]*.
2 a) +5 *[1 mark]*
   b) (i) $MnO_4^- + 8H^+ + 5e^- \rightarrow Mn^{2+} + 4H_2O$ *[1 mark]*
      (ii) $V^{3+} + 2H_2O \rightarrow VO_2^+ + 4H^+ + 2e^-$ *[1 mark]*
   c) $2MnO_4^- + 5V^{3+} + 2H_2O \rightarrow 2Mn^{2+} + 5VO_2^+ + 4H^+$ *[1 mark for correct reactants and products, 1 mark for correct balancing]*
   Remember to check that the charges balance when you combine two half-equations, you'll often need to multiply everything in one of them.

Page 167 — Complex Ions

1 a) (i) A species (atom, ion or molecule) that donates a lone pair of electrons to form a coordinate bond with a metal atom or ion *[1 mark]*. NH_3 is a ligand in $[Ag(NH_3)_2]^+$ *[1 mark]*.
      (ii) A covalent bond in which both electrons come from the same species *[1 mark]*. Both electrons come from a nitrogen *[1 mark]* in each coordinate bond in $[Ag(NH_3)_2]^+$.
      (iii) The number of coordinate bonds formed with the central metal atom or ion *[1 mark]*. In $[Ag(NH_3)_2]^+$, the coordination number is 2 because two NH_3 ligands are bonded to Ag^+ *[1 mark]*.
   b) Linear *[1 mark]*
2 a) Coordination number: 6 *[1 mark]* shape: octahedral *[1 mark]*
   b) Coordination number: 4 *[1 mark]* shape: tetrahedral *[1 mark]* formula: $[CuCl_4]^{2-}$ *[1 mark]*
   c) Cl^- ligands are larger than water ligands *[1 mark]*, so only 4 Cl^- ligands can fit around the Cu^{2+} ion *[1 mark]*.
3 a) $[Co(H_2O)_6]^{2+} + 3NH_2CH_2CH_2NH_2$
   $\rightarrow [Co(NH_2CH_2CH_2NH_2)_3]^{2+} + 6H_2O$
   [1 mark for the correct formula of ethane-1,2-diamine, and 1 mark for the correct equation]
   b)

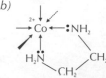

   [1 mark for showing two bonds between the nitrogens on the ligand and the Co ion, and 1 mark for indicating clearly that they are coordinate bonds.]

Page 169 — Formation of Coloured Ions

1 a) Energy is absorbed from visible light *[1 mark]* when electrons move from the ground state to a higher energy level *[1 mark]*.
   b) Change in oxidation state *[1 mark]*, ligand *[1 mark]* or coordination number *[1 mark]*
2 Prepare a range of dilutions of known concentrations *[1 mark]*. Measure the absorbance of the solutions *[1 mark]*. Plot a graph of concentration versus absorbance *[1 mark]*.
3 a) ΔE is the energy absorbed *[1 mark]* when an electron moves from the ground state to a higher energy level/excited state *[1 mark]*.
   OR ΔE is the difference between the ground state energy *[1 mark]* and the energy of an excited electron *[1 mark]*.
   b) (i) Cu^+ $[Ar]$ $3d^{10}$ *[1 mark]*
      (ii) Cu^{2+} $[Ar]$ $3d^9$ *[1 mark]*
   c) Cu^{2+} *[1 mark]* because it has an incomplete d-subshell *[1 mark]*.

Page 171 — Variable Oxidation States

1 a) $Cr_2O_7^{2-} + 14H^+ + 6e^- \rightarrow 2Cr^{3+} + 7H_2O$ *[1 mark]*
   b) (i) The two half-equations are:
      $Cr_2O_7^{2-} + 14H^+ + 16e^- \rightarrow 2Cr^{3+} 7H_2O$
      $CH_3CH_2OH \rightarrow CH_3CHO + 2H^+ + 2e^-$
      So the full equation is:
      $Cr_2O_7^{2-} + 3CH_3CH_2OH + 8H^+ \rightarrow$
      $2Cr^{3+} + 3CH_3CHO + 7H_2O$
      [1 mark for correct reactants and products, 1 mark for correct balancing]
      Don't forget to balance half-equations by adding H_2O, H^+ and e^- — see page 157 for more.
      (ii) orange $\rightarrow$ green *[1 mark]*
   c) (i) $SO_2 + 2H_2O \rightarrow SO_4^{2-} + 4H^+ + 2e^-$ *[1 mark]*
      (ii) $Cr_2O_7^{2-} + 3SO_2 + 2H^+ \rightarrow 2Cr^{3+} + 3SO_4^{2-} + H_2O$ *[1 mark]*
      For this one, you needed to triple everything in the half-equation from c(i) so that it has the same number of electrons as the half-equation in a).
2 a) $Cr^{3+}_{(aq)}$ or $[Cr(H_2O)_6]^{3+}$ *[1 mark]*
   b) -1 *[1 mark]*
   c) $2CrO_4^{2-} + H^+ \rightarrow Cr_2O_7^{2-} + OH^-$ *[1 mark]*
      orange *[1 mark]*

Page 173 — Titrations with Transition Metals

1 a) Moles of $Cr_2O_7^{2-}$ added $= \dfrac{0.0100 \times 29.40}{1000} = 2.94 \times 10^{-4}$
   [1 mark for the correct formula, 1 mark for the correct answer.]
   b) 6 moles of Fe^{2+} react with 1 moles of $Cr_2O_7^{2-}$
   So moles of Fe^{2+} reacted $= 2.94 \times 10^{-4} \times 6 = 1.76 \times 10^{-3}$ *[1 mark]*
   c) Mass of Fe $= A_r \times$ moles $= 55.8 \times 1.76 \times 10^{-3} = 0.0984$ g *[1 mark for the correct formula, 1 mark for the correct answer.]*
   d) % Fe $= \dfrac{0.0984}{0.100} \times 100 = 98.4\%$ *[1 mark]*

A2 Answers

2   Moles of MnO_4^- added $= \dfrac{0.0200 \times 18.30}{1000} = 3.66 \times 10^{-4}$

[1 mark for the correct formula, 1 mark for the correct answer.]

5 moles of $C_2O_4^{2-}$ react with 2 moles of MnO_4^-

so moles of $C_2O_4^{2-}$ added $= \dfrac{3.66 \times 10^{-4} \times 5}{2} = 9.15 \times 10^{-4}$

[1 mark for the correct formula, 1 mark for the correct answer.]

M_r of $Na_2C_2O_4 = (23 \times 2) + (12 \times 2) + (16 \times 4) = 134$

[1 mark]

Mass of $Na_2C_2O_4 = M_r \times$ moles $= 134 \times 9.15 \times 10^{-4}$
$= 0.123\ g$

[1 mark for the correct formula, 1 mark for the correct answer.]

3   Add dilute sulfuric acid to the iron(II) sulfate solution **[1 mark]**. Titrate with potassium manganate(VII) **[1 mark]**. The remaining Fe^{2+} will be oxidised to Fe^{3+} **[1 mark]**. Calculate the number of moles of potassium manganate(VII) that react and use this to calculate the number of moles of iron(II), and hence the concentration **[1 mark]**.

Page 176 — Uses of Transition Metals

1 a) Vanadium(V) is reduced **[1 mark]** to vanadium(IV)
    $V_2O_5 + SO_2 \rightarrow V_2O_4 + SO_3$ **[1 mark]**
    Vanadium(IV) is oxidised **[1 mark]** to vanadium(V)
    $V_2O_4 + \frac{1}{2}O_2 \rightarrow V_2O_5$ **[1 mark]**
 b) (i) impurities **[1 mark]** are adsorbed onto the surface of the catalyst **[1 mark]**.
        Example: lead in petrol poisons catalytic converter
        or: sulfur in hydrogen poisons the iron catalyst in the Haber Process **[1 mark]**
    (ii) Reduced efficiency **[1 mark]**, Increased cost **[1 mark]**

2 a) Homogeneous — reactants and products in the same phase **[1 mark]**
    Heterogeneous — reactants and products in different phases **[1 mark]**
 b) (i) homogeneous **[1 mark]**
    (ii)

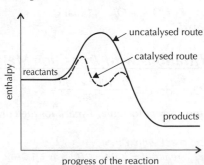

[1 mark for enthalpy level of products lower than that of reactants, 1 mark for uncatalysed reaction having the greater activation energy, 1 mark for a catalysed reaction curve with two humps.]

3 a) Aldehydes and ketones **[1 mark]**.
 b) silver mirror/precipitate **[1 mark]**
 c) (i) $[Ag(NH_3)_2]^+$ **[1 mark]**
    (ii) linear **[1 mark]**
    (iii) colourless **[1 mark]**

Unit 5: Section 4 — Inorganic Reactions

Page 179 — Metal-Aqua Ions

1   Fe^{3+} has a higher charge density than Fe^{2+} **[1 mark]**. This means Fe^{3+} polarises water molecules more **[1 mark]**, weakening the O-H bond more **[1 mark]** and making it more likely that H^+ ions are released into the solution **[1 mark]**.

2   A blue precipitate **[1 mark]** of copper hydroxide forms in the blue solution **[1 mark]** of copper sulfate. On addition of excess ammonia, the precipitate dissolves **[1 mark]** to give a deep blue solution **[1 mark]**.
    $[Cu(H_2O)_6]^{2+} + 2OH^- \rightleftharpoons Cu(H_2O)_4(OH)_2$ **[1 mark]** $+ 2H_2O$
    $Cu(H_2O)_4(OH)_2 + 4NH_3 \rightleftharpoons$
        $[Cu(NH_3)_4(H_2O)_2]^{2+}$ **[1 mark]** $+ 2H_2O + 2OH^-$
    [1 additional mark for each balanced equation.]

3 a) (i) $[Cr(H_2O)_6]^{3+}$ **[1 mark]**
    (ii) $Cr(H_2O)_3(OH)_3$ **[1 mark]**
    (iii) $[Cr(NH_3)_6]^{3+}$ **[1 mark]**
 b) $Cr(H_2O)_3(OH)_3 + 6NH_3 \rightarrow [Cr(NH_3)_6]^{3+} + 3H_2O + 3OH^-$
    [1 mark]

4 a) (i) Formation of a brown precipitate **[1 mark]**
    (ii) Formation of a green precipitate **[1 mark]**
 b) $[Fe(H_2O)_6]^{2+}{}_{(aq)} + CO_3{}^{2-}{}_{(aq)} \rightarrow FeCO_{3(aq)} + 6H_2O_{(aq)}$ **[1 mark]**
 c) (i) Formation of a brown precipitate **[1 mark]**
    (ii) Fe^{2+} has been oxidised to Fe^{3+} by the oxygen in the air **[1 mark]**

Page 181 — Substitution Reactions

1 a) $[Cu(H_2O)_6]^{2+} + 4Cl^- \rightarrow CuCl_4^{2-}$ **[1 mark]** $+ 6H_2O$
    [1 mark for balanced equation.]
    $CuCl_4^{2-}$ is yellow (allow green, which is the colour you would see when you actually do the experiment) **[1 mark]**
 b) $[Cu(H_2O)_6]^{2+} + 3C_2O_4^{2-} \rightarrow [Cu(C_2O_4)_3]^{4-}$ **[1 mark]** $+ 6H_2O$
    [1 mark for balanced equation.]
 c) $Cu(H_2O)_4(OH)_2$ **[1 mark]**, blue **[1 mark]**

2 a) $[Fe(H_2O)_6]^{3+} + EDTA^{4-} \rightarrow [FeEDTA]^-$ **[1 mark]** $+ 6H_2O$
    [1 mark for balanced equation.]
 b) The formation of $[FeEDTA]^-$ results in an increase in entropy **[1 mark]**, because the number of particles increases **[1 mark]**.
 c) $Fe(H_2O)_3(OH)_3$ **[1 mark]**

3 a) $[Co(H_2O)_6]^{2+} + 6\ NH_3 \rightarrow [Co(NH_3)_6]^{2+}$ **[1 mark]** $+ 6H_2O$
    [1 mark for balanced equation.]
    You can fit the same number of H_2O and NH_3 ligands around a Co^{2+} ion because they are a similar size.
 b) $[Co(H_2O)_6]^{2+} + 3NH_2CH_2CH_2NH_2 \rightarrow$
        $[Co(NH_2CH_2CH_2NH_2)_3]^{2+}$ **[1 mark]** $+ 6H_2O$
    [1 mark for balanced equation.]
 c) $[Co(H_2O)_6]^{2+} + 4Cl^- \rightarrow CoCl_4^{2-}$ **[1 mark]** $+ 6H_2O$
    [1 mark for balanced equation.]

Index

Index

Index

Index

Some Useful Stuff

Useful Formulas...

Here's a handy reference guide to the formulas you'll need. You'll find them all inside the book too.

Mole equations:
$$\text{Number of moles} = \frac{\text{mass of substance}}{\text{molar mass}}$$

$$\text{Number of moles} = \frac{\text{concentration} \times \text{volume (cm}^3)}{1000}$$

Ideal gas equation: $pV = nRT$ p = pressure (Pa), V = volume (cm³), n = number of moles, $R = 8.31$ J K⁻¹ mol⁻¹, T = temperature (K)

Rate equation: $\text{Rate} = k[A]^m[B]^n$

Equilibrium constant: $K_c = \dfrac{[D]^d[E]^e}{[A]^a[B]^b}$

pH formula: $pH = -\log_{10}[H^+]$

Ionic product of water: $K_w = [H^+][OH^-]$

Acid dissociation constant: $K_a = \dfrac{[H^+][A^-]}{[HA]}$

pK_a formula: $pK_a = -\log_{10}K_a$

Entropy change of a system: $\Delta S_{\text{system}} = \Delta S_{\text{products}} - \Delta S_{\text{reactants}}$

Entropy change of the surroundings: $\Delta S_{\text{surroundings}} = \dfrac{-\Delta H}{T}$

Total entropy change: $\Delta S_{\text{total}} = \Delta S_{\text{system}} + \Delta S_{\text{surroundings}}$

Free energy change: $\Delta G = \Delta H - T\Delta S$

Energy absorbed by an electron moving between orbitals: $\Delta E = h\nu$

Organic Compounds...

These can be a real nightmare. You need to learn all the types in this table and practise naming them...

Homologous series	Prefix or Suffix	Examples	Functional Group
alkanes	-ane	Propane $CH_3CH_2CH_3$	n/a
branched alkanes	alkyl- (-yl)	methylpropane $CH_3CH(CH_3)CH_3$	n/a
alkenes	-ene	propene $CH_3CH=CH$	$C=C$
alcohols	-ol	ethanol CH_3CH_2OH	$-OH$
aldehydes	-al	ethanal CH_3CHO	(aldehyde group)
ketones	-one	propanone CH_3COCH_3	(ketone group)
esters	alkyl -oate	propyl ethanoate $CH_3COOCH_2CH_2CH_3$	(ester group)
carboxylic acids	-oic acid	ethanoic acid CH_3COOH	(carboxylic acid group)
cycloalkanes	cyclo- -ane	cyclohexane C_6H_{12}	n/a
arenes	phenyl- (-benzene)	ethylbenzene $C_6H_5C_2H_5$	(benzene ring)
amines	-amine (primary)	methylamine CH_3NH_2	$-NH_2$
	di- -amine (secondary)	dimethylamine $(CH_3)_2NH$	(secondary amine group)
	tri- -amine (tertiary)	trimethylamine $N(CH_3)_3$	(tertiary amine group)
acyl chlorides	-oyl chloride	ethanoyl chloride CH_3COCl	(acyl chloride group)

The Periodic Table

Key

Relative Atomic Mass →	1.0	
	H	
Atomic number →	Hydrogen	
	1	

Periods	Group 1	Group 2													Group 3	Group 4	Group 5	Group 6	Group 7	Group 0
1																				4.0 **He** Helium 2
2	6.9 **Li** Lithium 3	9.0 **Be** Beryllium 4													10.8 **B** Boron 5	12.0 **C** Carbon 6	14.0 **N** Nitrogen 7	16.0 **O** Oxygen 8	19.0 **F** Fluorine 9	20.2 **Ne** Neon 10
3	23.0 **Na** Sodium 11	24.3 **Mg** Magnesium 12													27.0 **Al** Aluminium 13	28.1 **Si** Silicon 14	31.0 **P** Phosphorus 15	32.1 **S** Sulfur 16	35.5 **Cl** Chlorine 17	39.9 **Ar** Argon 18
4	39.1 **K** Potassium 19	40.1 **Ca** Calcium 20	45.0 **Sc** Scandium 21	47.9 **Ti** Titanium 22	50.9 **V** Vanadium 23	52.0 **Cr** Chromium 24	54.9 **Mn** Manganese 25	55.8 **Fe** Iron 26	58.9 **Co** Cobalt 27	58.7 **Ni** Nickel 28	63.5 **Cu** Copper 29	65.4 **Zn** Zinc 30	69.7 **Ga** Gallium 31	72.6 **Ge** Germanium 32	74.9 **As** Arsenic 33	79.0 **Se** Selenium 34	79.9 **Br** Bromine 35	83.8 **Kr** Krypton 36
5	85.5 **Rb** Rubidium 37	87.6 **Sr** Strontium 38	88.9 **Y** Yttrium 39	91.2 **Zr** Zirconium 40	92.9 **Nb** Niobium 41	95.9 **Mo** Molybdenum 42	98.9 **Tc** Technetium 43	101.1 **Ru** Ruthenium 44	102.9 **Rh** Rhodium 45	106.4 **Pd** Palladium 46	107.9 **Ag** Silver 47	112.4 **Cd** Cadmium 48	114.8 **In** Indium 49	118.7 **Sn** Tin 50	121.8 **Sb** Antimony 51	127.6 **Te** Tellurium 52	126.9 **I** Iodine 53	131.3 **Xe** Xenon 54
6	132.9 **Cs** Caesium 55	137.3 **Ba** Barium 56	138.9 **La** Lanthanum 57	178.5 **Hf** Hafnium 72	180.9 **Ta** Tantalum 73	183.9 **W** Tungsten 74	186.2 **Re** Rhenium 75	190.2 **Os** Osmium 76	192.2 **Ir** Iridium 77	195.1 **Pt** Platinum 78	197.0 **Au** Gold 79	200.6 **Hg** Mercury 80	204.4 **Tl** Thallium 81	207.2 **Pb** Lead 82	209.0 **Bi** Bismuth 83	210.0 **Po** Polonium 84	210.0 **At** Astatine 85	222.0 **Rn** Radon 86
7	223.0 **Fr** Francium 87	226.0 **Ra** Radium 88	227.0 **Ac** Actinium 89	261 **Rf** Rutherfordium 104	262 **Db** Dubnium 105	266 **Sg** Seaborgium 106	264 **Bh** Bohrium 107	277 **Hs** Hassium 108	268 **Mt** Meitnerium 109	271 **Ds** Darmstadtium 110	272 **Rg** Roentgenium 111							

The Lanthanides

140.1 **Ce** Cerium 58	140.9 **Pr** Praseodymium 59	144.2 **Nd** Neodymium 60	144.9 **Pm** Promethium 61	150.4 **Sm** Samarium 62	152.0 **Eu** Europium 63	157.3 **Gd** Gadolinium 64	158.9 **Tb** Terbium 65	162.5 **Dy** Dysprosium 66	164.9 **Ho** Holmium 67	167.3 **Er** Erbium 68	168.9 **Tm** Thulium 69	173.0 **Yb** Ytterbium 70	175.0 **Lu** Lutetium 71

The Actinides

232.0 **Th** Thorium 90	231.0 **Pa** Protactinium 91	238.0 **U** Uranium 92	237.0 **Np** Neptunium 93	239.1 **Pu** Plutonium 94	243.1 **Am** Americium 95	247.1 **Cm** Curium 96	247.1 **Bk** Berkelium 97	252.1 **Cf** Californium 98	252 **Es** Einsteinium 99	257 **Fm** Fermium 100	258 **Md** Mendelevium 101	259 **No** Nobelium 102	260 **Lr** Lawrencium 103